Origins of the Earth, Moon, and Life

Origins of the Earth, Moon, and Life

An Interdisciplinary Approach

Akio Makishima
Professor, Institute for Planetary Materials,
Okayama University, Japan

elsevier.com

Elsevier
Radarweg 29, PO Box 211, 1000 AE Amsterdam, Netherlands
The Boulevard, Langford Lane, Kidlington, Oxford OX5 1GB, United Kingdom
50 Hampshire Street, 5th Floor, Cambridge, MA 02139, United States

Library of Congress Cataloging-in-Publication Data
A catalog record for this book is available from the Library of Congress

British Library Cataloguing-in-Publication Data
A catalogue record for this book is available from the British Library

ISBN: 978-0-12-812058-3

For information on all Elsevier publications
visit our website at https://www.elsevier.com/books-and-journals

ELSEVIER
Book Aid
International
Working together
to grow libraries in
developing countries
www.elsevier.com • www.bookaid.org

Publisher: Candice Janco
Acquisition Editor: Marisa LaFleur
Editorial Project Manager: Marisa LaFleur
Production Project Manager: Mohanapriyan Rajendran
Designer: Christian J. Bilbow

Typeset by TNQ Books and Journals

To my wife Mayumi and to my children, Jun, Kei, and Yu'u.

Contents

Preface

This book takes an interdisciplinary approach to the study of the origins of the Earth, the Moon, and life by basing its 11 chapters on the five astrosciences: astronomy, astrobiology, astrogeology, astrophysics, and cosmochemistry. As the theories, experiments, observations, calculations, and analytical data presented are based on research from these astrosciences, deep knowledge of them is a prerequisite to follow and understand this up-to-date discussion. Unfortunately, this kind of interdisciplinary science does not yet exist, which is why the author talks ironically about having used an interdisciplinary approach.

In each science, masterpieces exist. For example, one masterpiece in cosmochemistry is *Cosmochemistry* by McSween and Huss (2010). For professional molecular biologists, *Molecular Biology of the Cell* by Alberts et al. (2014) is a masterpiece. For professional astrophysicists or astrobiologists, *Origin of the Earth and Moon* by Canup and Righter (2000) and *From Suns to Life* by Gargaud et al. (2006) are already published.

The largest difference between this book and such previously-published masterpieces are:

1. The author wants to explain and present the most accepted ideas in simpler words. To understand this book, only an advanced high school level of literacy is required, so undergraduate students or even curious high school students should be able to tackle this book.
2. Although the five astrosciences are required to describe the many problems regarding the origin of the Earth, Moon, or life, each professional science uses specific terminology that often is utterly incomprehensible to those in other disciplines. In this book, the author tries to break or destroy such boundaries. The author hopes that this book could work as a pipeline or synapse that connect the disciplines.
3. Nowadays, there are many professional papers and books. The speed of research has increased dramatically, with new results in each discipline issued daily, making it very difficult to keep up with the latest ideas. Here, the author wants to present generally accepted up-to-date information on the universe, stars, planets, moons, and life.

The readers are imagined to be undergraduate to graduate students in astrosciences who want to increase their astroliteracy. Of course, students in non-astrosciences are welcome too, especially pure biologists, chemists, geologists and physicists, because this book presents only qualitative results, which allows non-astroscience students to increase their astroliteracy without the advanced mathematics used in astronomy and astrophysics. Therefore, this book is suitable as a textbook for both undergraduate and graduate students. In this book, deeper knowledge and explanations are found in boxes and references.

The author hopes that readers will enjoy this interdisciplinary approach to the origins of the Earth, Moon, and life. For graduate students who already have interdisciplinary knowledge, this book will be a handy reference to get acquainted with other astroscience disciplines.

The framework of this book is as follows: Chapter 1 begins with the Big Bang. There is no physical evidence that the Big Bang actually occurred. Remember that the Big Bang is a purely theoretical phenomenon. Then we learn about the life of stars. The stars produce energy by burning hydrogen, and heavier elements up to nickel are synthesized. Stars die quietly or flamboyantly. The latter case is called a supernova, in which elements heavier than nickel are synthesized.

In Chapter 2, we learn about the solar nebula, from which planets from Mercury to Neptune, including the proto-Earth, were formed. We describe not "the Earth" but "the proto-Earth," because the "proto-Earth" is considered to have been destroyed by a Mars-sized planetesimal named "Theia" in a giant impact, forming the present Earth and Moon system (Chapter 3).

The abundance of the highly siderophile elements (HSEs) in the Earth's mantle is much higher than those calculated by elemental partitioning between molten iron and silicate melt. To increase the HSE abundance in the mantle, it is theorized that a "late veneer" occurred after the giant impact from Theia (Chapter 4).

As there is no water on the Moon, alteration or weathering by water on the Moon is considered not to have occurred. Therefore, remnants of the giant impact should remain on the Moon. However, the volcanism of the Moon, and the bombardment of meteorites forming craters make it difficult to obtain the date of the giant impact, or even evidence of the giant impact itself (Chapter 5).

In Chapter 6, the geological evidence for the oldest sample of the Earth's crust is discussed. In addition, the search for the oldest fossils on Earth is described.

At the end of the 20th century, the news that a record of life was found in a Martian meterorite traveled around the world (Chapter 7). It was eventually proved to be incorrect, but news of the discovery advanced the astrosciences, especially astrobiology.

In Chapter 8, trials to find life, which have now become graphite, began strenuously. To discuss life, the evolution of atmosphere is also required. In this discussion, the origin of the ocean is also important. Along with water, the origin of organic materials, the building blocks of life, are also discussed. Then life started on Earth. The geological evidence of life is also presented, though it can sometimes be erroneous.

In Chapter 9, laboratory experiments for making life-related materials, especially nucleic acids, are presented.

Chapter 10 describes how life began using life-related materials on Earth. The protocell and RNA world are explained. The cell cycle is successfully copied. How the eukaryotic cell was evolved from the archaea cell is discussed.

Finally in Chapter 11, many recent observations by spacecraft for evidence of liquid water on other planets and moons in our solar system, which implies life-friendly conditions, are summarized. It is surprising that there is liquid water in the moons of Jupiter and Saturn. Furthermore, Pluto is not a dead world but an active world of moving ice.

The author hopes the readers will enjoy these discussions on the origin of the Earth, Moon, and life based on interdisciplinary studies of the astrosciences. Furthermore, it is hoped that some readers find careers as scientists or science-related professionals based on their own curiosity.

Akio Makishima
Misasa, Tottori, Japan
July 8, 2016

References

Alberts B, Johnson A, Lewis J, et al. Molecular biology of the cell. 6th ed. New York: Garland Science; 2014.

Canup RM, Righter K. Origin of the Earth and Moon. Tucson: The University of Arizona Press; 2000.

Gargaud M, Claeys P, Lopez-Garcia P, et al. From suns to life: a chronological approach to the history of life on Earth. New York: Springer; 2006.

McSween Jr HY, Huss GR. Cosmochemistry. Cambridge: Cambridge University Press; 2010.

CHAPTER 1

Origin of Elements

Chapter Outline

1.1 ORIGIN OF PROTONS (HYDROGEN ATOMS)

The origins of Earth, the Moon, and Life start with the synthesis of elements, which date back to the formation of the universe. The universe was formed in the so-called Big Bang 13.8billion years ago. The inflation theory (Sato, 1981) predicts that the universe was formed from nothing to almost its present size in 1 s. After the inflation, the universe became cool enough for the electrons and protons to bind each other, forming hydrogen atoms (recombination) (Spergel and Keating, 2015).

The ^{1}H and neutron (^{1}n) produced deuterium (^{2}H); deuterium and neutron produced tritium (^{3}H); tritium decayed into 3-helium (^{3}He); and ^{3}He and neutron produced ^{4}He. The number in the superscript position indicates the mass number, which is the number of protons plus neutrons (see Box 1.1).

BOX 1.1 ATOM AND ISOTOPE

Everything is composed of atoms. The atom is made of the nucleus and electron(s), which are equal to the number of protons in the nucleus. The nucleus is made of protons and neutrons, and the mass number is the number of protons plus neutrons. The structure of the atom is schematically depicted in Fig. Box 1.1. An atom with the same proton number but a different neutron number is called an isotope. A list of naturally existing isotopes is shown in Table Box 1.1. Details for the elemental abundance of Earth is written in Box 1.2.

In general, when we want to indicate the element and isotope, for example, uranium-235, and uranium-238 (235 and 238 being mass numbers), we write them as ^{235}U and ^{238}U, respectively. The elemental abundance of Earth is explained in Box 1.2. ppm or ppb are explained in Box 1.3.

Origins of the Earth, Moon, and Life. http://dx.doi.org/10.1016/B978-0-12-812058-3.00001-6

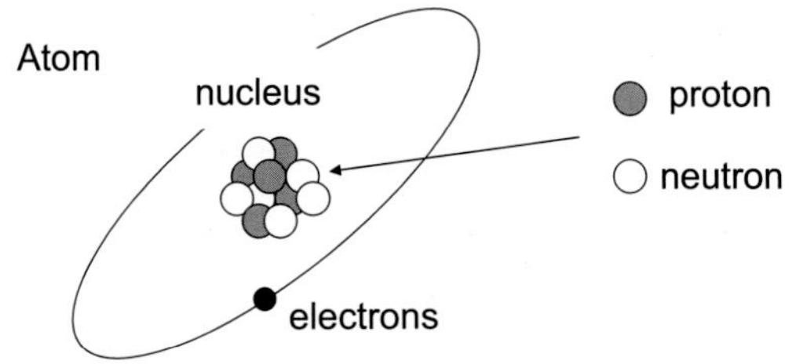

FIGURE BOX 1.1
The structure of the atom.

Table Box 1.1 A Table of Naturally Existing Isotopes (the Number Is the Percentage of Isotopic Abundance), Elemental Concentration in the CI Chondrite (CI Conc.), Atomic Weight, and Each Elemental Abundance Compared to Si × 10⁶ (X/Si × 10⁶). The elemental abundances of the CI chondrite are from McDonough and Sun (1995)

Atomic Number	Element	1	2	3	4	5	6	7	8	9	10	11	12
1	H	99.2	0.02										
2	He			Trace	100								
3	Li						7.5	92.5					
4	Be									100			
5	B										19.9	80.1	
6	C												98.9
7	N												
8	O												
9	F												
10	Ne												
		21	22	23	24	25	26	27	28	29	30	31	32
10	Ne	0.27	9.22										
11	Na			100									
12	Mg				79	10	11						
13	Al							100					
14	Si								92.2	4.67	3.1		
15	P											100	
16	S												95
17	Cl												
18	Ar												
19	K												
20	Ca												
		41	42	43	44	45	46	47	48	49	50	51	52
19	K	6.73											
20	Ca		0.65	0.14	2.09		Trace		0.19				
21	Sc					100							
22	Ti						8	7.3	73.8	5.5	5.4		
23	V										0.25	99.8	
24	Cr										4.35		83.8
25	Mn												
26	Fe												
27	Co												

Table Box 1.1 A Table of Naturally Existing Isotopes (the Number Is the Percentage of Isotopic Abundance), Elemental Concentration in the CI Chondrite (CI Conc.), Atomic Weight, and Each Elemental Abundance Compared to Si × 10^6 (X/Si × 10^6). The elemental abundances of the CI chondrite are from McDonough and Sun (1995)

13	14	15	16	17	18	19	20	CI Conc.	Unit	Atomic Weight	X/Si × 1e6
										1.0	
										4.0	
								1.5	ppm	6.9	5.7E+01
								0.025	ppm	9.0	7.3E−01
								0.9	ppm	10.8	2.2E+01
1.1								3.5	%	12.0	7.7E+05
	99.6	0.37						3180	ppm	14.0	6.0E+04
			99.8	0.04	0.2					16.0	
						100		60	ppm	19.0	8.3E+02
							90.5			20.2	
33	34	35	36	37	38	39	40				
								5100	ppm	33.0	4.1E+04
								9.65	%	24.3	1.0E+06
								0.86	%	27.0	8.4E+04
								10.65	%	28.1	3.8E−03
								1080	ppm	31.0	9.2E+03
0.75	4.21		0.02					5.4	%	32.1	4.4E+05
		75.8		24.2				680	ppm	35.5	5.1E+03
			0.34		0.06		99.6			40.0	
						93.3	0.01	550	ppm	39.1	3.7E+03
							96.9	0.925	%	40.1	6.1E+04
53	54	55	56	57	58	59	60				
								5.92	ppm	45.0	3.5E+01
								440	ppm	47.9	2.4E+03
								56	ppm	50.9	2.9E+02
9.5	2.37							2650	ppm	52.0	1.3E+04
		100						1920	ppm	54.9	9.2E+03
	5.8		91.7	2.2	0.28			18.1	%	55.9	8.5E+05
						100		500	ppm	58.9	2.2E+03

Continued...

Table Box 1.1 A Table of Naturally Existing Isotopes (the Number Is the Percentage of Isotopic Abundance), Elemental Concentration in the CI Chondrite (CI Conc.), Atomic Weight, and Each Elemental Abundance Compared to Si × 10^6 (X/Si × 10^6). The elemental abundances of the CI chondrite are from McDonough and Sun (1995) continued

Atomic Number	Element	1	2	3	4	5	6	7	8	9	10	11	12
28	Ni												
		61	62	63	64	65	66	67	68	69	70	71	72
28	Ni	1.13	3.59		0.91								
29	Cu			69.2		30.8							
30	Zn				48.6		27.9	4.1	18.8		0.6		
31	Ga									60.1		39.9	
32	Ge										20.5		27.4
33	As												
34	Se												
35	Br												
36	Kr												
		81	82	83	84	85	86	87	88	89	90	91	92
34	Se		9.2										
35	Br	49.3											
36	Kr		11.6	11.5	57		17.3						
37	Rb					72.2		27.8					
38	Sr				0.56		9.86	7	82.6				
39	Y									100			
40	Zr										51.5	11.2	17.2
41	Nb												
42	Mo												14.8
44	Ru												
		101	102	103	104	105	106	107	108	109	110	111	112
44	Ru	17	31.6		18.7								
45	Rh			100									
46	Pd		1.02		11.1	22.3	27.3		26.5		11.7		
47	Ag							51.8		48.2			
48	Cd						1.25		0.89		12.5	12.8	24.1
49	In												
50	Sn												0.97
51	Sb												
52	Te												
		121	122	123	124	125	126	127	128	129	130	131	132
50	Sn		4.63		5.79								
51	Sb	57.3		42.7									
52	Te		2.6	0.91	4.82	7.14	19		31.7		33.8		
53	I							100					
54	Xe				0.1		0.09		1.91	26.4	4.1	21.2	26.9
55	Cs												
56	Ba										0.11		0.1
57	La												
58	Ce												

Table Box 1.1 **A Table of Naturally Existing Isotopes (the Number Is the Percentage of Isotopic Abundance), Elemental Concentration in the CI Chondrite (CI Conc.), Atomic Weight, and Each Elemental Abundance Compared to Si × 10⁶ (X/Si × 10⁶). The elemental abundances of the CI chondrite are from McDonough and Sun (1995) continued**

13	14	15	16	17	18	19	20	CI Conc.	Unit	Atomic Weight	X/Si × 1e6
					68.3		26.1	10,500	ppm	58.7	4.7E+04
73	74	75	76	77	78	79	80				
								120	ppm	63.6	5.0E+02
								310	ppm	65.4	1.3E+03
								9.2	ppm	69.7	3.5E+01
7.8	36.5		7.8					31	ppm	72.6	1.1E+02
		100						1.85	ppm	74.9	6.5E+00
	0.9		9	7.6	23.6		49.7	21	ppm	79.0	7.0E+01
						50.7		3.57	ppm	79.9	1.2E+01
					0.35		2.25			83.8	
93	94	95	96	97	98	99	100				
								2.3	ppm	85.5	7.1E+00
								7.25	ppm	87.6	2.2E+01
								1.57	ppm	88.9	4.7E+00
	17.4		2.8					3.82	ppm	91.2	1.1E+01
100								240	ppb	92.9	6.8E−01
	9.25	15.9	16.7	9.55	24.1		9.63	900	ppb	95.9	2.5E+00
			5.52		1.88	12.7	12.6	710	ppb	101.1	1.9E+00
113	114	115	116	117	118	119	120				
								130	ppb	102.9	3.3E−01
								550	ppb	106.4	1.4E+00
								200	ppb	107.9	4.9E−01
12.2	28.7		7.49					710	ppb	112.4	1.7E+00
4.3		95.7						80	ppb	114.8	1.8E−01
	0.65	0.36	14.5	7.68	24.2	8.58	32.6	1650	ppb	118.7	3.7E+00
							0.1	140	ppb	121.8	3.0E−01
								2330	ppb	127.6	4.8E+00
133	134	135	136	137	138	139	140				
								450	ppb	126.9	9.4E−01
	10.4		8.9							131.3	
100								190	ppb	132.9	3.8E−01
	2.42	6.59	7.85	11.2	71.7			2410	ppb	137.3	4.6E+00
					0.09	99.9		237	ppb	138.9	4.5E−01
			0.19		0.25		88.8	613	ppb	140.1	1.2E+00

Continued...

Table Box 1.1 **A Table of Naturally Existing Isotopes (the Number Is the Percentage of Isotopic Abundance), Elemental Concentration in the CI Chondrite (CI Conc.), Atomic Weight, and Each Elemental Abundance Compared to Si × 10^6 (X/Si × 10^6). The elemental abundances of the CI chondrite are from McDonough and Sun (1995) continued**

Atomic Number	Element	1	2	3	4	5	6	7	8	9	10	11	12
		141	142	143	144	145	146	147	148	149	150	151	152
58	Ce		11.1										
59	Pr	100											
60	Nd		27.1	12.2	23.8	8.3	17.2		5.76		5.64		
62	Sm				3.1			15	11.3	13.8	7.4		26.7
63	Eu											47.8	
64	Gd												0.2
65	Tb												
66	Dy												
		161	162	163	164	165	166	167	168	169	170	171	172
66	Dy	18.9	25.5	24.9	28.2								
67	Ho					100							
68	Er		0.14		1.61		33.6	23	26.8		14.9		
69	Tm									100			
70	Yb								0.13		3.05	14.3	21.9
71	Lu												
72	Hf												
73	Ta												
74	W												
		181	182	183	184	185	186	187	188	189	190	191	192
73	Ta	99.9											
74	W		26.3	14.3	30.7		28.6						
75	Re					37.4		62.6					
76	Os				0.02		1.58	1.6	13.3	16.1	26.4		41
77	Ir											37.3	
78	Pt										0.01		0.79
79	Au												
80	Hg												
		201	202	203	204	205	206	207	208	209			232
80	Hg	13.2	29.8		6.85								
81	Tl			29.5		70.5							
82	Pb				1.4		24.1	22.1	52.4				
83	Bi									100			
90	Th												100
92	U												

Table Box 1.1 A Table of Naturally Existing Isotopes (the Number Is the Percentage of Isotopic Abundance), Elemental Concentration in the CI Chondrite (CI Conc.), Atomic Weight, and Each Elemental Abundance Compared to Si × 10⁶ (X/Si × 10⁶). The elemental abundances of the CI chondrite are from McDonough and Sun (1995) continued

13	14	15	16	17	18	19	20	CI Conc.	Unit	Atomic Weight	X/Si × 1e6
153	154	155	156	157	158	159	160				
								92.8	ppb	140.9	1.7E−01
								457	ppb	144.2	8.4E−01
	22.7							148	ppb	150.4	2.6E−01
52.2								56.3	ppb	152.0	9.8E−02
	2.18	14.8	20.5	15.7	24.8		21.9	199	ppb	157.3	3.3E−01
						100		36.1	ppb	158.9	6.0E−02
			0.06		0.1		2.34	246	ppb	162.5	4.0E−01
173	174	175	176	177	178	179	180				
								54.6	ppb	164.9	8.7E−02
								160	ppb	167.3	2.5E−01
								24.7	ppb	168.9	3.9E−02
16.1	31.8		12.7					161	ppb	173.0	2.5E−01
		97.4	2.59					24.6	ppb	175.0	3.7E−02
	0.16		5.21	18.6	27.3	13.6	35.1	103	ppb	178.5	1.5E−01
							0.01	13.6	ppb	180.9	2.0E−02
							0.13	93	ppb	183.8	1.3E−01
193	194	195	196	197	198	199	200				
								40	ppb	186.2	5.7E−02
								490	ppb	190.2	6.8E−01
62.7								455	ppb	192.2	6.2E−01
	32.9	33.8	25.3		7.2			1010	ppb	195.1	1.4E+00
				100				140	ppb	197.0	1.9E−01
			0.14		10	16.8	23.1	300	ppb	200.6	3.9E−01
233	234	235	236	237	238						
								140	ppb	204.4	1.8E−01
								2470	ppb	207.2	3.1E+00
								110	ppb	209.0	1.4E−01
								29	ppb	232.0	3.3E−02
	0.01	0.72			99.3			7.4	ppb	238.0	8.2E−03

BOX 1.2 THE ELEMENTAL ABUNDANCE OF EARTH

The elemental abundance of Earth's refractory elements is assumed to be the same as in carbonaceous chondrite (CI; see Box 1.7). This is supported by the similarity of the composition of the carbonaceous chondrite to that of the sun's photosphere. The elemental abundances of the CI chondrite are shown in Table Box 1.1. The semivolatile and volatile elements of Earth, such as C, N, O, S, halogens, etc., are difficult to estimate. This is because some are in the core, and some were lost or evaporated during the formation of Earth and the Moon. McDonough and Sun (1995) estimated the elemental abundance of Earth is 2.75 times that of CI chondrite for highly refractory elements.

Elemental abundance of CI chondrite compared to Si is shown in Fig. Box 1.3. The relative low abundance of Li, Be, and B is clearly observed and discussed in Box 1.4. The even atomic-number elements are more abundant than the odd atomic-number elements, which are clearly observed as the zig-zag pattern in heavy elements in Fig. Box 1.3. This is called as the "Oddo–Harkins" rule and is caused by the relative stability of the nuclei of even atomic-number elements compared to those of odd atomic-number elements of both sides.

BOX 1.3 SI PREFIX

When the SI unit is too large or small, SI prefixes such as kilo (k), mega (M), giga (G), and tera (T) are added to the SI unit. These prefixes are commonly used to express amounts of memories in computers and hard disks. Empirical expressions are used mainly when concentrations of elements or chemical compounds are indicated. The use of empirical expressions such as ppm, ppb, and so on is not recommended.

10^{18}	exa	E	
10^{15}	peta	P	
10^{12}	tera	T	
10^{9}	giga	G	
10^{6}	mega	M	
10^{3}	kilo	k	
10^{-3}	milli	m	Empirical expression
10^{-6}	micro	μ	$\mu g\,g^{-1} = ppm$
10^{-9}	nano	n	$ng\,g^{-1} = ppb$
10^{-12}	pico	p	$pg\,g^{-1} = ppt$
10^{-15}	femto	f	$fg\,g^{-1} = ppq$
10^{-18}	atto	a	

In this book, Gyr (1 billion years), Myr (1 million years), Ga (1 billion years ago), Ma (1 million years ago), km (kilometer), Mm (megameter), and Gm (gigameter) are often used.

The number of protons defines the characteristics of the element. In the periodic table (see Box 1.4), elements are plotted according to their atomic number horizontally and periodically, but the similar characteristics of elements appear vertically.

BOX 1.4 V.M. GOLDSCHMIDT (1888–1947), FATHER OF GEOCHEMISTRY

Modern geochemistry was started by V.M. Goldschmidt, who is called the father of geochemistry. He categorized elements into lithophile, chalcophile, siderophile, and atmophile elements (see Fig. Box 1.2).

The lithophile elements reside in silicate parts of Earth such as the mantle and the continental crust. The chalcophile elements are preferentially combined with sulfur. The siderophile elements prefer metallic iron and therefore reside in the core. The atmophile elements are volatile elements or noble gases that reside in the atmosphere. This classic classification is empirical and is still used in modern geochemistry.

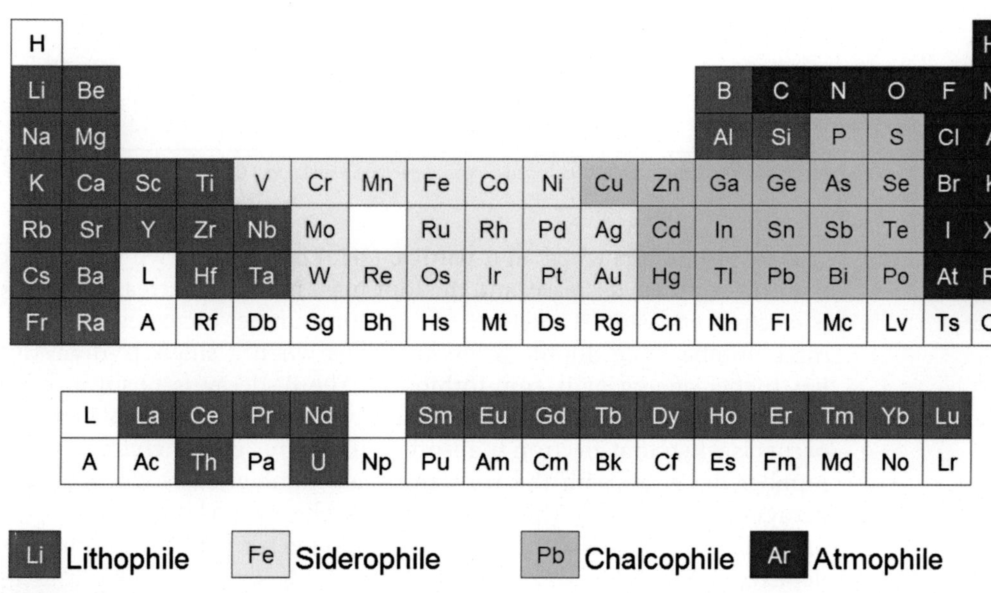

FIGURE BOX 1.2

Four geochemical classifications of elements by Goldschmidt: lithophile, siderophile, chalcophile and atmophile.
In June 2016, International Union of Pure and Applied Chemistry (IUPAC) announced the names and symbols of four elements. Elements 113, 115, 117, and 118 are nihonium (Nh), moscovium (Mc), tennessine (Ts), and oganesson (Og), respectively. Japan's RIKEN research institution had the honor of selecting the name and symbol of element 113. This is the first time that Asian researchers have named an element.
Nh is after "Nihon", which is "Japan" in Japanese. European-American collaborations involving Russia's Joint Institute for Nuclear Research and the U.S. Lawrence Livermore and Oak Ridge National Laboratories had rights to name the other three elements (115, 117, and 118). Mc and Ts are after the geographical regions of Moscow and Tennessee, where Joint Institute for Nuclear Research, Dubna (Russia), Oak Ridge National Laboratory (USA), Vanderbilt University (USA), and Lawrence Livermore National Laboratory (USA) are located.
Og honors Professor Yuri Oganessian for his pioneering contributions.

1.2 TYPES OF RADIOACTIVE DECAYS

Unstable isotopes are called radioactive isotopes, and they break up into more stable isotopes. This is called radioactive decay (see Table 1.1 for radioactive isotopes used in cosmochemistry).There are six main radioactive decay types: α-decay, β^--decay, double β^--decay, β^+-decay, electron capture (EC), and spontaneous fission (SF).

Table 1.1 **Radioactive Isotopes and Half-Lives Used in Cosmochemistry**

Radioactive Isotope	Decay Scheme	Daughter Isotope	Half-Life Y
^{40}K	EC	^{40}Ar	1.19×10^{10}
	β^-	^{40}Ca	1.40×10^{9}
^{87}Rb	β^-	^{87}Sr	4.88×10^{10}
^{138}La	EC	^{138}Ba	1.56×10^{11}
	β^-	^{138}Ce	3.03×10^{11}
^{147}Sm	α	^{143}Nd	1.06×10^{11}
^{176}Lu	β^-	^{176}Hf	3.71×10^{10}
^{187}Re	β^-	^{187}Os	4.16×10^{10}
^{190}Pt	β^-	^{190}Os	4.69×10^{11}
^{232}Th	Decay chain	^{208}Pb	1.40×10^{10}
^{235}U	Decay chain	^{207}Pb	7.04×10^{8}
^{238}U	Decay chain	^{206}Pb	4.47×10^{9}

Half-lives are from Dickin AP. Radiogenic isotope geology. Cambridge: Cambridge University Press; 2005.

The α-decay occurs when an α-particle (^{4}He) is emitted. Two proton and two neutron numbers decrease, and the four mass numbers decrease. The β^--decay is electron emission. Therefore, one proton number increases with the same mass number. The double β^--decay occurs when a single β^--decay nucleus has higher energy or is spin-forbidden. The β^+-decay is positron emission. Therefore, one proton number decreases with the constant mass number. EC occurs is when that electron is absorbed in the nucleus. Therefore, one proton number decreases with the constant mass number. SF is when the nucleus decays into two nuclei. This occurs only in a heavy nucleus such as ^{235}U, ^{238}U, or ^{240}Pu. The changes in the N (neutron number)–Z (proton number) plot are depicted in Fig. 1.1.

1.3 THREE FORCES SUSTAIN A STAR IN A FINE BALANCE

The three forces sustaining a star are depicted in Fig. 1.2. The three forces are: (1) energy produced by nuclear reaction, which makes the star swell; (2) gravity produced by the huge body, which makes the star contract; and (3) repulsion produced by electron degeneration, which prohibits a proton and an electron from being a neutron at the center of the star. It is easy to imagine (1) and (2). However, (3) may be difficult to imagine. If we suppose that there is no (3), the core of the star becomes a neutron star at once and the star ceases to glare.

1.4 ELEMENT SYNTHESIS IN STARS

The first stars appeared after 0.55 Gyr of the Big Bang; the heavier elements were then formed in stars. The Big Bang basically produced only ^{1}H and ^{4}He; ^{1}H

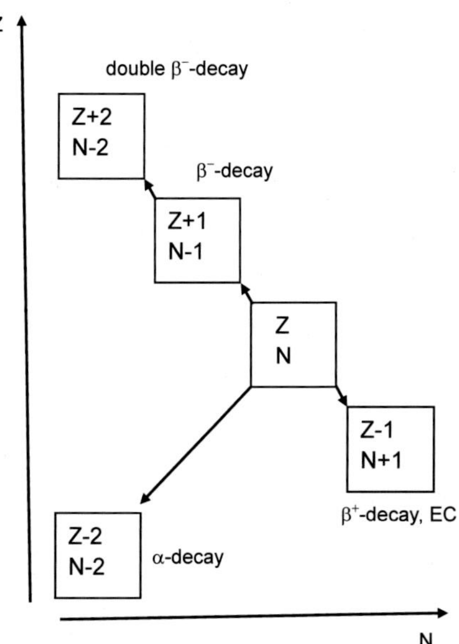

FIGURE 1.1
The radioactive decays. The changes of N (neutron number; the horizontal scale) and Z (proton number; the vertical scale) by the radioactive decays.

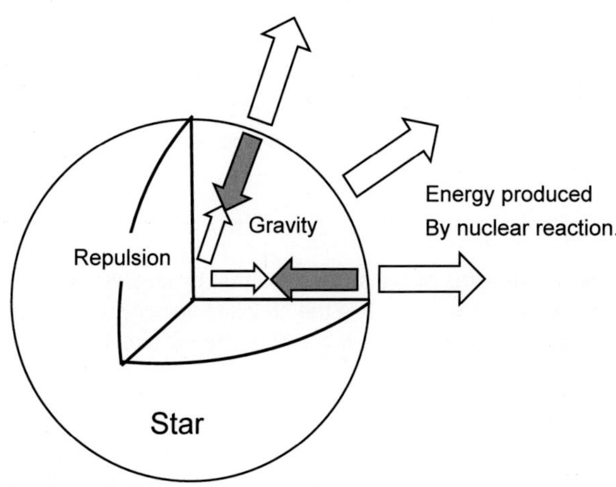

FIGURE 1.2
Three forces sustaining a star.

gas gathered and formed the star (the first-generation star; see Box 1.5). This star died when forming the elements as described in this section. Then ^{1}H gas including small amounts of the heavier elements formed the star again, and the star died when forming the elements. This process was repeated again and again until the present universe, our galaxy, and our solar system was formed.

BOX 1.5 THE FIRST STAR OR GALAXY

The first star or galaxy is made of the "first" material just after the Big Bang. It is the most distant galaxy with the highest–redshifted light (the more distant, the more redshifted; see Box 1.6). Their ultraviolet (UV) radiation, named Lyman-α, should be observable. The Lyman-α is the strongest UV spectral line emitted from the highly redshifted system. However, it is easily attenuated by absorption of interstellar dusts and neutral hydrogen atoms. Astronomers are looking for such luminous Lyman-α emitters.

BOX 1.6 DOPPLER EFFECT AND REDSHIFT

The siren of an ambulance car becomes higher than the original tone when the car is approaching and lower when the car is getting farther. This is called the Doppler effect. The Doppler effect is given by the following equation:

$$\lambda' = \lambda_0 \times (V - v_s) / (V - v_r) \qquad \text{(B1.5.1)}$$

where λ_0 and λ' are wavelengths at the source and the receiver, respectively; V is the velocity of the sound in medium (air); and v_s and v_r are the velocities of the source and the receiver, respectively (the direction from the receiver to the source is positive). Thus, when the source overtakes the fixed receiver ($v_r=0$), $v_s>0$ and $\lambda'<\lambda_0$ before the overtake. After the overtake, $v_s<0$ and $\lambda'>\lambda_0$, resulting in stretching the wavelength or lowering the frequency, because $V=\lambda\times\nu$ (V is the velocity; λ and ν are the wavelength and frequency, respectively).

The Doppler effect also occurs for light. However, the Doppler effect of starlight is different from that of sound, because the velocity of the light is constant and the time dilation occurs in the special relativity theory. The Doppler effect of light is expressed as:

$$\lambda' = \lambda_0 \times [1 - (v/c)\cos\theta] / \sqrt{\left[1 - (v/c)^2\right]} = \lambda_0 \times (1 + z) \qquad \text{(B1.5.2)}$$

where λ_0 and λ' are wavelengths of the source and that by an observer, respectively; c is the velocity of the light in vacuum; v is the absolute velocity of the light source; θ is the direction of the light source (when the light is coming directly to the observer, $\theta=0$); and z is redshift (see Eq. B1.5.3).

It is interesting that even if the light source has no component part of its velocity to the observer ($\theta=90$ degrees), the light is affected by the Doppler effect. As is clear from Eq. (B1.5.2), when the light source is approaching the observer, the wavelength of the light becomes shorter (blue shift), and when the light source is moving away, the wavelength becomes longer (redshift).

Generally, the wavelength of the absorption lines by hydrogen in the emitted light by a star becomes longer as the light-emitting star moves farther from us. This is based on the fact that the universe is still swelling uniformly. Actually, as the star exists farther from us, the escaping velocity of the star from us becomes faster. When the escaping velocity is faster, the wavelength of light becomes longer (more reddish) by the Doppler effect.

BOX 1.6 DOPPLER EFFECT AND REDSHIFT continued

The redshift is characterized by z, which is:

$$z = (\lambda_{obs}/\lambda_{emit} - 1) \quad \text{(B1.5.3)}$$

where λ_{obs} and λ_{emit} are wavelengths of the observed and emitted lights, respectively. The cosmic microwave background has a redshift of z=1089, corresponding to an age of 379,000 years after the Big Bang. The yet-observed oldest first light comes from Population III stars, which are very metal-poor stars with z=7–6. To find such old stars or galaxies in deep space is one of the big themes of astroscience.

BOX 1.7 CLASSIFICATION OF METEORITES

Meteorites are characterized as either meteorite falls or meteorite finds. A meteorite fall is a meteorite that was observed falling from the sky. The Chelyabinsk meteorite, which fell in Russia in 2013, and the Allende meteorite, which fell in Mexico in 1969, are typical meteorite falls. A meteorite find is not the same as a meteorite fall. Initially it was just a stone, which was later appraised by specialists as a meteorite. There are meteorite hunters who collect meteorites from deserts in Africa or Australia, as well as all over the world. Many meteorites are found in glaciers in Antarctica. A fusion crust is often proof of a meteorite. The fusion crust that surrounds the meteorite is melted due to the compaction of air at supersonic velocity during falling (not the friction with the air!).

Meteorites are finally characterized by chemical composition and textures observed by an optical microscope. As the appraisement of the meteorite is not easy, there are lots of fakes for sale on the Internet.

Meteorites are divided into stony meteorites, stony-iron meteorites, and iron meteorites. The stony meteorite is further divided into chondrite and achondrite (see Table Box 1.2). Chondrite is composed of chondrules and matrix. Chondrules are small silicate spherules (0.1 mm–a few centimeters), once melted at high temperature (>1800K) and quenched (~1000 K h^{-1}; Hewins, 1997). The matrix is the low-temperature component. Chondrites are also divided into carbonaceous chondrites (CI, CO, CV, CK, CR, etc.), which contain carbon compounds, ordinary chondrites (L, LL, and H), and enstatite chondrites (EL, EH). Chondrites are numbered from 1 to 7 based on heating (metamorphic) grades. Therefore, chondrites are called, such as CV3, LL6, etc. The composition of carbonaceous chondrites includes 5–20% water. Carbonaceous chondrites are the least affected by thermal events and are therefore considered to be the most primitive material in meteorites.

The achondrite has evolved-composition like aubrite (origin unknown), eucrite (basaltic), diogenite (peridotitic), and howardite (mixture of eucrites and diogenite). The howardite, eucrites, and diogenites are called HED (the combination of the first letters of howardite, eucrite, and diogenite) meteorites. These meteorites are believed to come from 4 Vesta, an evolved asteroid with a core and mantle (see Section 11.3). The other meteorite categories (iron and stony-iron) are also considered to come from the core and core-mantle boundary in evolved asteroids.

There are some special meteorites that are considered to have come from the Moon (e.g., ALHA A81005), from Mars (SNC meteorites; shergottites, nakhlites, and chassignites), and even from Mercury (e.g., NWA 7325).

Table Box 1.2 Characterization of Meteorites

Stony Meteorite			
	Chondrite	Carbonaceous chondrite (*1~*3)	CI, CO, CV, CK, CR
		Ordinary chondrite (*3~*7)	L, LL, H
		Enstatite chondrite (*3~*6)	EL, EH
	Achondrite	HED meteorite	Howardite, Eucrite, Diogenite
		SNC meteorite	Shergottite, Nakhlite, Chassignite
		Angrite, Aubrite, Brachinite, Acapulcoite, Ureilite	
Stony Iron Meteorite			
	Mesosiderite, Pallasite		
Iron Meteorite			
	IAB, IC IIAB, IIC, IID, IIE, IIG IIIAB, IIICD, IIIE, IIIF IVA, IVB		

At the beginning of a star, ^3He and ^1H burn into ^4He. There are two paths for this reaction, the proton–proton chain reaction and the CNO cycle. In the proton–proton reaction,

$$^1H + {}^1H \rightarrow {}^2H + e^+ + \nu + \gamma$$

occurs, and the reaction

$$e^+ + e^- \rightarrow 1.0\,\text{MeV}$$

occurs and energy is produced. Then the reaction

$$^2H\ (^1H, \gamma)\ ^3He$$

occurs. For this type of presentation of the nuclear reaction, see Box 1.8. γ is a positive energetic photon. Thus;

1. $^3He + {}^3He \rightarrow {}^4He + 2{}^1H + \gamma$
2. $^3He\ (^4He, \gamma)\ ^7Be$
   $^7Be\ (e^-, \nu)\ ^7Li$
   $^7Li\ (^1H, {}^4He)\ ^4He$
3. $^3He\ (^4He, \gamma)\ ^7Be$
   $^7Be\ (^1H, \gamma)\ ^8B$
   $^8B \rightarrow {}^8Be + e^+ + \nu + \gamma$
   $^8Be \rightarrow 2\ {}^4He + \gamma$

BOX 1.8 DESCRIPTION OF NUCLEAR REACTION

When we hit a ^{2}H atom with a proton and it becomes a ^{3}He atom with emission of the gamma ray, we describe it as:

$^{2}H\ (p, \gamma)\ ^{3}He$

This description is very useful when successive reactions occur, for example, $^{2}H \rightarrow {}^{3}He + p + \gamma \rightarrow {}^{3}He + {}^{4}He \rightarrow {}^{7}Be + \gamma$ and $^{7}Be + e^{-} \rightarrow {}^{7}Li + \nu$ (ν is neutrino), which can be described as:

$^{2}H\ (p, \gamma)\ ^{3}He\ (^{4}He, \gamma)\ ^{7}Be\ (e^{-}, \nu)\ ^{7}Li$.

Another reaction path is the CNO cycle,

$$^{12}C\ (p, \gamma)\ ^{13}N$$
$$^{13}N \rightarrow {}^{13}C + e^{+} + \nu + \gamma$$
$$^{13}C\ (p, \gamma)\ ^{14}N$$
$$^{14}N\ (p, \gamma)\ ^{15}O$$
$$^{15}O \rightarrow {}^{15}N + e^{+} + \nu + \gamma$$
$$^{15}N + p \rightarrow {}^{12}C + {}^{4}He + \gamma$$
$$\text{Total: } 4p \rightarrow {}^{4}He + 2e^{+} + 2\nu + 25.10\,\text{MeV}$$

However, these reactions cannot make heavier elements. In the core of the old star, ^{4}He becomes rich, and the temperature is >100 million K. In this condition, the triple-α process can occur:

$$^{4}He + {}^{4}He \rightarrow {}^{8}Be + \gamma$$
$$^{8}Be + {}^{4}He \rightarrow {}^{12}C + \gamma$$

In total, 7.275 MeV is produced. ^{8}Be is very unstable and returns to 2 ^{4}He in 2.6×10^{-16} s. However, in the condition where He burning is occurring, this reaction becomes an equilibrium reaction. In addition, the energy of the second reaction is almost the same with the exited state of ^{12}C. Therefore, these two rare reactions occur. At the same time, oxygen atoms are formed:

$$^{12}C + {}^{4}He \rightarrow {}^{16}O + \gamma$$

However, the next reaction forming ^{20}Ne does not occur because of the restriction from the nuclear spin. In 1957, Burbidge, Burbidge, Fowler, and Hoyle theorized that heavier elements are made by C-burning, Ne-burning, O-burning, Si-burning, s-process, r-process, and p-process.

In the C-burning, the following reactions occur, with the first two being main reactions. (E is energy):

$$^{12}C + {}^{12}C \rightarrow {}^{20}Ne + {}^{4}He + E$$
$$^{12}C + {}^{12}C \rightarrow {}^{23}Na + {}^{1}H + E$$
$$^{12}C + {}^{12}C \rightarrow {}^{23}Mg + n - E$$

$$^{12}C + ^{12}C \rightarrow ^{24}Mg + E$$
$$^{12}C + ^{12}C \rightarrow ^{16}O + 2\ ^{4}He - E$$

After the C-burning and C is consumed, the Ne-burning occurs. Higher pressure and temperature are required for the Ne-burning. In such high temperature, photodisintegration cannot be ignored (the first reaction), and part of the Ne nuclei decays. In a few years Ne is consumed in the core of the star, and the core becomes unstable, making O and Mg.

$$^{20}Ne + \gamma \rightarrow ^{16}O + ^{4}He$$
$$^{20}Ne + ^{4}He \rightarrow ^{24}Mg + E$$

The O-burning process starts after Ne is consumed.

$$^{16}O + ^{16}O \rightarrow ^{28}Si + ^{4}He + E$$
$$^{16}O + ^{16}O \rightarrow ^{31}P + ^{1}H + E$$
$$^{16}O + ^{16}O \rightarrow ^{31}S + n + E$$
$$^{16}O + ^{16}O \rightarrow ^{30}Si + 2\ ^{1}H + E$$
$$^{16}O + ^{16}O \rightarrow ^{30}P + ^{2}H - E$$

In this process, the core of the star becomes Si-rich; however, the temperature is not high enough to ignite Si. At this time, O-burning shell, Ne-shell, C-shell, He-shell, and H-shell exist from outside to inside like onion shells.

After all the O is consumed, Si-burning starts. The Si-burning process occurs only in a massive giant star with more than 8–11 the mass of the solar mass. This process is a 2-week-long process, and then the star becomes a type II supernova. The star with fewer than three solar masses finishes just after all H becomes He. The star with 3–8 solar masses dies after He-burning leaving a C core. The star with 8–11 solar masses can burn C, and finally Si.

In the Si-burning processes, new elements are produced in sequence. Within a day, ^{56}Ni is made and the process ends.

$$^{28}Si + ^{4}He \rightarrow ^{32}S$$
$$^{32}S + ^{4}He \rightarrow ^{36}Ar$$
$$^{36}Ar + ^{4}He \rightarrow ^{40}Ca$$
$$^{40}Ca + ^{4}He \rightarrow ^{44}Ti$$
$$^{44}Ti + ^{4}He \rightarrow ^{48}Cr$$
$$^{48}Cr + ^{4}He \rightarrow ^{52}Fe$$
$$^{52}Fe + ^{4}He \rightarrow ^{56}Ni$$

This reaction stops and does not go to ^{60}Zn because ^{56}Ni is the most stable nucleus. ^{56}Ni decays to ^{56}Co by β^+-decay, and ^{56}Co decays to ^{56}Fe by β^+-decay. Then gravity suddenly collapses the core of the star because the following reaction occurs:

$$^{56}Fe \rightarrow 13\ ^{4}He + 4\ n - E$$

This reaction is endothermic; therefore, the core collapses at once. Then, due to the shock wave of the collapse, the star explodes. This is the type II supernova (see Fig. 1.3). In the explosion, when the repulsion of electrons is overwhelmed by the shock of explosion, protons in ^{4}He become neutrons, and the core

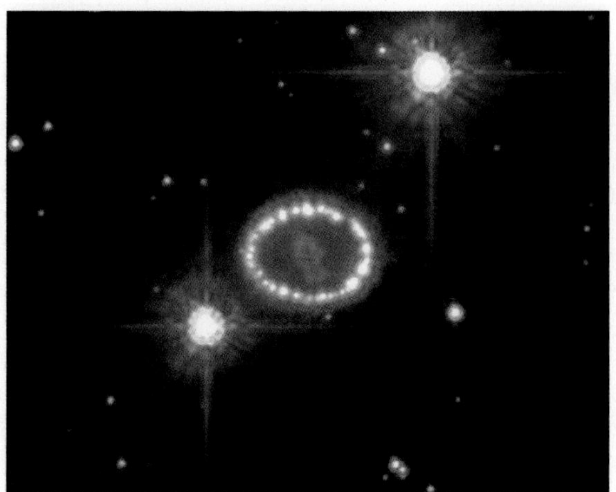

FIGURE 1.3
A remnant of SN 1987A, a type II supernova in the large Magellanic cloud. The exploded gas is seen as small rings. Two faint large rings including the bright ring (the shock waves) can be seen.
Image credit: NASA, ESA, P. Challis, and R. Kirshner (Harvard-Smithsonian Center for Astrophysics), http://www.nasa.gov/sites/default/files/images/170938main_image_feature_773_ys_full.jpg.

becomes neutron (a neutron star). When the mass is large, the core becomes not a neutron star but a black hole.

The story is easy to explain, but theoretical treatments are beyond the ability of the author. For those who are interested in the theories, the author suggests two classic books by Chandrasekhar (1967) and Clayton (1983).

1.5 ELEMENT SYNTHESES IN OTHER PROCESSES

r-process: As discussed in the previous section, elements lighter than Fe are formed. In the type II supernova, r-process occurs. "r" means "rapid," and it is believed that when the type II supernova occurs, a very neutron-rich condition is achieved before the element decays by β^--decay. Thus neutron-rich elements are formed at once. Then unstable neutron-rich isotopes decay to stable isotope through β^--decay.

p-process: In the type II supernova, very strong lights (photons) are also generated, and they break the nuclei. This photodisintegration process is called p-process. The idea of p-process was required to make proton-rich isotopes in the B^2FH theory, which cannot be generated by either r-process or s-process. In the early theory "p" meant "proton-rich," but now it means "photodisintegration." The p-isotopes are not very common.

s-process: The s-process is considered to form neutron-rich nuclei "slowly" in thousands of years and occurs in asymptotic giant branch stars (AGB stars; see Box 1.9) or stars with low metallicity. The seed is Fe, and neutrons are supplied by the following reactions:

$$^{13}C + {}^{4}He \rightarrow {}^{16}O + n$$
$$^{12}C + {}^{12}C \rightarrow {}^{23}Mg + n$$

BOX 1.9 HERTZSPRUNG–RUSSELL DIAGRAM (H–R DIAGRAM)

The H–R diagram is a classic plot of the effective surface temperature of stars versus the luminosity of stars, which is shown in Fig. Box 1.4. The effective surface temperature is color indexed from B to V or spectral classes. The luminosity is absolute bolometric magnitude, or L/L⊙ where ⊙ means the sun. Most stars are plotted along the main sequence. The AGB stars also appear in this diagram.

The neutron-rich isotopes of proton number Z are formed by the s-process (N + 1, N + 2, N + 3, ...) in Fig. 1.1, and finally a short half-life isotope appears, ends by β^--decay, and then goes up to the next element, Z + 1. This process goes up to ^{209}Bi.

x-process: In the B^{2}FH theory, Li, Be, and B were not synthesized. When the elements are plotted by the elemental abundances compared to Si according to the atomic number, Li, Be, and B show large depletion in the plot (see Fig. Box 1.3). Thus, they are named and formed by x-processes. They are considered to be synthesized by the

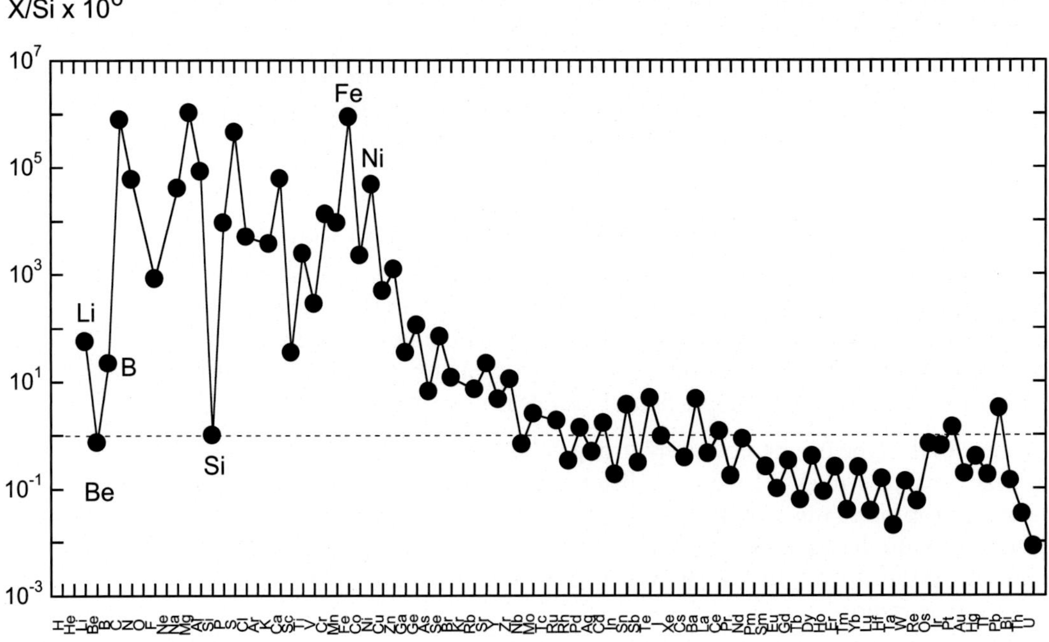

FIGURE BOX 1.3
Elemental abundances of CI chondrite compared to Si. Each elemental abundance in the CI chondrite is divided by the Si abundance times 10^6. The horizontal dotted line shows X/Si × 10^6 = 1.
The data are from McDonough WF, Sun S-S. The composition of the Earth. Chem Geol 1995;120:223–53, shown in Table 1.1.

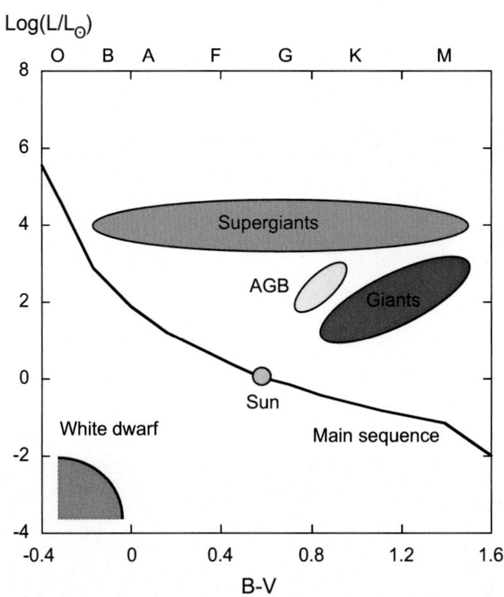

FIGURE BOX 1.4
The Hertzsprung-Russel diagram. The horizontal axis is the color index, B-V, or the spectral type from O to M (at the top). The vertical axis is the logarithms of luminosity against the Sun. AGB means the asymptotic giant branch.

spallation by high-energy galactic cosmic ray (Viola and Mathews, 1987). This is why the solar abundances of Li, Be, and B are very low. Recently, Tajitsu et al. (2015) proposed that ^{7}Li can be produced in a supernova (Hernanz, 2015).

1.6 TYPE IA SUPERNOVA

S. Chandrasekhar calculated the limit of the mass of a white dwarf (Chandrasekhar, 1931). He concluded that if the mass of the white dwarf becomes 1.26 masses of the sun (called the Chandrasekhar limit), the electron degeneration cannot hold the gravity, the white dwarf collapses, explodes as a supernova Type Ia, and could become a neutron star (Woosley et al., 2008).

In Section 1.4, it was said that the star with mass of fewer than three solar masses dies after H and He fuels are consumed. The gravity balances the repulsion of electron degeneration because of the small masses. The star becomes a white dwarf and gets cooler and cooler. There is a possibility—not low—that the white dwarf is twinned. In such a case, one star becomes a white dwarf, and the other star becomes a red giant (see Fig. 1.4). The gases of the red giant star with low gravity are attracted by the white dwarf with high gravity. The gases from the red giant gather on the white dwarf (see Fig. 1.4), and explosive nuclear fusion can occur on the white dwarf. This is considered to be a real identity of the AGB star. If the white dwarf continues to absorb gases from the companion giant star, the mass of the white dwarf can reach the Chandrasekhar limit. Then the white dwarf explodes as Type Ia supernova (see Fig. 1.5) (Box 1.10 and 1.11).

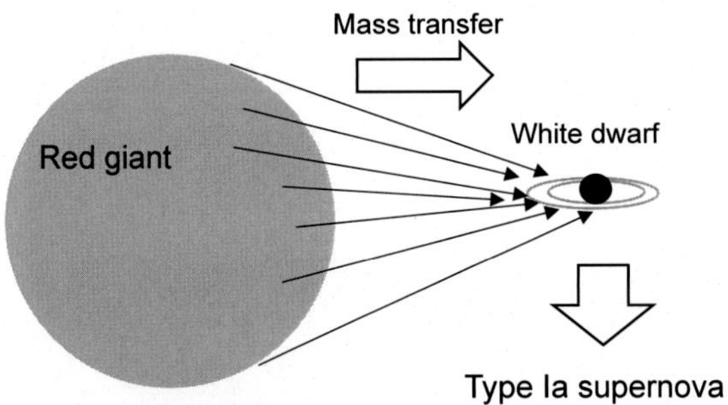

FIGURE 1.4
The fate of the twin stars.

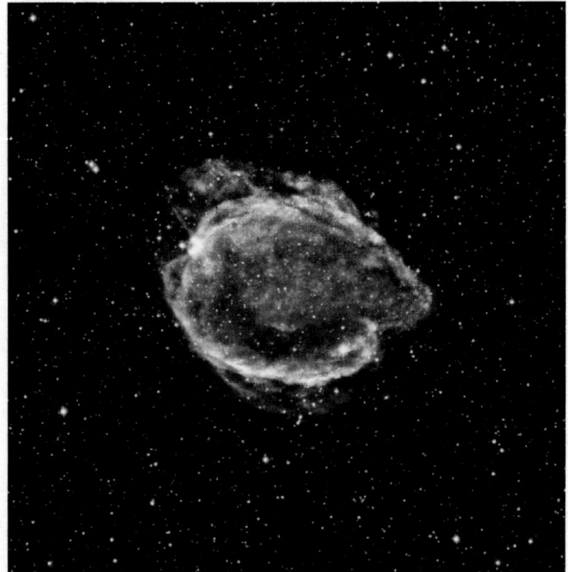

FIGURE 1.5
G229, a remnant of type Ia supernova (Post et al., 2014).
The image was taken by the Chandra X-ray Observatory (see Box 1.10). Image credit: NASA, https://www.nasa.gov/sites/default/files/thumbnails/image/g299.jpg.

BOX 1.10 THE CHANDRA X-RAY OBSERVATORY

The Chandra X-ray Observatory (CXO) was launched in 1999. CXO is 100 times more sensitive than any X-ray telescope on Earth's surface, because Earth's atmosphere absorbs most X-rays. CXO was named after S. Chandrasekhar. X-rays are produced when matter is heated to extreme heat in extreme magnetic fields and gravity, which are related to neutron stars and black holes, and in explosive forces, which are related to supernova. The image of CXO is shown in Fig. Box 1.5.

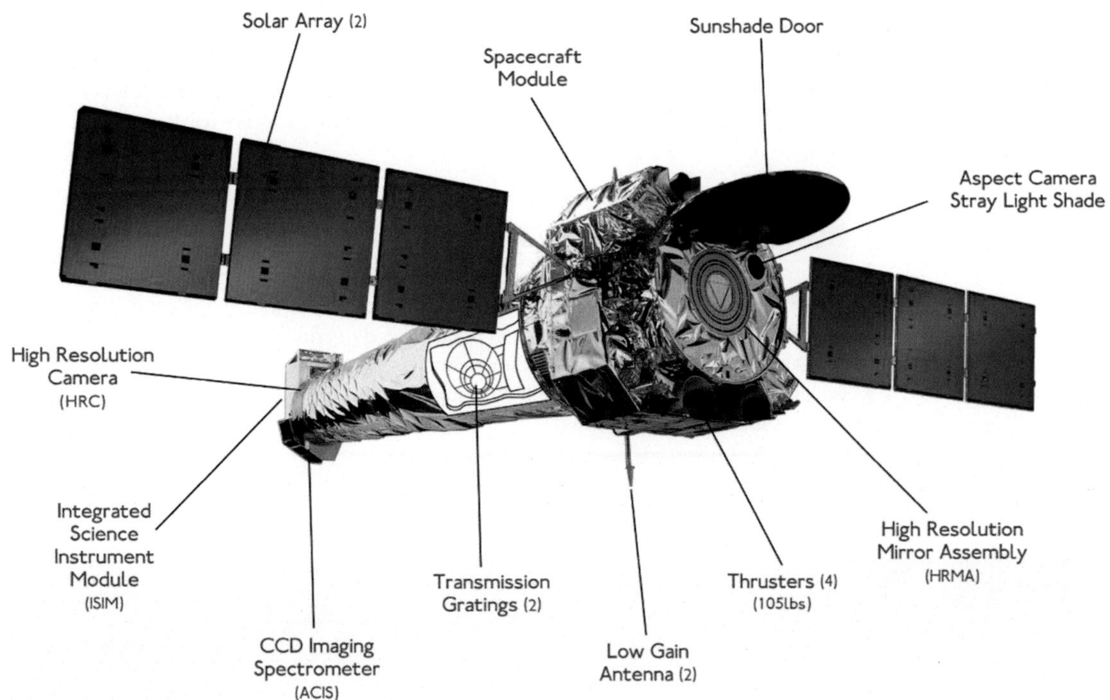

FIGURE BOX 1.5
The Chandra X-ray Observatory.
Image credit: NASA.
http://www.nasa.gov/sites/default/files/images/704250main_chandra-telescope_full.jpg.

BOX 1.11 THE SLOAN DIGITAL SKY SURVEY

The Sloan Digital Sky Survey (SDSS) is making the most detailed three-dimensional maps of the Universe, using deep multicolor images of one third of the sky, and spectra for more than 3 million astronomical objects. The Alfred P. Sloan Foundation has been supporting the SDSS. The Sloan Foundation has a 2.5-m telescope at Apache Point Observatory in New Mexico, USA. The southern hemisphere observation is the Irenee du Pont Telescope at Las Campagnas Observatory in Chile.

In its first 5 years of operation (SDSS-I/II; 2000–08), the SDSS carried out deep multicolor imaging over 8000 square degrees and measured spectra of more than 700,000 celestial objects. There are three measurements; legacy, supernova, and SEGUE (Sloan Extension for Galactic Understanding and Exploration)-1 surveys.

SDSS-III (2008–14) undertook a major upgrade of the venerable SDSS spectrographs and added new powerful instruments to execute interweaved sets of four surveys; APOGEE (Apache Point Observatory Galactic Evolution Experiment), BOSS (Baryon Oscillation Spectroscopic Survey), MARVELS (Multi-object APO Radial Velocity Exoplanet Large-area Survey), and SEGUE-2.

SDSS-IV: Current Surveys (2014–20): SDSS is now extending three main measurements: APOGEE-2, eBOSS (extended Baryon Oscillation Spectroscopic Survey), and MaNGA (Mapping Nearby Galaxies at APO).

1.7 THE OLDEST GALAXY EVER FOUND

The Hubble's Near Infrared Camera and Multi-Object Spectrometer (NICMOS) equipped on the Hubble Space Telescope (HST) and the Spitzer Space Telescope (SST; see Box 1.12) with the help of a natural "zoom lens" took detailed images

> ### BOX 1.12 THE SPITZER SPACE TELESCOPE
>
> The SST was launched in 2003 for the detection of infrared using an infrared array camera (IRAC), infrared spectrograph (IRS), and Multiband Imaging Photometer for Spitzer (MIPS). As the on-board liquid helium used to cool the infrared detector in 2009 was exhausted, the planned mission was finished. However, the two shortest-wavelength IRAC cameras are still available without liquid helium, so the SST could work as the Spitzer Warm Mission.

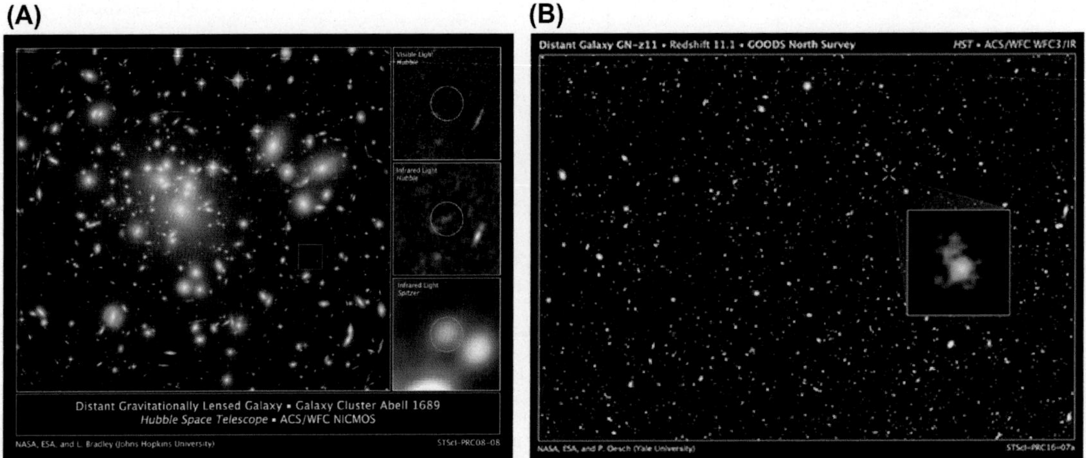

FIGURE 1.6
The oldest galaxy in the universe. (A) The left side large photograph: Galaxy cluster Abell 1689, which is a gravitationally lensed galaxy. The right three small photographs, enlarged of the white square area of the left photo (from top to bottom): a visible light image by the Hubble Space Telescope (HST) of the white square; an infrared image by HST; an infrared image by the Spitzer Telescope (see Box 1.12). (B) The oldest galaxy, GN-z11.
(A) Image credit: NASA, ESA and L. Bradley (Johns Hopkins University) http://www.nasa.gov/images/content/211728main_young_bright_lg.jpg. (B) Image credit: NASA, ESA, and P. Oesch (Yale University).

of an infant galaxy, Abell-zD1 (see Fig. 1.6A). The massive cluster of the galaxy named A1689 worked like the zoom lens (Bradley et al., 2008). The redshift of the galaxy was 7.5 (Watson et al., 2015), indicating that the galaxy was formed just after 680–850 Myr (Michalowski, 2015; 13.1–13.0 Gyr from the present). The galaxy contained cosmic dust like the Milky Way, indicating the galaxy was evolved even when it was formed in the early time of the Universe (Watson et al., 2015).

In 2016, a combination of the HST and SST found a surprisingly bright, infant galaxy, named GN-z11, which was 13.4 Gyr in the past, just 400 Myr after the Big Bang (see Fig. 1.6B). Astronomers measured the redshift of the galaxy to be 11.1. It was surprising that a galaxy so massive existed only 200–300 Myr after the very first stars started to form. It meant that stars were produced at a rate huge enough to form a galaxy (Oesch et al., 2016).

1.8 THE OLDEST STAR FOUND SO FAR

The supernova of a star with element abundance ratios of extremely low metallicity (low abundance of elements heavier than helium) shows, for example, very little iron. It could be the first star as discussed in Section 1.4. The optical spectrum of SMSS J031300.36-670839.3 shows no evidence of iron (Keller et al., 2014). The supernova's original mass was ~60 times heavier than that of the Sun. From chemical and physical investigation, a star that afterward became this supernova could be a first-generation star from ~13.6 Gyr ago.

1.9 ELEMENT SYNTHESES BY A SHORT-DURATION γ-RAY BURST

A short-duration γ-ray burst (GRB) lasting less than 2 s was observed (e.g., GRB 130603B), but it was one of the enigmas of the Universe. The origin of GRB was supposed to be a collision of two neutron stars that produces huge amounts of neutron-rich elements and isotopes and works like the r-process (Tanvir et al., 2013). The Universe is not static but dynamic, and even the neutron star, which seems to be the end of the star and materials, breaks and recycles to produce elements.

References

Bradley LD, Bouwens RJ, Ford HC, et al. Discovery of a very bright strongly lensed galaxy candidate at z approximate to 7.6. Astrophys J 2008;678:647–54.

Burbidge EM, Burbidge GR, Fowler WA, Hoyle F. Synthesis of the elements in stars. Rev Mod Phys 1957;29:547–650.

Chandrasekhar S. The maximum mass of ideal white dwarfs. Astrophys J 1931;74:81–2.

Chandrasekhar S. An introduction to the study of stellar structure. New York: Dover Publications, Inc.; 1967.

Clayton DD. Principles of stellar evolution and nucleosynthesis. New York: McGraw-Hill; 1983.

Dickin AP. Radiogenic isotope geology. Cambridge: Cambridge University Press; 2005.

Hernanz M. A lithium-rich stellar explosion. Nature 2015;518:307–8.

Hewins RH. Chondrules. Annu Rev Earth Planet Sci 1997;25:61–83.

Keller SC, Bessell MS, Frebel A, et al. A single low-energy, iron-poor supernova as the source of metals in the star SMSS J031300.36-670839.3. Nature 2014;506:463–6.

McDonough WF, Sun S-S. The composition of the Earth. Chem Geol 1995;120:223–53.

Michalowski MJ. Dust production 680–850 million years after the Big Bang. Astron Astrophys 2015;577. UNSP A80.

Oesch PA, Brammer G, van Dokkum PG, et al. A remarkably luminous galaxy at z = 11.1 measured with Hubble Space Telescope Grism Spectroscopy. 2016. arXiv:1603.00461.

Post S, Park S, Badenes C, Burrows DN, et al. Asymmetry in the observed metal-rich ejecta of galactic type Ia supernova remnant G299.2-2.9. Astrophys J Lett 2014;792(L20):1–6.

Sato K. Cosmological baryon number domain structure and the first order phase transition of a vacuum. Phys Lett B 1981;33:66–70.

Spergel D, Keating B. Cosmology: the oldest cosmic light. Nature 2015;518:170–1.

Tajitsu A, Sadakane K, Naito H, Arai A, Aoki W. Explosive lithium production in the classical nova V339 Del (Nova Delphini 2013). Nature 2015;518:381–4.

Tanvir NR, Levan AJ, Fruchter AS, Hjorth J, Hounsell RA, Wiersema K, et al. A 'kilonova' associated with the short-duration g-ray burst GRB130603B. Nature 2013;500:547–9.

Viola VE, Mathews GJ. The cosmic synthesis of lithium, beryllium and boron. Sci Am 1987;256:38–45.

Watson D, Christensen L, Knudsen KK, Richard J, Gallazzi A, Michalowski MJ. A dusty, normal galaxy in the epoch of reionization. Nature 2015;519:327–30.

Woosley S, Kasen D, Ma H, et al. Proc Sci 2008;047 (NIC X).

CHAPTER 2

Formation of the Proto-Earth in the Solar Nebula

Chapter Outline

2.1 EVOLUTION OF MOLECULAR CLOUDS TO THE SOLAR NEBULA

The evolution of an early star and its surrounding molecular cloud (nebula) are controlled by four components: (1) the centrifugal force; (2) mass ejection (mass and angular momentum loss as ejection jet); (3) mass accretion (mass

Origins of the Earth, Moon, and Life. http://dx.doi.org/10.1016/B978-0-12-812058-3.00002-8

gain from the disk surrounding the star); (4) a magnetic field (the mass ejection and the accretion follow the magnetic field). These four components are schematically depicted in Fig. 2.1 (Montmerle et al., 2006).

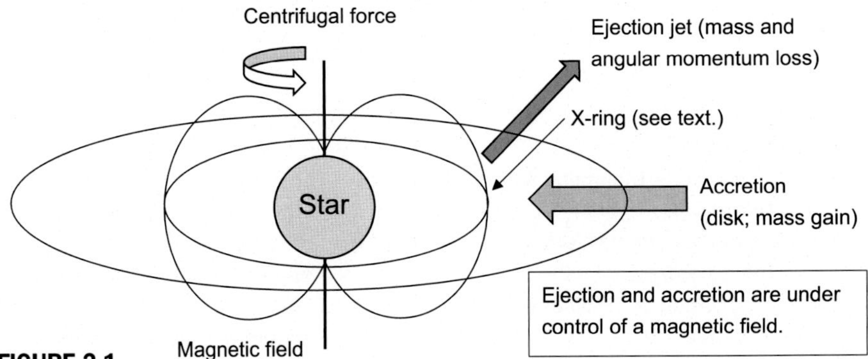

FIGURE 2.1
The four components for evolution of the early star and surrounding disk (each component is not in scale).

The star has two strong magnetic fields (B). One magnetic field extends vertically along the rotation axis of the star. Another magnetic field stretches horizontally. In the accretion-ejection model, these two magnetic fields meet along the X-ring (or at the X-point if plotted two dimensionally). The accreting materials from the horizontal direction are ejected at the X-ring.

Our solar system is considered to have been formed from cosmic gas clouds as shown in Fig. 2.2. The Fig. 2.2 was obtained by the Hubble Space Telescope

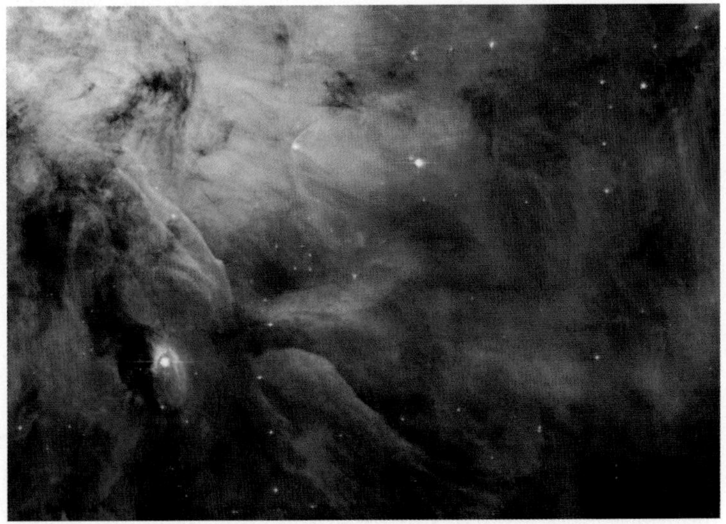

FIGURE 2.2
The close-up of cosmic clouds and stellar winds at the Orion Nebula.
Image credit: NASA, European Space Agency, and the Hubble Heritage Team http://www.nasa.gov/sites/default/files/images/724329main_hubble_feature_full_full.jpg.

(HST) (see Box 2.1). These gas clouds are made of hydrogen, helium, and heavy elements that are the remnants of a supernova (see Chapter 1). The shockwave from a supernova produced inhomogeneity of the gas cloud near the location where our Sun would appear. The gas cloud about 3 light years began to gather and form a dense gas disk ($10^3 \sim 10^4$ AU; AU is the present distance between the Sun and Earth). At the center of the gas disk was a huge gas ball (protostar) mainly composed of hydrogen.

The various stellar phases with timescales and sizes are shown in Fig. 2.3 after Feigelson and Montmerle (1999). These models were established due to the advancement of observation techniques using infrared and X-ray telescopes. The initial stage, which is previously explained, is called the infalling protostar stage (see the top figure of Fig. 2.3). The accretion and ejection of the molecular gases occur. Although the gas ball is very large, it cannot be seen from outside because of the surrounding clouds. This stage lasts fewer than 10^4 years.

The next stage is the evolved protostar stage. The accretion and ejection of the gases are still vigorously occurring. However, the size of the gas ball becomes 10 times smaller. This stage lasts for 0.1 Myr. The gas disk also glows with accretion.

Then the classic T Tauri stage begins. This stage is also the ejection stage. The gas clouds are vigorously ejected, and the ejection jet is clearly observed. This stage lasts for 1–10 Myr, and the thick disk shrinks to about 100 AU. In this stage, the planetary system would be formed.

The next stage is the weak-lined T Tauri stage. "Weak-lined" means that a "weak-Hα" line of ionized hydrogen is observed in the spectrum of the (proto)star.

Finally, the clouds dissipate and the Sun can be seen. The Sun becomes a main sequence star in the H-R diagram (see Fig. Box 1.4).

The evolution model described here is for the evolution of the Sun (star) based on actual observation of stars and theories. However, when the masses of the presolar nebula or the planetary system are too small compared to the Sun, the evolution of the presolar nebula requires different views and theories.

BOX 2.1 THE HUBBLE SPACE TELESCOPE (HST)

The HST was launched into Earth orbit in 1990 (see Fig. Box 2.1A and B). Its 2.4-m mirror and main instruments can observe the near ultraviolet, visible, and near infrared spectra. The telescope was named after astronomer Edwin Hubble. The HST can take high-resolution images without atmospheric distortion and the background light of the atmosphere. The HST was built by the National Aeronautics and Space Administration (NASA, the U.S. space agency), with contributions from the European Space Agency.

Continued...

BOX 2.1 THE HUBBLE SPACE TELESCOPE (HST)—continued

(A)

(B)

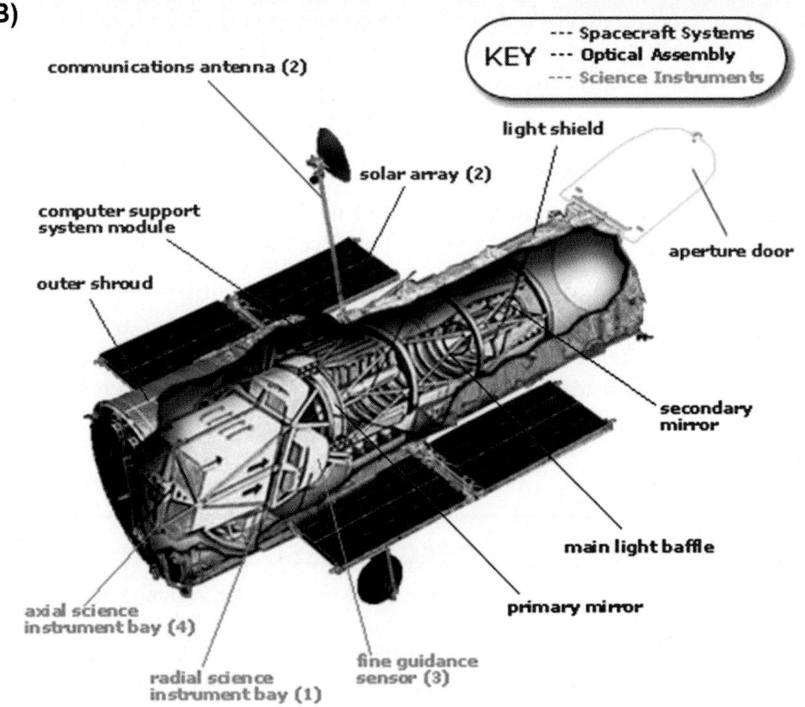

FIGURE BOX 2.1
(A) The photograph of the Hubble Space Telescope (HST) taken in 1997. (B) The schematic diagram of the HST.
(A) Credits: NASA. http://www.nasa.gov/sites/default/files/thumbnails/image/345535main_hubble1997_hi.jpg.
(B) Credits: NASA. http://asd.gsfc.nasa.gov/archive/sm4/art/technology/parts/newCutawayHST.jpg.

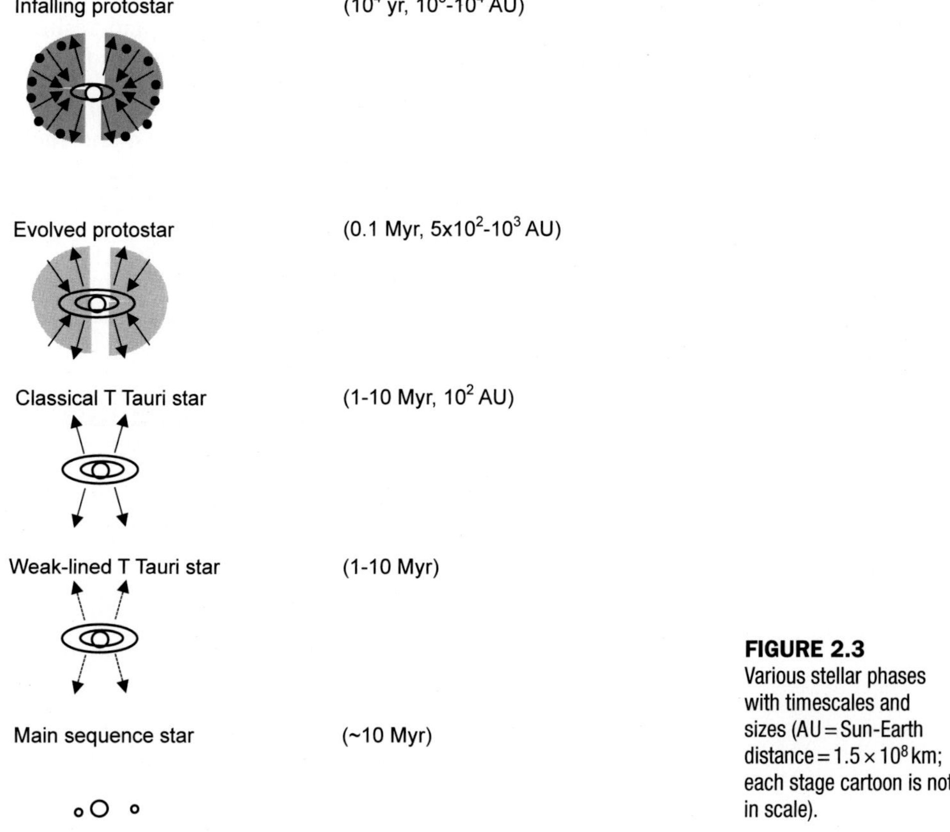

FIGURE 2.3
Various stellar phases with timescales and sizes (AU = Sun-Earth distance = 1.5×10^8 km; each stage cartoon is not in scale).

2.2 FROM THE PRESOLAR NEBULA TO THE SOLAR SYSTEM

The surrounding gas of the proto-Sun becomes hotter and flatter by the effects of gravity, magnetic field, and angular momentum. This disk is called the nebula or presolar disk. As discussed in Section 2.1, the presolar nebula is made up of gas clouds. As the temperature of the gas clouds decreases, silicates, iron, iron oxides, and carbon compounds form that are sub-micrometers to micrometers in size. These dust grains collide and build up larger particles in the turbulent circumstellar disk, and finally in most models of planet formation, planets grow by the accumulation of smaller bodies called planetesimals about 10–100 km in diameter (e.g., Chambers, 2004).

The theories seems to predict that the nebular gas evolves into planets without problems. However, the theories cannot solve two problems (Montmerle et al., 2006). Thus, the presolar nebula cannot evolve smoothly and continuously into the Solar System. One problem is what was the condition of the collision velocity and the density of the gas. If the dust grains have typical velocity in theories, they do not accrete, but disrupt each other. Thus, the dust grains do not grow to the size of pebbles. Another problem is that radial migration of planetesimals is unknown during growth. The gases easily drag the planetesimals into the central star within 100–1000 years. Therefore, all planetesimals fall into the star.

Investigation by the author revealed that the first problem is not yet solved. To make the meter-sized particles, physical sticking and inelastic collision, in which the kinetic energy is changed into other energy such as heat, are required. However, these conditions have not yet been determined.

The second problem seems to be solved by the model calculation produced by Johansen et al. (2007), for example. They started with meter-sized boulders and could create asteroids such as 900-km Ceres by considering magnetorotational instability in the model calculation.

2.3 AGE DATING BY RADIOACTIVE ISOTOPES

Using dating methods using radioactive isotopes, time-criteria can be given in theoretical models. Therefore, age dating is very important in astronomy. Radioactive decay is expressed as the following equation, when N is the number of a radioactive isotope:

$$dN/dt = -\lambda N \tag{2.3.1}$$

λ is called as the decay constant. The half-life, $T_{1/2}$ has the relation with λ as:

$$T_{1/2} = \ln 2/\lambda = 0.693/\lambda \tag{2.3.2}$$

When the Eq. (2.3.1) is integrated,

$$N_t = N_0 e^{-\lambda t} \tag{2.3.3}$$

N_0 is the initial number of the radioactive isotope, and t is the elapsed time from the start (the element synthesis). The number of the daughter isotope is:

$$D_t = D_0 + (N_0 - N_t) = D_0 + N_0 \left(1 - e^{-\lambda t}\right) \tag{2.3.4}$$

Generally, in cosmochemistry, T = 0 is present, and T is the elapsed time, that is the age of the sample. Thus the above equation changes as follows:

$$D_P = D_0 + N_0 \left(1 - e^{-\lambda T}\right) = D_0 + N_P \left(e^{\lambda T} - 1\right) \tag{2.3.5}$$

where "P" indicates the present, and T is the elapsed time from the element synthesis. For example, we consider the decay of ^{87}Rb to ^{87}Sr. Sr has a stable isotope, ^{86}Sr, whose number does not change over time (T). Then Eq. (2.3.5) is expressed as:

$$\left(^{87}Sr/^{86}Sr\right)_P = \left(^{87}Sr/^{86}Sr\right)_0 + \left(^{87}Rb/^{86}Sr\right)_P \left(e^{\lambda T} - 1\right) \tag{2.3.6}$$

This equation is the basic equation for age dating.

In actual applications, we use the isochron plot (see Fig. 2.4). We measure $\left(^{87}Sr/^{86}Sr\right)_P$ and $\left(^{87}Rb/^{86}Sr\right)_P$ of several mineral phases. If all the phases became

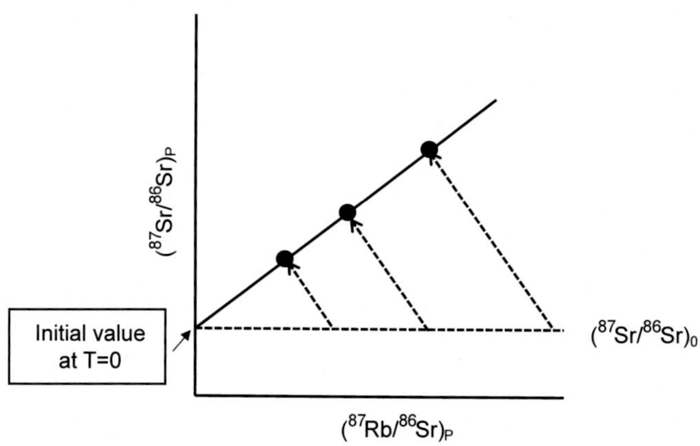

FIGURE 2.4
The isochron plot for the ^{87}Rb–^{87}Sr system.

isotopic equilibria at age T and did not suffer any disturbances from T to present, the data should form a line called an isochron, as shown in Fig. 2.4. If we calculate the y-intercept, it has an initial value of $\left(^{87}Sr/^{86}Sr\right)_0$. The slope corresponds to the age. As the age becomes older and older, the slope becomes steeper and steeper. The age calculation can be done by the famous ISOPLOT by Ludwig (1999), which can be obtained freely from their web site.

2.4 MODEL AGE

Here the model age is explained using the ^{147}Sm–^{143}Nd isotope system and Fig. 2.5. In Fig. 2.5A, the horizontal axis is the time. T_0 is the initial point when the solar system was formed. $T_{present}$ is the present time. Thus the time goes from right to left. The vertical axis is the ^{143}Nd:^{144}Nd ratio. For example, A and B are the neodymium isotope ratios of samples, and the concentration ratio of ^{147}Sm:^{144}Nd is expressed as $f^{i}_{Sm/Nd}$, where i is A, B, or chondritic uniform reservoir (CHUR), respectively. CHUR is the average value of the carbonaceous chondrites, and both $f_{Sm/Nd}^{CHUR}$ and $\left(^{143}Nd/^{144}Nd\right)_{present}^{CHUR}$ values are assumed to be constant. The time, T_M is called the model age, at which CHUR evolved at the point C, and the CHUR separated into the two reservoirs A and B by partial melting. The value of $f^{i}_{Sm/Nd}$ is the value of $^{147}Sm/^{144}Nd$ of the reservoir i (i = A, B, or CHUR).

Sample A has a higher $f_{Sm/Nd}^{A}$ value than CHUR (it is called "depleted reservoir"), resulting in a higher $\left(^{143}Nd/^{144}Nd\right)_{present}^{A}$ value than CHUR. In contrast, $f^{B}_{Sm/Nd}$ becomes lower than CHUR, which is called "enriched reservoir." The evolution curve of sample A crosses the evolution curve of CHUR at point C at the time T_M as the time goes back. In contrast, because sample B has a lower $f^{B}_{Sm/Nd}$ value than CHUR, a $\left(^{143}Nd/^{144}Nd\right)_{present}^{B}$ value becomes lower than that of CHUR; although the isotope ratio of B at the time TM (the point C) is identical with that of CHUR. Of course, this is an overly simplified two-stage model; however, it is very useful to treat the radiogenic isotope data at the beginning.

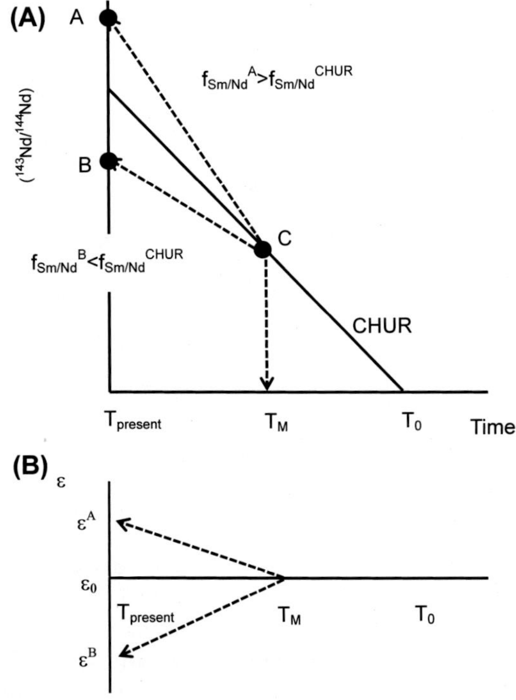

FIGURE 2.5
Model age in ^{147}Sm–^{143}Nd system. (A) Evolution curve of ($^{143}Nd/^{144}Nd$). (B) Evolution curve in ε-notation.

The $\left(^{143}\text{Nd}/^{144}\text{Nd}\right)_{\text{present}}^{\text{CHUR}}$ value is also used to calculate ε value, where

$$\varepsilon_{\text{present}}^{\text{i}} = \left[\left(^{143}\text{Nd}/^{144}\text{Nd}\right)_{\text{present}}^{\text{i}} / \left(^{143}\text{Nd}/^{144}\text{Nd}\right)_{\text{present}}^{\text{CHUR}} - 1\right] \times 10^4 \qquad (2.4.1)$$

As shown in Fig. 2.5B, when ε-notation is used, the evolution curve of CHUR becomes a horizontal linear line, and the evolution curves of A and B split into ε^A and ε^B at present.

The model age, T_M, where the present isotope ratio with the parent daughter ratio crosses with the CHUR evolution curve, are very convenient, and can be applied to most isotopic systems, such as Rb–Sr, La–Ce, Sm–Nd, Lu–Hf, Re–Os, and Pt–Os isotope decay systems.

2.5 EXTINCT NUCLIDES

Some nuclides have relatively short half-lives of 0.1–100 Myr (see Table 2.1). The half-lives of these elements are so short that they are already extinct. Therefore, these elements are called extinct nuclides. However, their inhomogeneous existences remain as isotopic anomalies of daughter nuclides in the materials formed in the early solar system.

Table 2.1 Important Extinct Nuclides

Radioactive Parent	Decay Scheme	Daughter Isotope	Horizontal Isotope Ratio	Half Life Myr
^{26}Al	β^-	^{26}Mg	$^{27}Al/^{24}Mg$	0.7
^{60}Fe	$2\beta^-$	^{60}Ni	$^{56}Fe/^{58}Ni$	1.5
^{53}Mn	β^-	^{53}Cr	$^{55}Mn/^{52}Cr$	3.7
^{107}Pd	β^-	^{107}Ag	$^{108}Pd/^{109}Ag$	6.5
^{182}Hf	$2\beta^-$	^{182}W	$^{180}Hf/^{184}W$	9.0
^{129}I	β^-	^{129}Xe	$^{127}I/^{132}Xe$	16
^{146}Sm	α	^{142}Nd	$^{149}Sm/^{144}Nd$	103

The horizontal isotope ratio means the horizontal axis in the fossil isochron.
Decay constants are from Dickin AP. Radiogenic isotope geology. 2nd ed. Cambridge: Cambridge Univ Press; 2008.

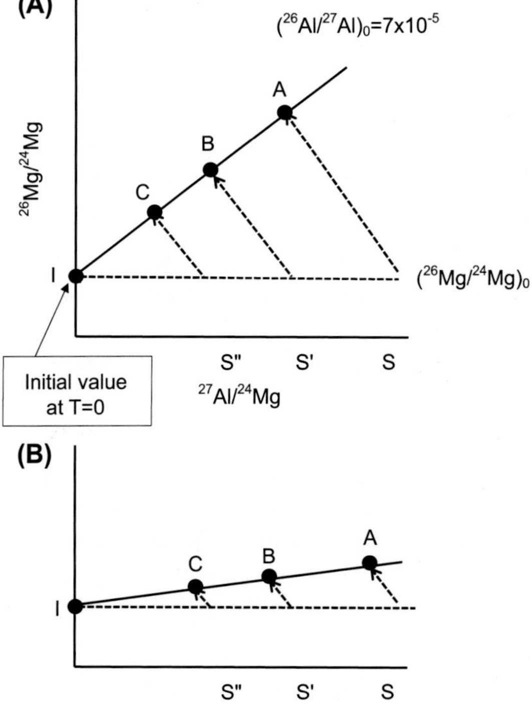

FIGURE 2.6
The fossil isochron plot for the ^{26}Al–^{26}Mg system. (A) Elemental fractionation (S-S″) occurred just after formation of ^{26}Al synthesis. (B) The same elemental fractionation occurred 2.1 Myr ($3 \times T_{1/2}$). The vertical axis of (B) becomes eight times smaller than that of (A).

The existence of these elements in the early solar system is proved in the fossil isochron, as shown in Fig. 2.6A. Here, we use ^{26}Al as an example. The differences from the isochron in Fig. 2.5A are as follows: As ^{26}Al is already extinct, we cannot use ^{26}Al in the horizontal axis in the isochron. The daughter isotope is ^{26}Mg, and Mg has nonradiogenic isotopes of ^{24}Mg and ^{25}Mg. Thus, the horizontal axis can be Al/Mg or $^{27}Al/^{24}Mg$. ($^{27}Al/^{25}Mg$ can be also applicable.)

The case that the elemental fractionation between Al and Mg occurred into three fractions (the points S, S′, and S″) just after the elemental synthesis is shown in Fig. 2.6A. From the slope, the initial value of $^{26}Al/^{27}Al$ of 7×10^{-5} is obtained. Or, if we assume the initial value of $^{26}Al/^{27}Al$, the age at the initial separation can be determined.

Fig. 2.6B presents the case that the elemental fractionation between Al and Mg occurred 2.1 Myr after the initial elemental synthesis, which is three times of the half-life of ^{26}Al (0.7 Myr). The vertical variation becomes 8 times smaller, because the amount of ^{26}Al is $(1/2)^{(2.1/0.7)} = 1/8$ compared to Fig. 2.6A.

BOX 2.2 THE CALCIUM-ALUMINUM–RICH INCLUSIONS (CAIS) IN THE CARBONACEOUS CHONDRITES: REMNANTS OF THE EARLY SOLAR SYSTEM (SOLAR NEBULA)

The CV3 chondrite Allende contains not only chondrules, but also white inclusions called CAIs. These white inclusions have compositions so refractory that they are considered to have formed in the early cooling stage of the solar system and survived homogenization and heat in the solar nebula. The age of CAIs is believed to be the oldest age in the solar system determined so far, from 4567.4 ± 1.1 to 4567.17 ± 0.70 Myr (Amelin et al., 2002; these ages are determined by the Pb-Pb method; see Box 2.5). The formation of the solar nebula is believed to have begun within 1 Myr of the shockwave of the supernova, because the CAIs contain evidence of ^{26}Al, which has a half-life of 0.7 Myr (see Section 2.5).

2.6 THE ^{182}HF–^{182}W ISOTOPE SYSTEM

^{182}Hf decays to ^{182}W through the double β^--decay, with a short half-life of 8.9 Myr. The ^{182}Hf–^{182}W isotope system has a special character compared to other extinct nuclides. Hafnium is a lithophile element and never becomes a siderophile element; therefore, Hf prefers silicates (mantle) in the mantle-core segregation. However, tungsten is a moderately siderophile element but also a moderate lithophile element. Therefore, elemental fractionation between Hf and W are expected to occur, when silicate (Hf-rich) and metal phases (W-rich) are separated, especially by melt-silicate separation or core-formation.

After all the ^{182}Hf decayed, W isotope ratios are frozen, and Hf–W chemical fractionation should be dated if the fractionation occurred early enough in solar system history. Generally, 5 times a half-life is a limit for detection of isotopic ratio anomalies, so that 45 Myr from the formation of the solar nebula is a limit for application of the ^{182}Hf–^{182}W isotope system. Harper and Jacobsen (1994) and Halliday and Lee (1999) successfully determined the isotopic anomaly of ^{182}W using multicollector inductively coupled plasma mass spectrometry (MC-ICP-MS; see Section 2.7.4).

The schematic fossil isochron for ^{182}Hf–^{182}W is shown in Fig. 2.7. For the denominator isotope, a stable isotope, ^{184}W generally is used. Most terrestrial samples are plotted at A in Fig. 2.7. However, early differentiated silicate

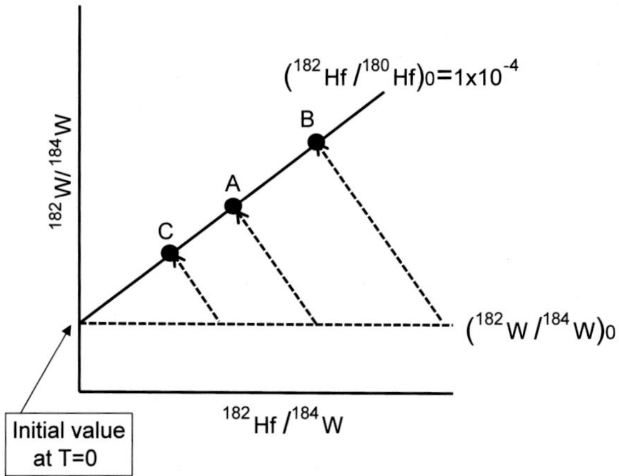

FIGURE 2.7
The fossil isochron plot for the ^{182}Hf–^{182}W system.

phases are Hf-rich and plotted at B. In contrast, the metal phase (core) is plotted at C. The W isotope ratios are presented in a ε-unit such as Eq. (2.4.1) defined as:

$$\varepsilon^{182}\mathrm{W} = \left[\left(^{182}\mathrm{W}/^{184}\mathrm{W}\right)_{\mathrm{sample}}/\left(^{182}\mathrm{W}/^{184}\mathrm{W}\right)_{\mathrm{standard}} - 1\right] \times 10^4 \tag{2.6.1}$$

and this ε-notation is used for presentation of degree of fractionation between Hf and W.

Because Hf and W isotope ratio determination was difficult to determine by using conventional thermal ionization mass spectrometry (TIMS; see Section 2.7.1), the development of MC-ICP-MS greatly contributed to easing the calculation (Harper and Jacobsen, 1994; Halliday and Lee, 1999).

The W isotope ratios are presented in an ε-unit (Eq. 2.6.1), but also in μ-units defined as:

$$\mu^{182}\mathrm{W} = \left[\left(^{182}\mathrm{W}/^{184}\mathrm{W}\right)_{\mathrm{sample}}/\left(^{182}\mathrm{W}/^{184}\mathrm{W}\right)_{\mathrm{standard}} - 1\right] \times 10^6 \tag{2.6.2}$$

2.7 MASS SPECTROMETRY

The weight of 1 mol of ^{12}C is defined to be exactly 12 u (unified atomic mass unit) or 12 Da (dalton). As the weights of electrons are approximately 1/1800 and negligible compared to that of the atom, the weight of the atom is almost the same weight as that of the nucleus. The weight of 1 mol of the element is almost the same as that of the mass number.

Mass spectrometry is the method that uses the specific mass number of each element as evidence of the element. For example, when the mass number of 235 appears in the mass spectrum, uranium-235 is judged to exist.

BOX 2.3 WHY ARE CHONDRULES GLOBULAR SHAPED?

Chondrules are millimeter-scale, previously melted silicates and cooled-very-fast spherules in chondrites. The ages of the chondrules are from 4564.17 ± 0.7 to 4566.7 ± 1 Myr (Amelin et al., 2002; these ages are also determined by the Pb-Pb method; see Box 2.5). The ages of chondrules are significantly 2–3 Myr younger (after) than those of the CAIs (see Box 2.2). From Box 2.4, planetesimal formation and even metallic core segregation in meteorite parent bodies are already finished within 1 Myr, and chondrule formation is the subsequent event.

There are no accepted models for the formation of chondrules. Here, one of the recent ideas, impact jetting, is explained. Johnson et al. (2014) proposed the idea of impact jetting in which two spherical projectiles impact each other, forming the melts. Johnson et al. (2015) further reported that impacts can produce enough chondrules during the first 5 Myr of planetary accretion to explain their observed abundance (see Fig. Box 2.2). They simulated protoplanetary impacts, finding that material is melted and ejected at high speed when the impact velocity exceeds 2.5 km s^{-1}. The jetted melt will form millimeter-scale droplets and the cooling rates will be fast, 10–1000 K h^{-1}. An impact origin for chondrules implies that meteorites are a byproduct of planet formation rather than leftover building blocks of planets.

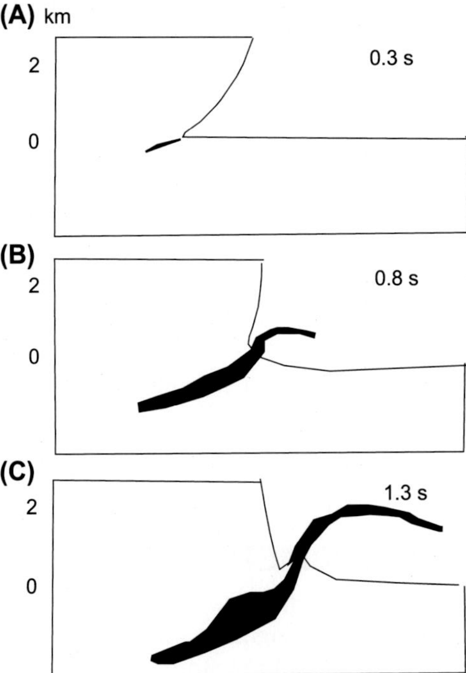

FIGURE BOX 2.2
Schematic diagrams of jetting of melted material during an accretionary impact (Johnson et al., 2015). A projectile of 10 km in diameter strikes the flat target at 3 km s^{-1}. The black parts show the melt jets with temperature of >1600 K.

BOX 2.4 THE EVIDENCE OF EARLY FORMATION AND DIFFERENTIATION OF PROTOPLANETS: CONSTRAINTS FROM IRON METEORITES

Iron meteorites are considered to be the core of the meteorite parent bodies, and good samples for constraining the protoplanet differentiation and accretion in the pre-solar disk. Kruijer et al. (2014) found variations of 5–20 ppm in ^{182}W, resulting from the decay of now-extinct ^{182}Hf in iron meteorite groups IIAB, IID, IIIAB, and IVA. These ^{182}W isotope ratios imply that core formation occurred after approximately 1 Myr, and the iron meteorite parent bodies probably accreted approximately 0.1–0.3 Myr after the formation of the CAIs.

The mass spectrum is obtained by a mass spectrometer. A mass spectrometer is simply composed of three parts: the ion source, the mass separator, and the detector. The ions are made at the ion source. Popular ion sources are thermal ionization (TI), ICP, and field emission (FE) ion sources. The mass separator separates ions according to m/z, where m and z are mass and charge, respectively. A sector electromagnet, a Q-pole mass filter (Q-pole) or time of flight (TOF) are mainly used. Faraday cups or the secondary electron multiplier (SEM) are used as detectors to count for the number of ions.

In observations of stars, mass spectrometers are useless, because only light comes from the target star to Earth. Thus, the available method is only photospectrometry (see Box 2.5). However, infrared spectra emitted by vibration of molecules become different when the constituent masses of the molecule are different. For example, ^{14}N–^{16}O and ^{15}N–^{6}O emit different vibration lights, enabling a rough estimate of the ^{15}N:^{14}N ratio.

BOX 2.5 PHOTO- AND MASS-SPECTROMETRY FOR BULK SAMPLE ANALYSIS

In cosmochemistry, the absorption of lights is used in analytical chemistry such as colorimetry, flame atomic absorption spectrometry (FAAS), and flameless atomic absorption spectrometry (FLAAS). The emission of lights is used in atomic emission spectrometry (AES) and ICP atomic emission spectrometry (ICP-AES) (see Fig. Box 2.3).

In astronomy, the same method is applied. The collected lights of the star by a telescope are analyzed by their wave length or color. The sharp absorption lines in the spectra are observed and used as evidence for the presence of some elements. (Actually, the astronomical method was used in cosmochemistry. It is very interesting that photospectrometry is applied for analysis from hand specimen to galaxy!)

Mass spectrometry is rarely used in astronomy, but generally used in cosmochemistry and astrophysics. TIMS was invented in the 1950s and still is used. Q-pole type ICP mass spectrometry (ICP-QMS) is derived from ICP-AES. The ICP ion source and the sector magnet of TIMS make sector-type ICP mass spectrometry (ICP-SFMS) or multicollector ICP-MS (MC-ICP-MS). The importance of hybridization is not limited to astrosciences (see Fig. Box 2.3).

Continued...

BOX 2.5 PHOTO- AND MASS-SPECTROMETRY FOR BULK SAMPLE ANALYSIS—continued

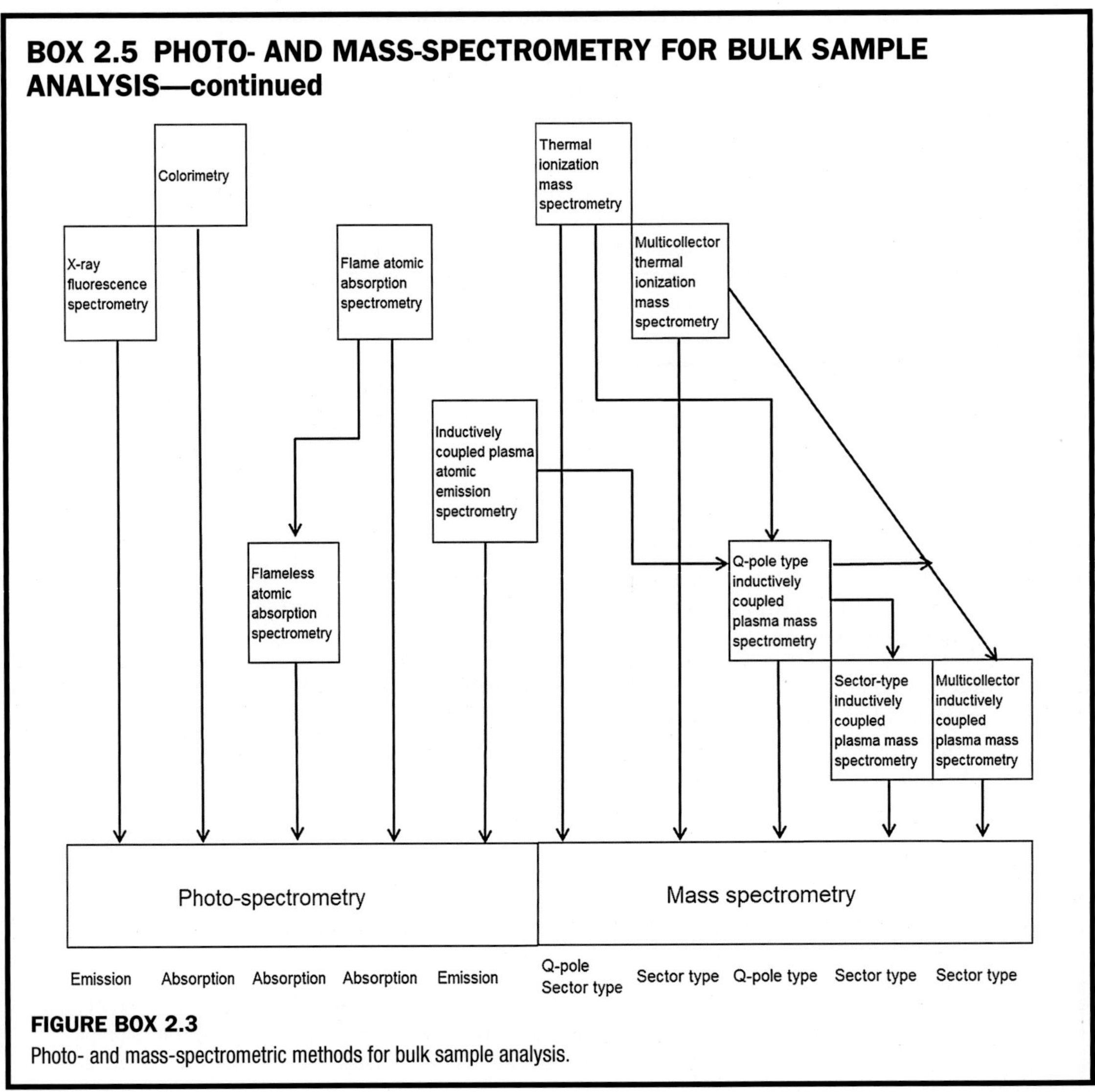

FIGURE BOX 2.3
Photo- and mass-spectrometric methods for bulk sample analysis.

On spacecraft, only photospectrometers are equipped for elemental determination. Mass spectrometers are generally heavy, because the magnet used in a mass spectrometer is a chunk of metals or metal oxides. However, spacecrafts are sometimes equipped with a mass spectrometer for determination of isotopic ratios, such as a D:H ratio, because the magnet for hydrogen is very light. For detection of life on a planet such as Mars, mass spectrometers for determination of isotopic ratios of hydrogen, carbon, nitrogen, and oxygen are loaded on the spacecraft.

Mass spectrometers generally are used for BULK isotopic and quantitative analyses of samples on Earth. For elemental quantification, TIMS (see Section 2.7.1), isotope dilution method (ID; see Section 2.7.2), ICP-QMS (see Section 2.7.3), and MC-ICP-MS (see Section 2.7.4) generally are used. The antonym of BULK analysis is SPOT analysis, which is performed by secondary ion mass spectrometers (SIMS) or laser ablation (LA)-ICP-MS (see Fig. Box 4.1).

When the isotope ratio analysis has an error rate greater than 0.3%, ICP-QMS is used. To obtain higher accuracy, TIMS and MC-ICP-MS are used. As the TIMS signal is not proportional to the sample amounts introduced into TIMS, ID must be used for quantitative analysis. In contrast, signal intensities of ICP-QMS are proportional to the abundance of target elements, ICP-QMS can be used directly with a combination of perfect dissolution of samples. For further explanation of sample dissolution and mass spectrometric techniques, a textbook by Makishima (2016) is a good reference.

2.7.1 Thermal Ionization Mass Spectrometry and Negative Thermal Ionization Mass Spectrometry

A schematic diagram of TIMS with a multicollector is shown in Fig. 2.8. The spectrometer is composed of three units: the thermal ionization ion source, the magnet, and the multicollector detection unit.

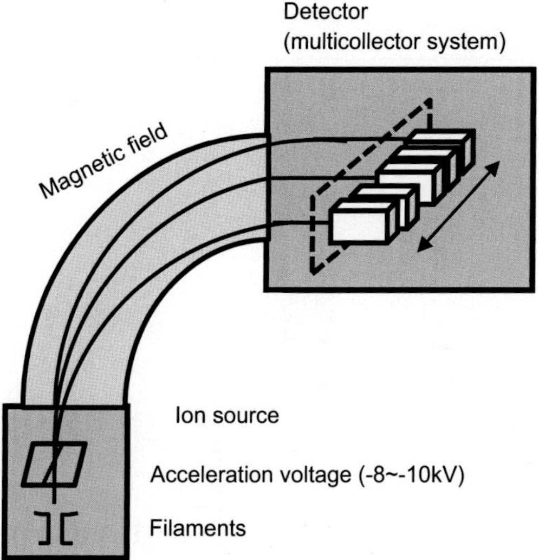

FIGURE 2.8
A schematic diagram of TIMS with multicollection. The dark area is kept in high vacuum by vacuum pumps. The magnetic field is supplied in the sector area by the sector magnet.

The ion source is composed of filaments, on which the purified element is loaded. The acceleration voltage in the ion source is supplied with 8–10 kV for positive-TIMS, and −8 to −10 kV for negative-TIMS. Accelerated ions go into the magnetic field produced by the electromagnet and are separated according to m/z. The flight tube and the detector unit are evacuated by the ion pumps ($\sim 1 \times 10^{-4}$ Pa). There is a focal plane along which ions of each m/z are focused. Along the focal plane, multiple Faraday cups are arrayed. Each Faraday cup can be moved from outside the vacuum along the focal plane. Generally, the center detector can be changed from a fixed Faraday cup to a SEM by changing the voltage supplied to the deflection plate. As the multicollector became an obvious option in TIMS, a TIMS machine means a "multicollector" TIMS machine.

N-TIMS, which uses negative ions formed on the filaments, is also widely used for isotope ratio determination of Os, Pt, W, Re, and so on. For N-TIMS application, positive acceleration voltage and a magnet for positive ions are required.

2.7.2 Isotope Dilution Method

The ID method is necessary for accurate elemental determination by TIMS. The drawback of the ID method is that it cannot be applied to monoisotopic elements, such as Na, P, Co, and Au.

Here ID is explained using Fig. 2.9. A target element needs to have more than two isotopes, shown as 1 and 2 in Fig. 2.9. We obtain a "spike," which is artificially

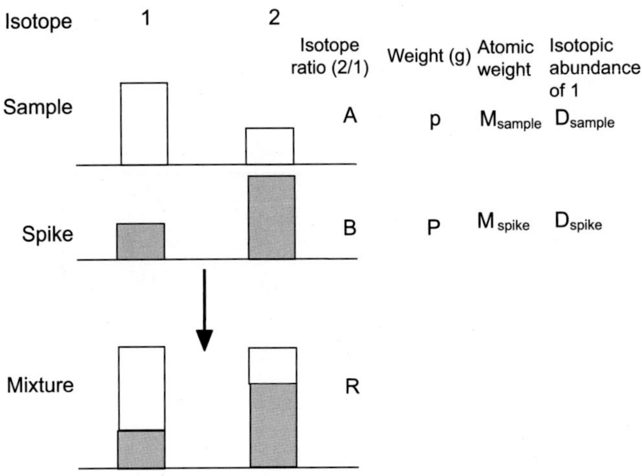

FIGURE 2.9
Isotope dilution method.

enriched in one isotope. Here, "2" is the enriched isotope in the spike. Usually, the spike is used in a solution form. p and P are the weights (g) of the target elements in the sample and spike; A, B, and R are isotope-2/isotope-1 ratios of the sample, spike, and sample-spike mixture; D_{sample} and D_{spike} are the isotopic abundances of the isotope-1 in the sample and spike; and M_{sample} and M_{spike} are the atomic weights of the target element in the sample and spike, respectively.

Then, the target element concentration $\left(C_T{}^{sample}\right)$ is expressed as:

$$C_T{}^{sample} = [(B-R)/(R-A)] \times Q' \times m_{spike}/m_{sample} \tag{2.7.1}$$

$$Q' = \left(D_{spike}/D_{sample}\right) \times \left(M_{sample}/M_{spike}\right) \times C_{spike} \tag{2.7.2}$$

where C_T^{sample} and C_{spike} are the concentrations of the target element in the sample and spike; m_{sample} and m_{spike} are the sample and spike weights; and Q′ in Eq. (2.7.2) is called the pseudo-concentration of the spike, because Q′ has the dimension of concentration. The pseudo-concentration of the spike solution should be calibrated beforehand.

After A, B, and Q′ are determined beforehand, and m_{sample} and m_{spike} are measured for each sample, the concentration of the target element C_T^{sample} is determined only by measurement of R, using Eq. (2.7.1). This equation stands as long as the isotope equilibrium for the target element between the sample and the spike is achieved.

The largest merit of ID is that once isotope equilibrium is achieved, losses of the target element by the ion exchange column chemistry, solvent extraction, or even by poor handling, do not affect the determination result or accuracy.

2.7.3 Q-Pole Type Inductively Coupled Plasma Mass Spectrometry

The sample solution is made into mist, and put into the Ar inductively coupled plasma (ICP) in air pressure (Fassel, 1978). As the plasma temperature is 6000–10,000K and very high, most elements are ionized in almost 100%. The air pressure pushes the plasma, including target element ions, into a small hole in the center of a sample cone. The back chamber of the sample cone is evacuated by a large rotary pump to be approximately 1 Pa. The flow of the plasma gas further hits the skimmer cone, which also has a hole in the center to lead the ions into the second chamber. The Ar gas flow is introduced into the first chamber as a supersonic jet flow through the sample cone, which is sampled again by the skimmer cone.

After passing the skimmer cone, the ions are accelerated by the extraction plates. The acceleration is 0~200V. The ions then pass through ion lenses, are curved and separated from strong UV lights. After separation from the lights, the ions go into the Q-pole mass filter.

Although various m/z ions are introduced, all ions except one m/z ion resonate, oscillate and hit the Q-poles, and thus cannot pass through the Q-pole mass filter. The merit of the Q-pole is that the mass scan can be performed very quickly because there is no magnet, which prevents fast change of the magnetic field by hysteresis. The disadvantage is that the peak shape has no flat peak top; therefore, the precision of isotopic ratios is limited to greater than 0.3%.

Finally, selected mass ions are detected by an SEM and counted by the ion counting system. Taking all these into account, ICP-QMS, which was invented by Houk et al. (1980) and Date and Gray (1981), is suitable not for precise isotope analysis but for quantitative elemental analysis with or without ID. However, its merit resides in its wide applicability to various elements using the calibration curve method (Makishima and Nakamura, 2006; Lu et al., 2007).

2.7.4 Multicollector Inductively Coupled Plasma Mass Spectrometry

MC-ICP-MS also appeared at the end of the 1980s to target precise isotope ratio measurement. The fluctuation of each isotope signal by the plasma was perfectly canceled out by simultaneous measurement by multicollectors, and precision drastically improved to the level of TIMS. Furthermore, MC-ICP-MS has an advantage over TIMS in that some elements like the Hf isotope ratio, which is very difficult to measure by TIMS, can be easily determined.

The ions produced by ICP pass through the sample and skimmer cones and are accelerated by −8 to approximately −10 kV. An entrance slit is placed in front of the electric field to change the resolution. Then the ions go into the electric field to make the ion energy the same, and then go into the magnet. The ions are separated according to m/z and detected by several Faraday cups or an ion counting system. The capability of MC-ICP-MS was first shown in the measurement of the Hf isotope ratios.

For the two-isotope elements, a standard-sample bracketing (SSB) method was developed. In the SSB method, the standard and sample solutions are alternately measured; thus, the mass discrimination can be canceled, resulting in high precision for isotopic analyses of the two-isotope element. Nowadays MC-ICP-MS is used in astrosciences for mass fractionation measurements of two-isotope elements such as V, Cu, and Tl, most of which could not be determined by TIMS.

MC-ICP-MS showed the possibility of determining the natural mass fractionation of most solid elements from Li to U. Applications of the natural mass fractionations of new elements are expanding from cosmochemistry o astrobiology.

2.8 FORMATION OF GAS GIANT PLANETS AND ASTEROIDS: THE NICE MODEL

Researchers in Observatoire de la Cote d'Azur in Nice, France (Tsiganis et al., 2005; Morbidelli et al., 2005; Gomes et al., 2005) established comprehensive models explaining the evolution of the solar system of 0–0.6 and 700–1000 Myr (the ages indicate the time from the beginning of the solar system) and origins of the giant planets, the main asteroid belt, the Trojan asteroids, and the Kuiper belt objects, based on the N-body simulation (see Box 2.8) calculation of 3500 particles in a solar nebula with 35 M_E (M_E is the mass of Earth) and truncated at 30 AU. In these two periods, large evolutionary events of the solar system are considered to have occurred. In Fig. 2.10, the schematic diagram for the evolution of the solar system is depicted after DeMeo and Carry (2014). These models are called the Nice model, after the location of the research observatory.

The model starts just after the giant planets (Jupiter, Saturn, Uranus, and Neptune) are formed and the surrounding gas is dissipated. The solar system is composed of the sun, the giant planets, and a debris disk of small planetesimals. The planets then start to erode the disk by accreting or scattering away the planetesimals. The

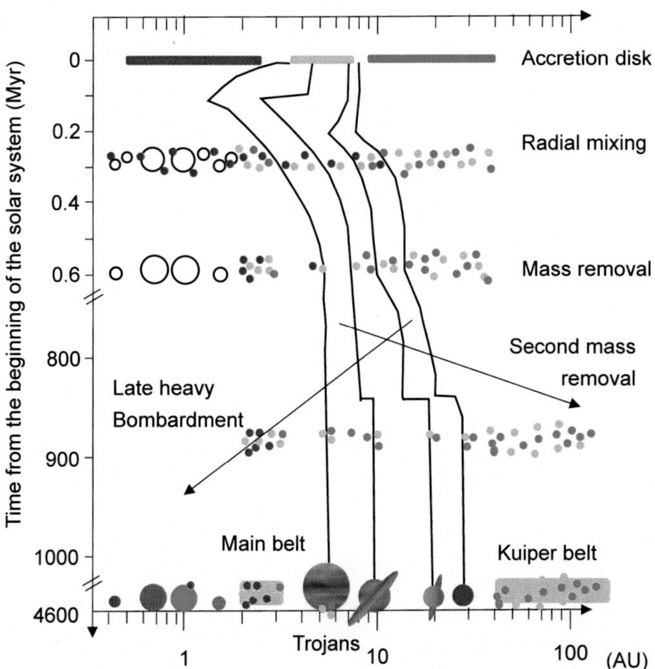

FIGURE 2.10
The schematic diagram for evolution of the solar system. The vertical axis is the time from the beginning of the solar system (Myr). Three scales of time (Myr), 0–0.6, 800–1000 and the present (4567) are shown in one axis. The horizontal axis is the semimajor axis in AU unit. The light and dark blue indicate C-type and outer-disk planetesimals, and the red indicates S-type (rocky) planetesimals, respectively. The blue planetesimals contain lots of water, which could be a source of Earth's water.

planets migrate because of the exchange of angular momentum with the disk particles in this process. The simulation shows that the fully formed Jupiter starts at 300 M_E at 3.5 AU, which is favorable for giant planet formation because it is outside the snow line. Saturn's core is initially 35 M_E and approximately 4.5 AU, and remains approximately 4.5 AU, with increasing the mass to 60 M_E. The cores of Uranus and Neptune begin at approximately 6 and 8 AU and grow from approximately 5 M_E. Their masses increase just slightly (see Fig. 2.10).

When the mass of Saturn reaches 60 M_E, Saturn suddenly begins migration inwards to 2 AU. Jupiter is forced to move inward to approximately 1.5 AU within 0.1 Myr. Uranus and Neptune migrate slightly inward with slow delay due to Jupiter's migration. Then on catching Jupiter, Saturn is trapped in a 1:2 resonance with Jupiter. The migration is reversed, and Saturn and Jupiter migrate outward together. They capture Uranus and Neptune in resonance, causing both to migrate outward as well. Saturn, Uranus, and Neptune reach their full mass at the end of the migration when Jupiter reaches 5.4 AU. The migration rate decreases exponentially as the gas disk dissipates. The final orbital configuration of the giant planets is consistent with their current orbits. The terrestrial planets are formed from this disk within 30–50 Myr after the migration of Jupiter and other planets ended.

BOX 2.6 A SINGLE ZIRCON DATING AND THE PB-PB DATING METHOD

Zircon ($ZrSiO_4$) is a common mineral in acidic rocks. This mineral is very refractory and stands against metamorphism, erosion, and weathering. As shown in Fig. Box 2.4, the zircon crystal is not so homogeneous, and contains inclusions (mineral, melt, and fluid), corroded rims and metamict (amorphous) parts, which are formed by radiation of radioactive

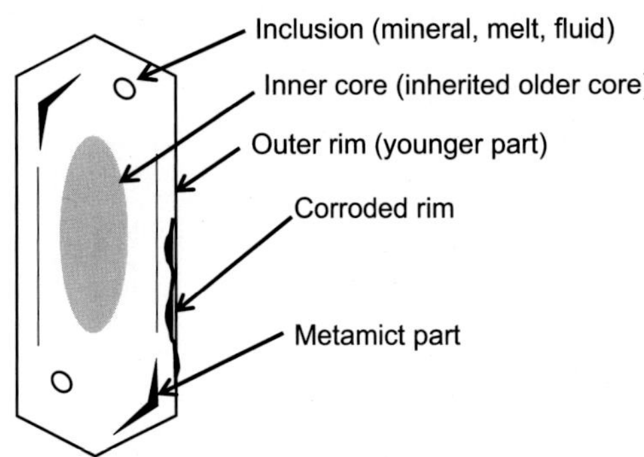

FIGURE BOX 2.4

A schematic diagram of a zircon crystal.

U and Th in the zircon itself. Old zircons are often made of inner core and outer rim. The inner core is made of the older zircon, which means that there were older source of the zircon such as granites. The outer rim forms the euhedral zircon with eroded rims. Our targets for age determination are pure inner core and outer rim as well. By establishing a high resolution secondary ion mass spectrometers (HR-SIMS; see Box 2.7), accurate ages of both inner and outer parts from a single zircon can be obtained.

The zircon crystal contains a high concentration of U but very low Pb in its structure; therefore, Pb in zircon is considered to be only radiogenic Pb. There are two decays of uranium isotopes:

$$^{238}U \rightarrow {}^{206}Pb^* \tag{B2.6.1}$$

$$\left({}^{206}Pb^*/{}^{204}Pb\right) = \left({}^{238}U/{}^{204}Pb\right) \times \left(e^{T\lambda 8} - 1\right) \tag{B2.6.2}$$

$$^{235}U \rightarrow {}^{207}Pb^* \tag{B2.6.3}$$

$$\left({}^{207}Pb^*/{}^{204}Pb\right) = \left({}^{235}U/{}^{204}Pb\right) \times \left(e^{T\lambda 5} - 1\right) \tag{B2.6.4}$$

The asterisks refer to the radiogenic lead; T is the crystalized age of the zircon; $\lambda 8$ and $\lambda 5$ are the decay constants of ^{238}U and ^{235}U, respectively. Note that there are no initial terms in this case. Sometimes a phrase of the "common lead" appears. The common lead means non-radiogenic Pb incorporated in zircon. From Eqs. (B2.6.2) and (B2.6.4),

$$\left({}^{206}Pb^*/{}^{207}Pb^*\right) = \left({}^{238}U/{}^{235}U\right) \times \left(e^{T\lambda 8} - 1\right) / \left(e^{T\lambda 5} - 1\right) \tag{B2.6.5}$$

where ($^{238}U/^{235}U$) is constant and 137.88. Thus if ($^{206}Pb^*/^{207}Pb^*$) is measured, T is obtained. This is called the Pb-Pb dating method. In order to determine T precisely, not only accurate determination of the isotope ratio of $^{206}Pb^*:^{207}Pb^*$, but also selection

BOX 2.6 A SINGLE ZIRCON DATING AND THE PB-PB DATING METHOD—continued

of zircons without common lead are prerequisites. In addition, choosing less disturbed zircons after formation is also important. "Disturb" means Pb or U are lost or added from outside of the zircon crystal by geological processes such as metamorphism or weathering.

If ($^{206}Pb^*/^{238}U$) is plotted against ($^{207}Pb^*/^{235}U$), one curve that starts from the origin is obtained. This is called a concordia curve, which is shown in Fig. Box 2.5. If the data is on this curve, the U–Pb of zircon has not been disturbed from formation of zircon to present. If the data is not on this line, the U–Pb of zircon has been disturbed. When there is only one disturbance and loss of Pb, the data form a line. One end shows the disturbing age (1 Ga in Fig. Box 2.5), and the other end shows the formation age (3.5 Ga in Fig. Box 2.5). This line is called a discordia line. This dating method is called the concordia method.

The age of CAIs (see Box 2.2) or chondrules (see Box 2.3) are determined by the Pb–Pb age dating method by assuming the initial Pb isotopic compositions.

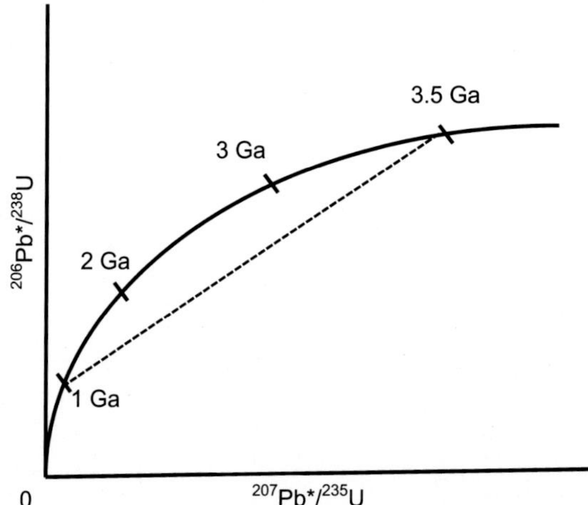

FIGURE BOX 2.5
The concordia diagram. The *dotted line* indicates a discordia line. One intercept of the discordia line to the concordia at 3.5 Ga shows the formation age of zircon. Another intercept of the discordia line at 1 Ga indicate the resetting age (the metamorphic age) of zircon.

BOX 2.7 HIGH RESOLUTION SECONDARY ION MASS SPECTROMETRY (HR-SIMS)

In the U–Pb dating of a single zircon (see Box 2.6) by SIMS, not only precise Pb isotope ratios but also a Pb:U concentration ratio is required. The largest problem is molecular interferences. In order to remove the molecular interferences, high mass resolution is required. By using conventional SIMS, operation at such high resolution is possible, but the count rates are not enough to date zircon precisely.

Continued...

BOX 2.7 HIGH RESOLUTION SECONDARY ION MASS SPECTROMETRY (HR-SIMS)—continued

To enable accurate dating, a high-resolution SIMS was designed and constructed by an Australian National University group. It was named Sensitive High Resolution Ion Microprobe (SHRIMP), but the size was huge (~3 m × ~7 m). The larger the size, the larger the number of secondary ions that can be collected. Cameca Company also build large ion probes, ims-1270 and -1280 (Fig. Box 2.6). Now the biggest problem is not the mass spectrometer but that homogeneous standard materials need to be prepared with precisely known amounts of sub-ppm levels of U and Pb.

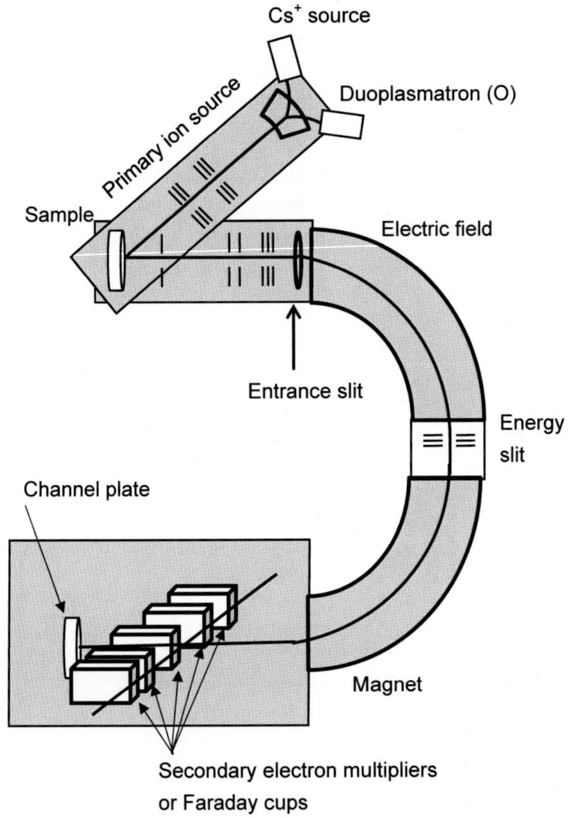

FIGURE BOX 2.6

A schematic diagram of a high-resolution multicollector secondary ion mass spectrometer (HR-SIMS).

BOX 2.8 N-BODY SIMULATION

In physics and astronomy, an N-body simulation often is used. In astrophysics especially, this simulation assumes multiple bodies (from two to a few million depending on the power of the computer) and gravity, and calculates the dynamic evolution of the multiple bodies depending on the particle-particle interactions. The calculation increases on the order of N^2; therefore, many computational techniques are needed and have been invented.

2.9 THE LATE HEAVY BOMBARDMENT IN THE NICE MODEL

The late heavy bombardment (LHB) is recorded in the Moon's craters. The frequency of crater-causing collisions on the Moon was strangely very high in 700–1000 Myr (the ages indicate the time from the beginning of the solar system). Gomes et al. (2005) explained LHB by the Nice model, in which Jupiter and Saturn become 1:2 resonant.

Prior to the 1:2 resonance, Saturn's orbital period was less than twice that of Jupiter. When the two planets crossed the 1:2 mean motion resonance, their orbits became eccentric. This abrupt transition destabilized the giant planets, leading to a short phase of close encounters among Saturn, Uranus, and Neptune. As a result of the interactions of the ice giants with the disk, Uranus and Neptune reached their current distances and Jupiter and Saturn evolved to their current orbital eccentricities. The main idea is that Jupiter and Saturn crossed the 1:2 resonance at approximately 700 Myr.

Initially, the giant planets migrated slowly owing to leakage of particles from the disk. This phase lasted 880 Myr. After the resonance crossing event, the orbits of the ice giants became unstable and they were scattered into the disk by Saturn. They disrupted the disk and scattered objects all over the Solar System, including the inner regions. A total of 9×10^{18} kg struck the Moon after the resonance crossing.

As Jupiter and Saturn moved from 1:2 resonance toward their current positions, secular resonances swept across the entire asteroid belt. These resonances can drive asteroids into orbits with eccentricities and inclinations large enough to allow them to evolve into the inner solar system and hit the Moon.

The asteroid objects slowly leak out of the asteroid belt and evolved into the inner solar system. Roughly the particles arrived in the first 10–150 Myr.

It was estimated that $(3–8) \times 10^{18}$ kg of asteroids hit the Moon. This amount is comparable to the amount of comets. So the Nice model predicts that the LHB impactors should have been a mixture of comets and asteroids. Within the first 30 Myr comets dominated, but the last impacters were asteroids. These results support a cataclysmic model for the lunar LHB. Although much about the LHB is not well known, their simulations reproduce two of the main characteristics attributed to this episode: (1) the 700 Myr delay between the LHB and terrestrial planet formation, and (2) the overall intensity of lunar impacts. Their model predicts a sharp increase in the impact rate at the beginning of the LHB.

2.10 THE GRAND TACK MODEL: THE RESONANCE OF JUPITER AND SATURN

The explanation of the large-scale mixing of reddish and bluish material in the asteroid belt (see Fig. 2.10) is lacking in the Nice model. To overcome this problem, the "Grand Tack" model (Walsh et al., 2011) appeared. "Tack" in the Grand Tack model means the migration of the planetesimals by Jupiter, using the sailing analogy. Although this model came out from the Nice group, the Jupiter and Saturn resonance changed from 1:2 to 2:3. The 2:3 resonance is proposed by Masset and Snellgrove (2001) as the type II migration, which is now widely accepted.

In this model, during the time of terrestrial planet formation (before the events of the Nice model), Jupiter could have migrated as close to the Sun as Mars is today. Jupiter would have moved through the primordial asteroid belt, emptying it and then repopulating it with scrambled material from both the inner and outer solar system. Then Jupiter reversed and headed back towards the outer Solar System. Once the details of the resulting distribution in the Grand Tack model have been compared to the emerging observational picture, it will become clear whether this model can crack the asteroid belt's compositional order. Planetary migration ends within the first billion years of our solar system's 4.5-billion-year history.

However, the asteroid belt is still dynamic today. Collisions between asteroids are continuously making the bodies smaller and smaller. Smaller asteroids from 10 cm to 10 km are then subject to the Yarkovsky effect, in which uneven heating and cooling of the asteroid alters its orbit (see Box 2.9). The Yarkovsky effect thoroughly mixes small bodies within each section of the main belt, but once they reach a major resonance such as the 3:1 and 5:2 mean motion resonances (MMR) at the locations, they are swiftly ejected from the main belt (see Table 2.3).

BOX 2.9 THE YARKOVSKY EFFECT

When the asteroid is rotating, it collects heat on the day surface, and emits heat (radiation) on the night surface. The transfer of heat or radiation can be inhomogeneous, and the radiation becomes a force that drifts the asteroid from its expected orbit. This is called as the Yarkovsky effect.

Current observations and models indicate that the strong resonances with Jupiter inhibit the crossing of material from one region to another. These processes continue to form the asteroid belt, erasing the past history and creating new structures in it. New observational evidence that reveals a greater mixing of bodies supports the idea of a solar system that is still evolving.

2.11 FORMATION OF INNER PLANETS IN THE GRAND TACK MODEL

In this section, we consider that the terrestrial planets can be produced by the Grand Tack model by Walsh et al. (2011). The terrestrial planets are best formed when the planetesimal disk is truncated with an outer edge at 1 AU. These conditions are created naturally if Jupiter tacked at approximately 1.5 AU.

However, before concluding that Jupiter tacked at approximately 1.5 AU, it is necessary to consider that the asteroid belts between 2 and 3.2 AU can survive the passage of Jupiter. Volatile-poor asteroids (S-types) are dominant in the inner asteroid belt, while volatile-rich asteroids (C-types) are dominant in the outer asteroid belt. These two main classes of asteroids have partly overlapping semi-major axis distributions, although there are more C-types than S-types beyond approximately 2.8 AU. The planetesimals from the inner disk are considered to be S-type and those from the outer regions C-type.

From hydrodynamic simulations, the inward migration of the giant planets shepherds much of the S-type material inward by resonant trapping, eccentricity excitation, and gas drag. The mass of the disk inside 1 AU doubles, reaching approximately 2 M_E. This reshaped inner disk constitutes the initial condition for terrestrial planet formation. However, a fraction of the inner disk (~14%) is scattered outward, ending up beyond 3 AU. During the subsequent outward migration of the giant planets, this scattered disk of S-type materials is encountered again. A small fraction (~0.5%) of this material is scattered inward and left decoupled from Jupiter into the asteroid belt region.

The giant planets then come across the materials in the Jupiter–Neptune formation region, some of which (~0.5%) is also scattered into the asteroid belt. Finally, the giant planets meet the disk of material beyond Neptune (within 13 AU) of which only approximately 0.025% reaches a final orbit in the asteroid belt. When the giant planets have finished their migration, the asteroid belt population is in place; however, the terrestrial planets require approximately 30 Myr to complete accretion.

The asteroid belt is composed of two separate populations: S-type planetesimals within 3.0 AU (left-overs from the giant planet accretion process: ~0.8 M_E of materials between the giant planets and ~16 M_E of planetesimals from the 8.0–13 AU region). The other is C-type material placed into orbits crossing the still-forming terrestrial planets. Every C-type planetesimal from beyond 8 AU was implanted in the outer asteroid belt. The 11–28 AU C-type planetesimals ended up in high-eccentricity orbits that enter the terrestrial-planet-forming region (<1.0–1.5 AU), and may represent a source of water for Earth.

For the Jupiter–Uranus region this ratio is 15–20, and for the Uranus–Neptune region it is 8–15. Thus, it is expected that (3–11) × 10^{-2} M_E of C-type material entered the terrestrial planet region. This exceeds by a factor of 6–22 the minimal mass required to bring the current amount of water to Earth (~5 × 10^{-4} M_E; Abe et al., 2000).

The migration of Jupiter creates a truncated inner disk matching initial conditions of previously successful simulations of terrestrial planet formation. As there is a slight build-up of dynamically excited planetary embryos at 1.0 AU, simulations of the accretion needed 150 Myr for Earth and Venus to grow within the 0.7–1 AU, accreting most of the mass, while Mars is formed from embryos scattered out beyond the edge of the truncated disk. The final distribution of planet mass versus distance quantitatively reproduces the large mass ratio existing between Earth and Mars, and also matches quantitative metrics of orbital excitation.

2.12 INCONSISTENCY BETWEEN OUR SOLAR SYSTEM AND EXOSOLAR SYSTEMS FOUND BY THE KEPLER PROGRAM

The Kepler program (see Box 2.10) found thousands of exoplanets in about four hundred stellar systems. Thus, more than half of the Sun-like stars have planet(s) in low-eccentricity orbits. However, the periods of the orbits were days to months, and the masses of the planets were 1 $M_E < M_p < 50$ M_E, where M_E and M_p are the masses of Earth and the planet, respectively (Mayor et al., 2011;

BOX 2.10 THE KEPLER SPACE OBSERVATORY

Kepler is a space observatory launched by NASA in 2009, named after a famous astronomer, Johannes Kepler. It was designed to search for Earth-sized extrasolar planets in Milky Way stars. The Kepler instrument is only a photometer that continuously monitors the brightness of more than 145,000 main sequence stars in a fixed view. Then data are analyzed to find periodic dimming of stars by crossing of extrasolar planets in front of the host stars (Box 2.11 and 2.12). The image of the Kepler is shown in Fig. Box 2.7.

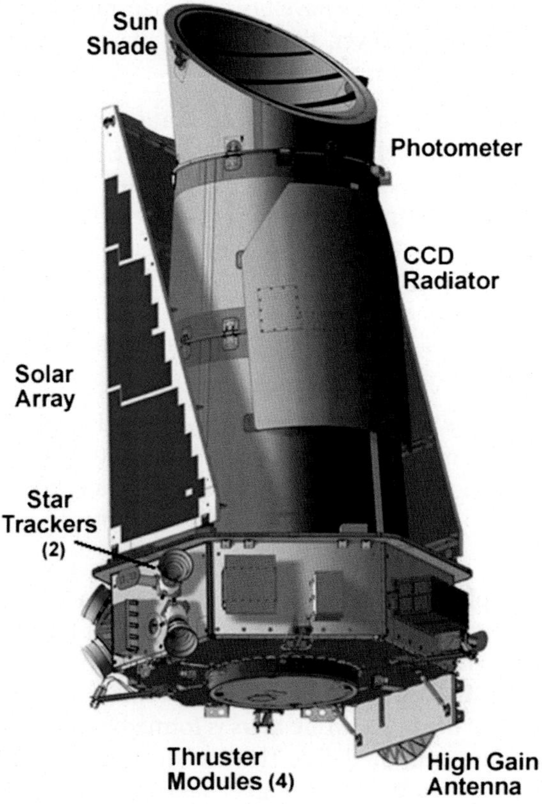

FIGURE BOX 2.7
The Kepler Space Observatory.
Credits: NASA. http://kepler.nasa.gov/Mission/QuickGuide/.

BOX 2.11 THE WIDE-FIELD INFRARED SURVEY EXPLORER AND NEOWISE

The wide-field infrared survey explorer (WISE) scans the entire sky in infrared light, picks up the glow of hundreds of millions of objects, and produces millions of images. The mission covers the search for the coolest stars, the most luminous galaxies, and the darkest near-Earth asteroids and comets. By measuring the infrared light of the object, a good estimate of the size distribution of the asteroid population was obtained. This information will identify hazardous asteroids. WISE can detect brown dwarfs, which are gas balls like Jupiter that failed to become a star. It was active from December 2009 to February 2011. The NEOWISE project succeeded the WISE project at the end of 2014. Fig. Box 2.8 shows a schematic diagram of WISE.

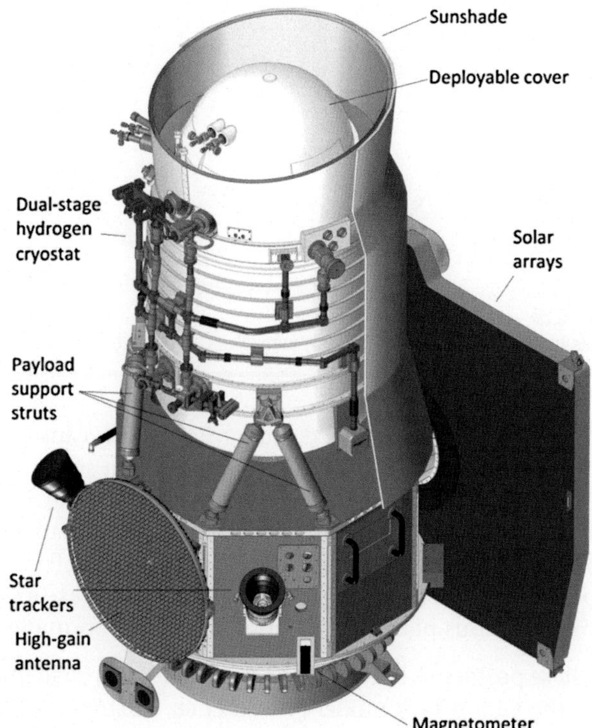

FIGURE BOX 2.8
The Wide-field Infrared Survey Explorer (WISE).
Credits: NASA, http://wise.ssl.berkeley.edu/documents/wise%20launch%202009-12-03.pdf.

BOX 2.12 LAGRANGE POINTS

The Lagrange points are positions in two large bodies, one of which revolves around another larger body. The two bodies are affected only by gravity, and small objects at the Lagrange points can be maintained at the stopped position. For example, the larger body is the Sun, and the smaller body is Earth; or the Sun and Jupiter.

Continued...

BOX 2.12 PISTON CYLINDER HIGH PRESSURE APPARATUS—continued

As shown in Fig. Box 2.9, there are five points (L1–L5). Three points—L1, L2, and L3—are on the line connecting two bodies. The points of L4 and L5 are ahead and behind on the orbit of the smaller revolving body. Many artificial satellites are placed at L1 and L2. Earth has one asteroid at L4, but no asteroids at L5 so far discovered. Jupiter has many small asteroids at L4 and L5 called Trojans.

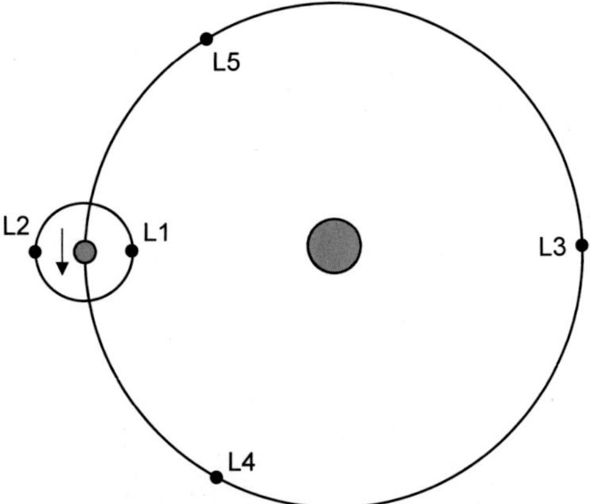

FIGURE BOX 2.9
Lagrange points (not to scale). The smaller body is revolves around the larger body.

Batalha et al., 2013). These are called "super-Earths", because the planets had larger mass but closer inner orbits than Earth's orbit. In summary, the exosolar planet studies revealed the peculiarity of our solar system. In addition, the first planetesimals have formed approximately 1 Myr after the Sun's birth, whereas final accretion of terrestrial planets were in the timescale of 100–200 Myr, well after the dispersion of the nebular gases. In contrast, the exosolar super-Earths are inferred to be gaseous planets, indicating their early formation.

To overcome these problems, Batygin and Laughlin (2015) challenged the calculation following the Grand Tack model (Walsh et al., 2011; Masset and Snellgrove, 2001), where Jupiter moves inward followed by outward migration. Their developed model based on calculation is summarized in Fig. 2.11.

In Fig. 2.11A, the left yellow half ball is the Sun. The super-Earths, which are common in the Kepler observation, are indicated blue balls near the Sun. In this case, two gas planets are depicted in the figure. These are formed in the gaseous disk, which is indicated by the trapezoid from the Sun. The small dark green circles indicate terrestrial planetesimals with a radius of 10–1000 km. The right side vertical arrow shows the time. Jupiter begins to move from approximately 3 AU after the Grand Tack model.

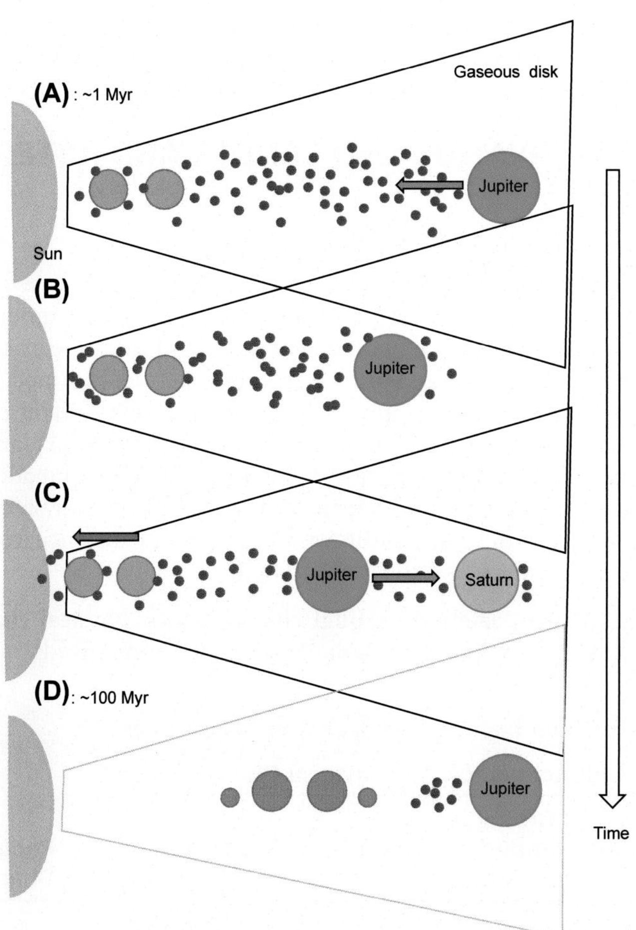

FIGURE 2.11
The model of Batygin and Laughlin (2015) of the early solar system evolution.
The figure is modified from Naoz S. Jupiter's role in sculpting the early Solar System. Nature 2015;112:4189–90.

As Jupiter migrates inward, the planetesimals collide with each other (Fig. 2.11B), and are broken and grinded into a size less than 100 m. Then in Fig. 2.11C, Jupiter migrates to approximately 1.5 AU and becomes resonant with Saturn, and begins to go on an outside orbit. However, the super-Earths are carried to the Sun by inward-drifting debris of planetesimals less than 100 m size. Once the planetesimals become smaller than 100 m, aerodynamic drag induces rapid orbital decay.

Thus the solar system is reset by the movement of Jupiter. After super-Earths are cleared up, the "second-generation planets" are formed from the surviving gas and materials (Fig. 2.11D). The terrestrial planets can be formed, instead of the observed gas-rich short-period planets.

This model clearly explains why the short-period super-Earths do not exist in our Solar System, and why it took a long time (~100 Myr) for the final formation of the terrestrial planets, compared to the early formation (~1 Myr) of the Sun.

2.13 OBSERVATION OF PLANETS AND ASTEROID BELTS

2.13.1 The Planetary System

As discussed in Box 2.4, the silicate planetesimals and planets with a metallic core were formed in the early stage of the solar nebula. The inner part of the solar nebula is considered to be hot, where water and methane exist as gas, and rocky planets (Mercury, Venus, Earth, Mars, and asteroids) formed. In contrast, the outer part of the nebula (outside the snow line where water exists as ice) is cold, so that ices of water and methane condensed, forming giant planets as Jupiter, Saturn, Uranus, and Neptune. Weidenschilling (1977) made a model of the distribution of mass in the solar nebula of the planetary system. He found the surface density of the planets follows as $r^{-3/2}$ (r is the distance from the Sun), except for Mercury, Mars, and the asteroid belt.

In Table 2.2, planets, satellites, orbital characteristics, physical characteristics, and peculiar characteristics in our Solar System are summarized.

Table 2.2 Sun, the Planets, and Their Major Satellites in the Solar System

		Orbital Characteristics		Physical Characteristics		
Mother Planet	**Famous Satellite**	**Semi-major Axis (AU)**	**Orbital Period (× Earth)**	**Radius (× Earth)**	**Mass (M_E)**	**Peculiar Characteristics**
Sun	–	–	–	109	3.33×10^5	
Mercury	0	0.387	87 d	0.384	0.0553	Refractory
Venus	0	0.723	224 d	0.952	0.815	CO_2 clouds
Earth	1	(1.49×10^8 km)	(365.25 d)	(6.357×10^3 km)	(5.97×10^{24} kg)	Ocean, life
	Moon	0.00257	27.321 d	0.273	0.0123	No atmosphere
Mars	2	1.52	1.88	0.532	0.107	CO_2 air
	Phobos	6.3×10^{-5}	7.66 h	0.0035	1.8×10^{-9}	D-type asteroid
	Deimos	1.6×10^{-4}	30.35 h	0.0020	3.4×10^{-10}	C-type asteroid
Jupiter	67	5.203	11.862	11.00	318	
	Io	0.00283	1.77 d	0.286	0.0150	Sulfur-volcanoes
	Europa	0.00450	3.55 d	0.246	0.0080	Ice-covered
	Ganymede	0.0995	7.15 d	0.414	0.0248	Magnetsphere
	Callisto	0.0722	16.7 d	0.379	0.0180	Ice-covered
Saturn	62	9.58	29.46	9.16	95.15	Rings
	Enceladus	0.00160	1.37 d	0.040	1.8×10^{-5}	Salty sea, life?
	Titan	0.0082	15.9 d	0.405	0.023	Atmosphere
Uranus	27	19.2	84.0	3.99	14.53	Tilted axis, rings
Neptune	14	30.07	165	3.86	17.1	Ice giant
	Triton		Retrograde orbit			Captured

2.13.2 Observation of the Asteroids

Gradie and Tedesco (1982) reported that the compositional types of asteroids in 1.8–5.2 AU systematically follow their heliocentric distance. The asteroids are classified as types E, R, S, M, F, C, P, and D, based on composition (spectral reflectance and albedo, i.e., surface brightness). This is the first comprehensive study of the compositional trends of the asteroid belt, which served for decades as the backbone for interpretation of the main belt.

The asteroid belts are summarized in Table 2.3. The types are not systematically rigid as proposed by Gradie and Tedesco (1982), and they are gradually changing as shown in Table 2.3.

Table 2.3 Asteroid Belts From 1.8 to 5.2 AU

Distance (AU)	Name	Type	Famous Asteroid
1.8–2.0	Hungaria	E, C, S	
4:1MMR			
2.1–2.5	Major asteroid belt Inner	 V, S, M, D	 Vesta
3:1 MMR			
2.5–2.8	Middle	C, B, P	Ceres
5:2 MMR			
2.8–3.3	Outer	C, B, P, M	Hygeia
2:1 MMR			
3.3–3.5	Cybele	P, C, D	
5:1 MMR			
3.9–4.0 5.1–5.3	Hilda Trojan	P, C, D D, P, C	

AU is the distance from the center of the Sun to Earth. MMR indicates mean motion resonance.

2.13.3 Observation of the Asteroids From Space

Observation methods have improved very much from those by Gradie and Tedesco (1982). Ivezić et al. (2001) measured asteroids by the Sloan Digital Sky Survey (SDSS; see Box 1.11), which provided multifilter photometry in the visible spectrum for more than 100,000 asteroids. Mothe-Diniz et al. (2003) carefully looked at asteroid compositional distributions of smaller sizes, which indicated that the distribution was different from that of Gradie and Tedesco (1982). Mainzer et al. (2011) reported measurements by WISE (Box 2.11), which provides diameters and albedos for more than 100,000 asteroids.

Carry (2012) conducted a detailed analysis of 994 mass estimates and 1500 volume determinations of 300 asteroids, demonstrating density trends per asteroids' taxonomic class. DeMeo and Carry (2013) created a new framework to quantify the compositional makeup of the asteroid belt by looking at the distribution by mass rather than numbers as done by Gradie and Tedesco (1982).

In Table 2.3, asteroid belts from 1.8 to 5.3 AU are shown with distance, names, and types. The major difference from those of the age of Gradie and Tedesco (1982) is that Type is mixed according to distance. For example, Trojan-type asteroids are observed in the inner belt. The bluish asteroids (C-type) and the reddish asteroids (S-type) are observed in many asteroid belts.

2.13.4 Sample Return Program From an Asteroid

The Hayabusa space mission was planned by the Japanese Aerospace Exploration Agency (JAXA) to collect samples from asteroid 25143 Itokawa and return to Earth (see a landing image of the Hayabusa spacecraft in Fig. 2.12). Samples from Earth, Mars, and Venus have undergone severe weathering by atmosphere and water. However, asteroids are considered to be in the same condition as when they were made 4.56 Gyr ago. Hayabusa was expected to reveal what kind of materials and processes were related in the formation of the solar system, and how the solar system looked just after the planets formed. It was launched in May 2003, and returned to Earth in June 2010.

FIGURE 2.12
An artistic image of asteroid explorer Hayabusa touching down and collecting samples from asteroid 25143 Itokawa.
Image credit: A. Ikeshita, JAXA, ISAS. http://www.jaxa.jp/projects/pr/brochure/pdf/04/sat14.pdf.

The size of 25143 Itokawa was 535 × 294 × 209 m, and its density was 1.9 ± 0.13 g cm^{-3}. The existence of large boulders suggests an early collisional breakup of a pre-existing parent asteroid followed by a reagglomeration into a rubble-pile object (Fujiwara et al., 2006). Although the sample collection system did not work properly, hundreds of mineral particles of 25143 Itokawa were collected. The mineral chemistry of the recovered sample was that of LL ordinary chondrite, which made a direct link between S-type asteroids and ordinary chondrites (Nakamura et al., 2011; see Box 1.8). On 50-μm micrograins, micrometer to submicrometer-size adhered melts and submicrometer-size craters were observed, indicating a hostile environment by impacts of micro-meteorites on the asteroid's surface (Nakamura et al., 2012).

In 2014 and 2016, JAXA and NASA launched Hayabusa 2 and OSIRIS-REx, respectively. Both spacecraft will visit carbonaceous asteroids (C-type asteroids 162173 Ryugu and 101955 Benuu) and return samples to Earth. Both projects will tell us more about the relationship between carbonaceous asteroids and carbonaceous chondrites. Carbonaceous chondrite is considered to be the most primitive meteorite, which means the most primitive material from the solar nebula (see Box 1.8).

References

Abe Y, Ohtani E, Okuchi T, Righter K, Drake M. Water in the early Earth. In: Canup RM, Righter K, editors. Origin of the Earth and Moon. Tucson: Univ Arizona Press; 2000. p. 413–33.

Amelin Y, Krot AN, Hutcheon ID, Ulyanov AA. Lead isotopic ages of chondrules and calcium-aluminum-rich inclusions. Science 2002;297:1678–83.

Batalha N, Rowe MJF, Bryson ST, et al. Planetary candidates observed by Kepler III. Analysis of the first 16 months of data. Astrophys J Suppl Ser 2013;204. Article Id: 24 (21pp.).

Batygin K, Laughlin G. Jupiter's decisive role in the inner Solar System's early evolution. Proc Natl Acad Sci USA 2015;112:4214–7.

Carry B. Density of asteroids. Planet Space Sci 2012;73:98–118.

Chambers JE. Planetary accretion in the inner solar system. Earth Planet Sci Lett 2004;223:241–52.

Date AR, Gray AL. Plasma source mass spectrometry using an inductively coupled plasma and a high resolution quadrupole mass filter. Anal Lond 1981;106:1255–67.

DeMeo FE, Carry B. The taxonomic distribution of asteroids from multi-filter all-sky photometric surveys. Icarus 2013;226:723–41.

DeMeo FE, Carry B. Solar System evolution from compositional mapping of the asteroid belt. Nature 2014;505:629–34.

Dickin AP. Radiogenic isotope geology. 2nd ed. Cambridge: Cambridge Univ Press; 2008.

Fassel VA. Quantitative elemental analyses by plasma emission spectroscopy. Science 1978;202:183–91.

Feigelson ED, Montmerle T. High-energy processes in young stellar objects. Annu Rev Astron Astrophys 1999;37:363–408.

Fujiwara A, Kawaguchi J, Yeomans DK, et al. The rubble-pile asteroid Itokawa as observed by Hayabusa. Science 2006;312:1330–4.

Gomes R, Levison HF, Tsiganis K, Morbidelli A. Origin of the cataclysmic Late Heavy Bombardment period of the terrestrial planets. Nature 2005;435:466–9.

Gradie J, Tedesco E. Compositional structure of the asteroid belt. Science 1982;216:1405–7.

Halliday AN, Lee DC. Tungsten isotopes and the early development of the Earth and Moon. Geochim Cosmochim Acta 1999;63:4157–79.

Harper CL, Jacobsen SB. ^{182}Hf-^{182}W: a new cosmochronometer and method for dating planetary core formation events. In: Sixth meeting of the European Union of Geosciences. 1994. p. 451.

Houk RS, Fassel VA, Flesh GD, Svec HJ, Gray AL, Taylor CE. Inductively coupled argon plasma as an ion source for mass spectrometric determination of trace elements. Anal Chem 1980;52:2283–9.

Ivezić Ž, Tabachnik S, Rafikov R, et al. Solar System objects observed in the Sloan Digital Sky Survey commissioning data. Astron J 2001;122:2749–84.

Johansen A, Oishi JS, Low M-MM, Klahr H, Henning T. Rapid planetesimal formation in turbulent circumstellar disks. Nature 2007;448:1022–5.

Johnson BC, Bowling TJ, Melosh HJ. Jetting during vertical impacts of spherical projectiles. Icarus 2014;238:13–22.

Johnson BC, Minton DA, Melosh HJ, Zuber MT. Impact jetting as the origin of chondrules. Nature 2015;517:339–41.

Kruijer TS, Touboul M, Gischer-Gödde M, Berminghan KR, Walker RJ, Kleine T. Protracted core formation and rapid accretion of protoplanets. Science 2014;344:1150–4.

Lu YH, Makishima A, Nakamura E. Coprecipitation of Ti, Mo, Sn and Sb with fluorides and application to determination of B, Ti, Zr, Nb, Mo, Sn, Sb, Hf and Ta by ICP-MS. Chem Geol 2007;236:13–26.

Ludwig KR. ISOPLOT/version 2.01. A geochronological tool kit for Microsoft Excel. Berkeley (CA): Geochronological Center; 1999.

Makishima A. Thermal ionization mass spectrometry (TIMS). Silicate digestion, separation, measurement. Weinheim: Wiley-VCH; 2016.

Makishima A, Nakamura E. Determination of major, minor and trace elements in silicate samples by ICP-QMS and ICP-SFMS applying isotope dilution-internal standardization (ID-IS) and multi-stage internal standardization. Geostand Geoanal Res 2006;30:245–71.

Masset F, Snellgrove M. Reversing type II migration: resonance trapping of a lighter giant protoplanet. Mon Not R Astron Soc 2001;320:L55–9.

Mainzer A, Bauer J, Grav T, et al. Preliminary results from NEOWISE: an enhancement to the Wide-field Infrared Survey Explorer for Solar System science. Astrophys J 2011;731:53–66.

Mayor M, Marmier M, Lovis C, et al. The HARPS search for southern extra-solar planets XXXIV. Occurrence, mass distribution and orbital properties of super-Earths and Neptune-mass planets. 2011. aiXiv:1109.2497.

Mothe-Diniz T, Carvano J, Lazzaro D. Distribution of taxonomic classes in the main belt of asteroids. Icarus 2003;162:10–21.

Montmerle T, Augereau J-C, Chaussidon M, et al. 3. Solar system formation and early evolution: the first 100 million years. Earth, Moon, Planets 2006;98:39–95.

Morbidelli A, Levison HF, Tsiganis K, Gomes R. Chaotic capture of Jupiter's Trojan asteroids in the early Solar System. Nature 2005;435:462–5.

Nakamura E, Makishima A, Moriguti T, et al. Space environment of an asteroid preserved on micrograins returned by the Hayabusa spacecraft. Proc Natl Acad Sci USA 2012;109:E624–9.

Nakamura T, Noguchi T, Tanaka M, et al. Itokawa dust particles: a direct link between S-type asteroids and ordinary chondrites. Science 2011;333:1113–6.

Naoz S. Jupiter's role in sculpting the early Solar System. Nature 2015;112:4189–90.

Tsiganis K, Gomes R, Morbidelli A, Levison HF. Origin of the orbital architecture of the giant planets of the Solar System. Nature 2005;435:459–61.

Walsh KJ, Morbidelli A, Raymond SN, O'Brien DP. Mandell: a low mass for Mars from Jupiter's early gas-driven migration. Nature 2011;475:206–9.

Weidenschilling SJ. The distribution of mass in the planetary system and solar nebula. Astrophys Space Sci 1977;51:153–8.

CHAPTER 3
The Giant Impact Made the Present Earth–Moon System*

Chapter Outline

* In October 2016, while this book was in proofreading process, Wang and Jacobsen (2016) found a new evidence for the giant impact. They proposed that the $d^{41}K$ value of the Moon was 0.4‰ higher than the BSE or chondrites (CI, L6, and H6). This means that the Moon was formed not by the low-energy disk equilibration model (e.g., Pahlevan and Stevenson, 2007), but by the high-energy, high-angular-momentum model (e.g., Lock et al., 2016), in which well-mixed continuous extended structure reached beyond the Roche limit, as proposed by Ringwood (1986). If this discovery is true, it is required to modify this chapter. Such new finding makes the book's author unhappy, but makes the science itself very exciting! This is why scientists cannot abandon the research.

Origins of the Earth, Moon, and Life. http://dx.doi.org/10.1016/B978-0-12-812058-3.00003-X

3.1 INTRODUCTION

In this chapter, the formation of inner silicate planets and the Earth–Moon system is discussed. It is generally accepted that Earth and the Moon were formed by the giant impact, in which a Martian-size object hit the proto-Earth, creating the Earth–Moon system. Chemical as well as physical simulations played important roles to establish the giant impact model. The important observations required for a giant impact having formed the Earth–Moon system is described in Chapter 3.

Fig. 3.1 is a cartoon showing the relation between the proto-Earth, the impacter named Theia, the giant impact and the late veneer, in chronological order, to help you understand these various events. We learn about the giant impact in Chapter 3, and the late veneer in Chapter 4.

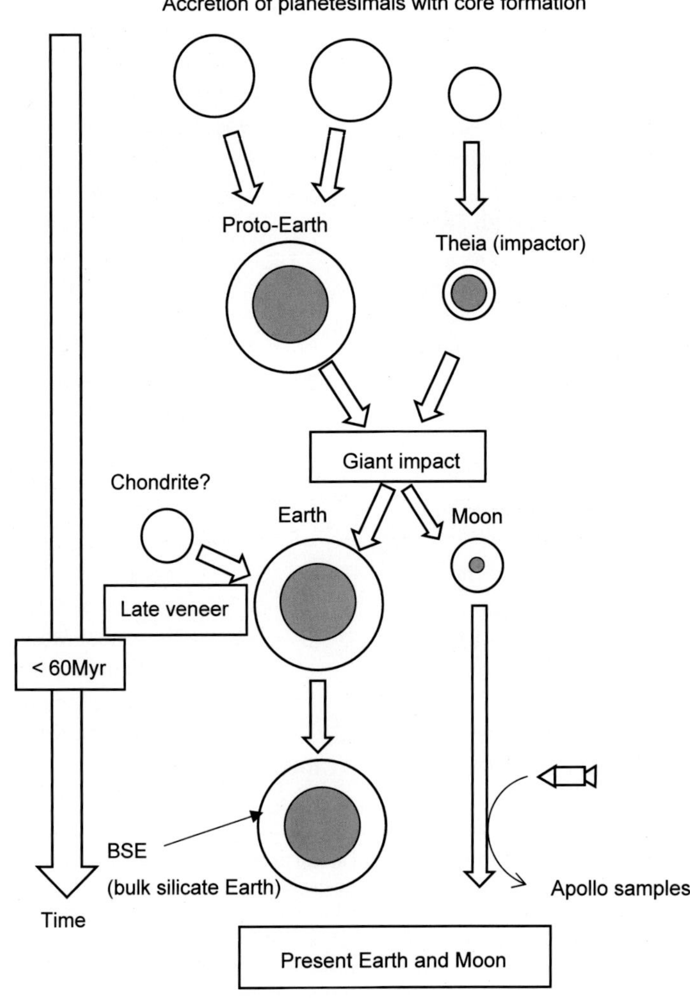

FIGURE 3.1
A cartoon showing the relation between the proto-Earth, Theia, the giant impact and the late veneer.

The author considers that history, process, and progress of discovery of ideas in the astro-sciences are important; therefore, history related to the giant impact models is shown in chronological order in Section 3.2.

3.2 HISTORY OF A GIANT IMPACT MODEL

Darwin (1879) discussed the origin of the Earth–Moon system and suggested that the Moon should be torn out from the molten Earth mantle by the tidally induced "fission process." This is like a fission process of a heavy nucleus, in which one sphere is torn into two spheres.

Later, Daly (1946) proposed that the Moon was created by an impact. However, this impact origin of the Moon was not given attention until Cameron and Ward (1976) brought out the idea of the giant impact again. In Daly's initial giant impact model, some important conclusions were derived:

1. The Martian size object (0.1 M_E) hit the proto-Earth (Cameron, 1985);
2. The collision made a silicate gas disk with low volatile elements;
3. The gas accreted and formed the Moon in a short time;
4. The strangely large angular momentum of the Earth–Moon system could be explained.

Hartmann and Davis (1975) are often referred to as other pioneers of the giant impact model as well; however, they showed that the planetesimal size could be the Martian size when following the model of Cameron and Ward (1976). Therefore, the author supposes the pioneer of the giant impact model to be Cameron and Ward (1976).

The impacting body was named "Theia", who gave birth to the Moon goddess, Selene, in the Greek myth, as proposed by Halliday (2000) (Box 3.1 and 3.2).

BOX 3.1 THE RARE EARTH ELEMENT PATTERN

The rare earth element (REE) pattern is the plot that the logarithms of sample REE concentrations normalized to the chondritic values (the vertical axis) are plotted in the order of the atomic number (the horizontal axis) (see Fig. Box 3.1; it is also called as a Masuda–Coryell diagram). The REE pattern utilizes the smooth decrease of ionic radii of each REE according to an increase of the atomic number.

As all data in Fig. Box 3.1 were obtained by isotope dilution-thermal ionization mass spectrometry (ID-TIMS), the mono-isotopic elements of Praseodymium (Pr), terbium (Tb), holmium (Ho), and thulium (Tm) could not be determined. Details of TIMS and ID are explained in Sections of 2.7.1 and 2.7.2, respectively.

Twenty years ago, the precise REE pattern was obtained only by the ID-TIMS method. Nowadays, however, the REE patterns can be easily determined by inductively coupled plasma quadrupole mass spectrometry (ICP-QMS) (see Section 2.7.3), which can determine all REEs including Pr, Tb, Ho and Tm, as well as other trace elements such as lithium (Li), boron (B), rubidium (Rb), strontium (Sr), yttrium (Y), zirconium (Zr), niobium (Nb), molybdenum (Mo), cesium (Cs), barium (Ba), tin (Sn), antimony (Sb),

Continued...

BOX 3.1 THE RARE EARTH ELEMENT PATTERN continued

hafnium (Hf), tantalum (Ta), lead (Pb), thorium (Th), and uranium (U) without difficulty (Makishima and Nakamura, 2006; Lu et al., 2007).

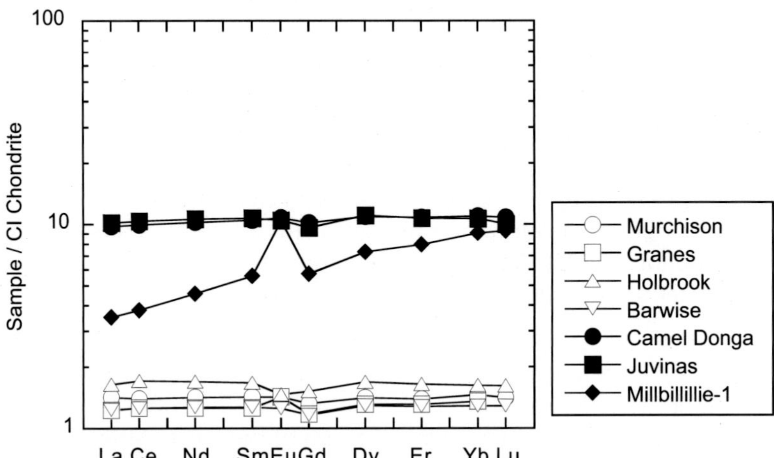

FIGURE BOX 3.1
Rare earth element (REE) patterns of meteorites. The horizontal axis is the REE elements. The vertical axis is the value of concentrations of samples divided by those of the CI chondrite (average or the Leedey chondrite). Murchison, Granes, Holbrook, Barwise are chondrites of CM2, L6, L6, and H5, respectively. Camel Donga, Juvinas, and Millbillillie are eucrites. For the classification of meteorites, see Box 1.8. The data were obtained by isotope dilution-thermal ionization mass spectrometry (ID-TIMS).
The data are after Makishima A, Masuda A. Primordial Ce isotopic composition of the solar system. Chem Geol 1993;106:197–205.

BOX 3.2 EUROPIUM ANOMALY (EU ANOMALY)

Rare earth elements [lanthanum (La), cerium (Ce), Pr, neodymium (Nd), samarium (Sm), europium (Eu), gadolinium (Gd), Tb, dysprosium (Dy), Ho, erbium (Er), Tm, ytterbium (Yb), and lutetium (Lu)] generally have an ionic charge of +3. However, Ce and Eu show anomalous values of +4 and +2, respectively. When Eu becomes +2 it can substitute for the site of Ca^{2+} in plagioclase when plagioclase crystallizes in magma. As a result, Eu shows a positive anomaly for the plagioclase in the REE pattern (see Fig. Box 3.1), and, in contrast, a negative anomaly appears in the crystallized residual magma. For example, the positive Eu anomaly is observed in the REE pattern of Millbillillie-1 (eucrite) in Fig. Box 3.1.

3.3 APOLLO PROGRAM: LANDING ON THE MOON AND SAMPLING MOON ROCKS

3.3.1 The Apollo Program

The Apollo program was an American human spaceflight program by the National Aeronautics and Space Administration (NASA). The program succeeded in landing on the Moon from 1969 to 1972 (see Fig. 3.2 for the landing sites), and collected a total of 380.95 kg of samples (see Table 3.1).

FIGURE 3.2
Apollo landing sites.
Image credit: NASA.
http://science.nasa.gov/media/medialibrary/2005/07/01/11jul_lroc_resources/landingsites_600.jpg.

Table 3.1 The Apollo Programs and Collected Sample Amounts

Program	Year	Amounts (kg)
Apollo 11	1969	21.55
Apollo 12	1969	34.30
Apollo 14	1971	42.80
Apollo 15	1971	76.70
Apollo 16	1972	95.20
Apollo 17	1972	110.40

3.3.2 Rocks From the Moon

The Moon rocks are classified into two main types: lunar highland rocks and mare rocks (summarized in Table 3.2). The highland rocks are mainly composed of mafic plutonic rocks and regolith breccias. The highland breccias are formed by impacts of highland igneous rocks. There are three types of highland igneous rocks: the ferroan anorthosite suite, the magnesian suite, and the alkali suite. Warren (1993) compiled non-mare Moon rocks. The ferroan anorthosite suite is composed of anorthosite (>90% calcic plagioclase) (see Fig. 3.3A). These rocks are considered to represent plagioclase cumulates floated during the lunar magma ocean. The plagioclase is extremely calcic (An 94–96), indicating the extreme depletion of alkalis [sodium (Na) and potassium (K)] and other volatile elements.

Table 3.2 Rocks of the Moon

Highland igneous rocks
Mafic plutonic rocks
Ferroan anorthosite (mafic minerals have low Mg:Fe ratios; e.g., Fig. 3.3A)
Anorthosite
(>90% calcic plagioclase; floated plagioclase cumulates)
Magnesian suits (relatively high Mg:Fe ratios)
Dunite
Troctolites (olivine-plagioclase)
Gabbro (plagioclase-pyroxene)
Alkali suits
Sodic plagioclase
Norites (plagioclase–orthopyroxene)
Gabbronorites (plagioclase–clinopyroxene–orthopyroxene)
Regolith breccias (made by impacts of the highland igneous rocks)
Mare rocks (basalts)
High-Ti basalts (e.g., Fig. 3.3B)
Low-Ti basalts (e.g., Fig. 3.3C)
Very–low-Ti (VLT) basalts
Very–high-K (VHK) basalts

The magnesian suite (Mg suite) is composed of dunites (>90% olivine), troctolites (olivine–plagioclase) and gabbros (plagioclase–pyroxene) with relatively high Mg:Fe ratios. The plagioclase is still calcic (An 86–93). The trace element contents in plagioclase require equilibrium with a KREEP-rich [K, REE, and phosphorus (P)-rich; so-called "incompatible" elements] magma, which is inconsistent with their Mg rich character.

The alkali suite has high alkali content compared to other Moon rocks. The alkali suite is composed of alkali anorthosites with sodic plagioclase (An 70–85), norites (plagioclase-orthopyroxene), and gabbronorites (plagioclase–clinopyroxene–orthopyroxene). The trace element contents in plagioclase also indicate a KREEP-rich parent magma.

The mare rocks are mainly composed of four types of basalts; high-Titanium (Ti) basalts (Fig. 3.3B), low-Ti basalts (Fig. 3.3C), very low-Ti (VLT) basalts and very high-K (VHK) basalts. They show a large negative Eu anomaly in the REE pattern. Extremely K rich content is found in the VHK basalt (Neal and Taylor, 1992).

3.4 SIMILARITY BETWEEN CHEMICAL COMPOSITIONS OF THE MOON AND EARTH'S MANTLE

Ringwood (1959, 1966, 1979) had a strong interest in the origin of the Moon. Ringwood first noticed the depletion of vanadium (V), chromium (Cr), and manganese (Mn) of the Moon to CI chondrites, but that they were similar to

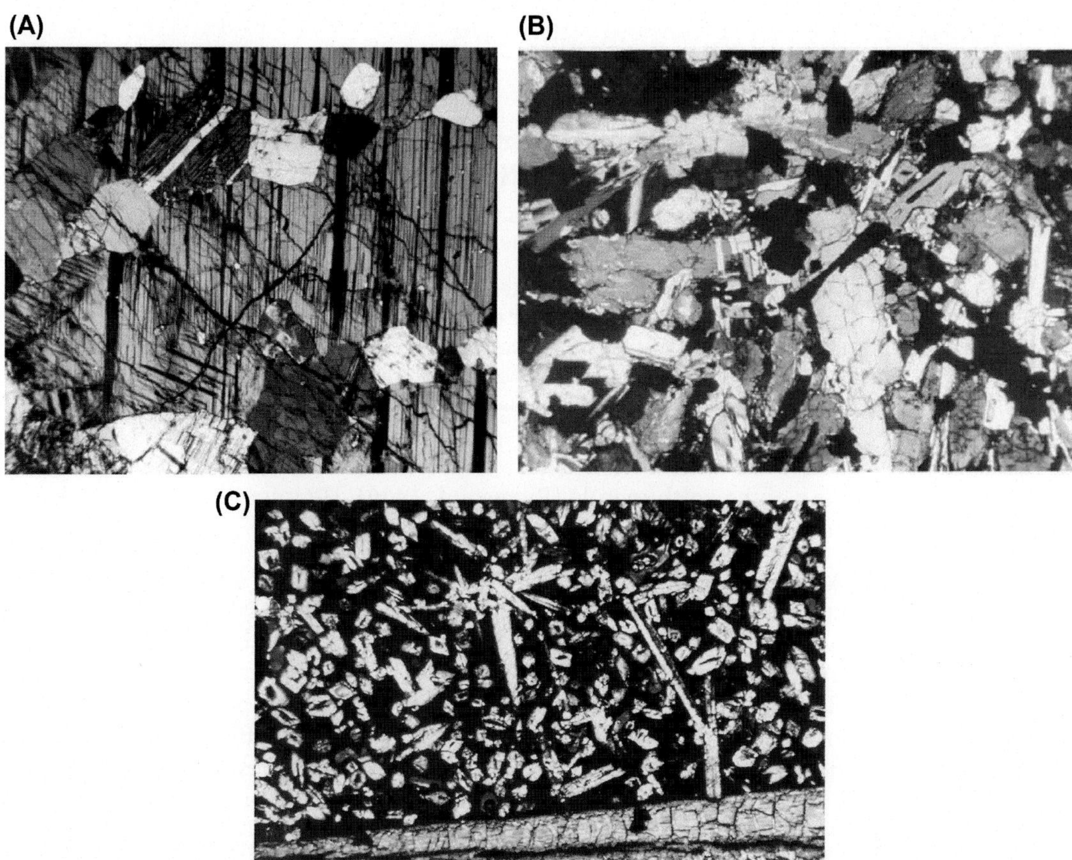

FIGURE 3.3

Photomicrographs of thin sections (see Box 3.3) of lunar rocks. (A) Ferroan anorthosite, 15415. Cross polarlized. Field of view is 3 mm. Twinning in mildly shocked plagioclase, which made up 98% of the rock. It is a unique lunar sample, which is the original crust of the Moon by plagioclase floatation in a magma ocean. The Rb–Sr and U–Th–Pb methods could not date 15415. However, the Ar–Ar dating and recalculation gave 4.08 Gyr with the old decay constant (Albarede, 1978). (B) Ilmenite basalt (high K), 10024. Cross polarlized. Field of view is 2.5 mm. Black crystals are ilmenite. Papanastassiou and Wasserburg (1971) obtained the ^{87}Rb–^{87}Sr isochron age to be 3.61 ± 0.07 Gyr. (C) Vitrophyric (phenocrysts are embedded in a glassy groundmass) pigeonite basalt, 15597. Cross polarlized. Field view is 2 mm × 1 mm. Black groundmass is glass and acicular minerals are pyroxenes. A large crystal at the bottom is pigeonite (Ca-poor pyroxene). It was quenched from high temperature giving insight to the crystallization history and cooling rate of lunar basalts. The bulk composition of 15597 was used in numerous experimental studies.

(A) Image Credit: Meyer C. 2011. NASA S71-52630. http://history.nasa.gov/alsj/a15/LunarSampleCompendium15415.pdf. (B) Image Credit: Meyer C. 2011. NASA S70-49978. http://www.history.nasa.gov/alsj/a11/a11-10024.pdf. (C) Image Credit: Meyer C. 2010. NASA S 71-51794. http://history.nasa.gov/alsj/a15/LunarSampleCompendium15597.pdf.

BOX 3.3 THIN SECTION

In mineralogy and petrology, thin sections of samples like rocks, minerals and soils are prepared. Then the thin section is observed using a polarizing petrographic microscope, an electron microscope, an electron probe micro analyzer (EPMA), and a secondary ion mass spectrometer (SIMS).

The sample chip is cut, polished, and pasted on a glass slide. Then the sample is cut and polished into approximately 30 μm-thick pieces of the sample. The observation of rock samples using a polarizing petrographic microscope is done in two steps. The first is observation by open Nicol, which is done by a single polarizing light.

Continued...

BOX 3.3 THIN SECTION continued

The other is observation by cross Nicol, in which the polarized light passed through the sample is further observed through the analyzer (the second polarized plate).

Each mineral has its own characteristic optical properties; thus, most rock-forming minerals are identified easily. For example, plagioclase is a clear mineral with twin crystals. Clinopyroxene and orthopyroxene are green minerals in open Nicol. Olivine is colorful when observed in cross Nicol.

those of the Earth mantle (Ringwood, 1966). He inferred that the origin of the Moon was related to that of the Earth's mantle.

Geochemistry of the Moon started to be discussed by geochemists after Moon samples were obtained by the Apollo spacecraft (see Section 3.3). Dreibus and Wänke (1979) noted that the moderately to highly siderophile elements (including Mn, V and Cr) are depleted in the Earth's mantle as well as the Moon relative to those of the CI chondrite composition. They also found that the compositions of the Earth's mantle and the Moon are different from those of the eucrite parent body (EPB), which is proposed as 4 Vesta. Therefore, the Moon should not have been formed from the CI chondrite or the EPB.

The Moon samples were depleted in volatile elements and metals compared to the Earth's mantle, with mean densities of 3.34 and 4.4 $g\,cm^{-3}$ for the Moon and the Earth's mantle, respectively. Drake (1983) discussed the origin of the Moon based on the siderophile element abundances [W, rhenium (Re), Mo, P, gallium (Ga), and germanium (Ge)].

Ringwood and Kesson (1977) concluded that the average abundances of iron (Fe), cobalt (Co), nickel (Ni), W, P, sulfur (S), and selenium (Se) of the Moon were similar to those of the Earth's mantle within a factor of approximately 2. Thus they concluded again that the similarity in siderophile elements between the Moon and the Earth's mantle implies that the Moon was derived from the Earth's mantle after the Earth's core had segregated. This model was consistent with the model of Wänke et al. (Dreibus and Wänke, 1979). Ringwood (1986) discussed the origin of the Moon again. He emphasized that high depletion of metallic iron is a remarkable feature of the Moon. The metallic core of the Moon is less than 2%, while the metallic core of the Earth is 32%.

Fig. 3.4 is a plot of abundance ratios of siderophile elements in the Moon versus those of the Earth's mantle against their metal/silicate partition coefficients (Ringwood, 1986) (see Box 3.4). Less siderophile elements than Ni (Mn, V, Cr, Fe, W, P, Co, and S) are similar in both bodies. Elements more siderophile than Ni [e.g., copper (Cu), Mo, Re, and gold (Au)] are depleted to degrees that correspond well with their metal/silicate partition coefficients. If a small metallic core of less than 1% of the lunar mass was separated after accretion of the Moon, depletion of highly siderophile elements in Fig. 3.4 can be explained. Note that there are no depletions of Mn, V, and Cr in Fig. 3.4. The vertical axis is the value of Moon/

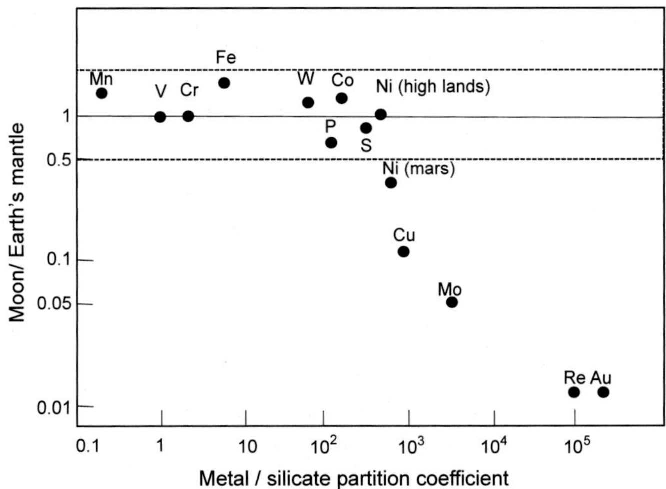

FIGURE 3.4
Abundance ratios of siderophile elements in the Moon and Earth's mantle versus their metal/silicate partition coefficients after Ringwood (1986). Both vertical and horizontal scales are logarithmic. The *horizontal line* indicates ×1 of the elemental abundance of the Earth's mantle, and the *dotted lines* show those of ×2 and ×0.5, respectively.

Earth's mantle. Therefore, although there are depletions in concentrations of Mn, V, and Cr compared to CI chondrite, they show no depletion in Fig. 3.4.

BOX 3.4 PARTITION COEFFICIENTS

The concepts of distribution coefficients and partial melting are used in cosmochemistry. We assume the system is partially molten. The weight of the liquid, the solids, and the total of major components are M_L, M_S, and M, respectively. One of the trace elements, T, is distributed based on the partition coefficient or distribution coefficient, K_d. The K_d value is defined as:

$$K_d = C_S/C_L \tag{B3.4.1}$$

where C_S and C_L are concentrations of the element T in solid and liquid, respectively. When the melt fraction, F, is introduced,

$$M_L = FM; \qquad M_S = (1-F)\,M \tag{B3.4.2}$$

$$m_L = fm; \qquad m_S = (1-f)\,m \tag{B3.4.3}$$

The weights of the trace element T in the liquid, solid, and total are m_L, m_S, and m, respectively. Then:

$$K_d = C_S/C_L = (m_S/M_S)\,/\,(m_L/M_L) = (1-f)\times F/\,[f\times(1-F)]$$

or

$$F = f\,K_d/\,(1 + f\,K_d - f) \tag{B3.4.4}$$

This equation indicates that if K_d is previously determined experimentally, and f is obtained from the sample, the fraction of the partial melting, F, can be estimated.

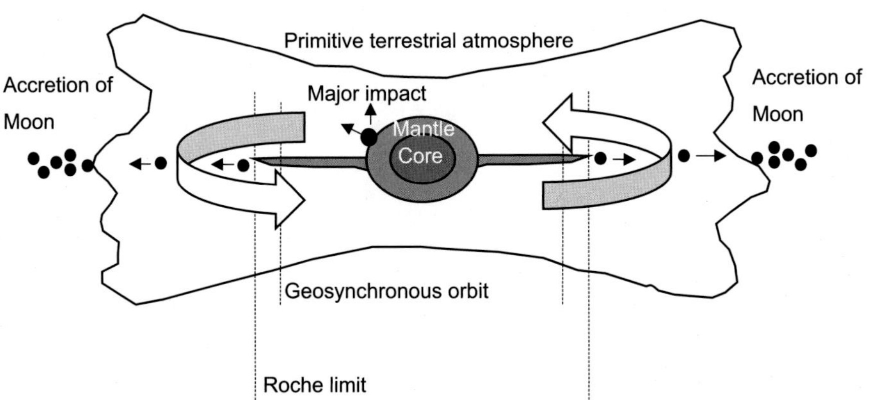

FIGURE 3.5
A fission model of the origin of the Moon by Ringwood (1986). The co-rotation speed of the primitive terrestrial atmosphere is approximately 5 h. The primitive terrestrial atmosphere is approximately 5 RE. Geosynchronous orbit is approximately 2.5 RE. The Roche limit (see Box 3.5) is approximately 3 RE. Planetesimals, which came into geosynchronous orbit, finally fell to Earth. The spray of melts formed by the major impact, which went outside of the Roche limit (see Box 3.5), eventually exited the primitive terrestrial atmosphere (~5 RE) and formed moonlets.

BOX 3.5 THE ROCHE LIMIT

Assume that one celestial body (B), which forms its shape by gravitational self-attraction, is approaching another larger celestial body (A). The Roche limit of A is the distance from A at which B will disintegrate by tidal forces of A. For example, consider that A and B are the Sun and a comet, respectively. If a comet approaches the Sun, and the comet goes into the Roche limit of the sun, the comet breaks up and forms a ring along the Roche limit around the Sun.

Ringwood (1986) made a cartoon for the proto-Earth based on a fission model (Fig. 3.5). The angular momentum was transferred from the rapidly rotating Earth to proto-lunar materials by corotating the primitive terrestrial atmosphere. Accretion of the Earth should occur before dissipation of the primitive gasses of the solar nebula. During the later stage of accretion, high temperature was produced by rapid accretion and thermal insulation. Thus materials from the mantle evaporated and spun out of the disk. The main problem of this model is that the Earth must be formed in a short timescale (~1 Myr). In addition, Earth's mantle should have completely melted and degassed for CO_2, N_2, chlorine (Cl), and noble gases, which do not suit the geochemical observations of Ringwood's model. Furthermore, it could be difficult to obtain fast internal spin of approximately 5 h. This model still exists and sometimes appears as a "fission model" (e.g., Ćuk and Stewart, 2012).

The key role of the Moon genesis in this model is the primitive terrestrial atmosphere of approximately 5 R_E. The higher FeO content of the Moon compared to the Earth's mantle should also be observed for Mn, Co, or W. The excess FeO should be derived from planetesimals involved in the major impacts. Thus, this explanation might be applied to Mn, Co, and W.

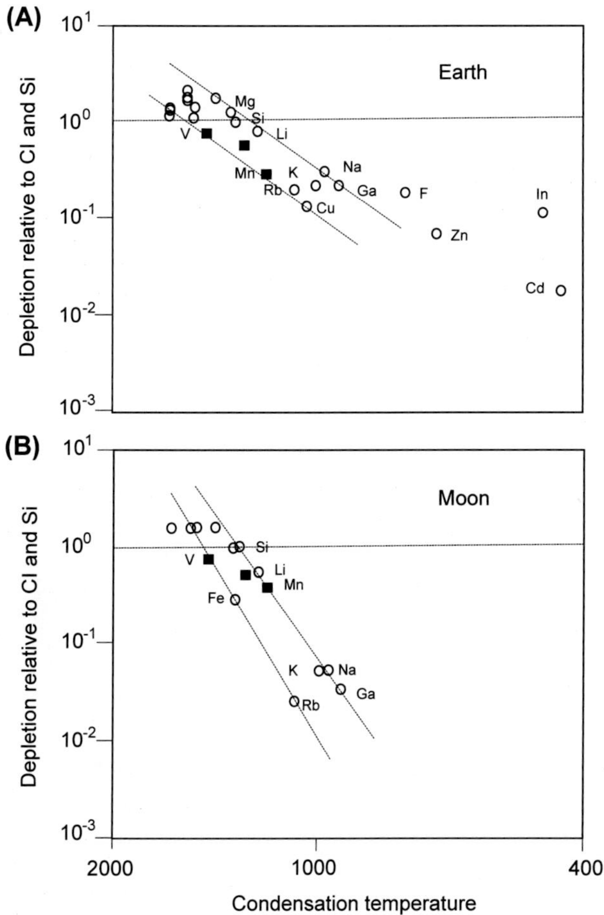

FIGURE 3.6

Depletion of elements in (A) Earth and (B) the Moon. Both horizontal and vertical scales are logarithmic, respectively. The horizontal axis is the condensation temperature. The vertical axis is the depletion normalized to CI and Si.

Data are after Wänke H, Baddenhausen, Blum K, et al. On the chemistry of lunar samples and achondrites. Primary matter in the lunar highlands: a re-evaluation. Proc Lunar Sci Conf 8th 1977:2191–213 and Drake MJ, Newsom CJ, Capobianco CJ. V, Cr and Mn in the Earth, Moon, EPB and SPB and the origin of the Moon: experimental studies. Geochim Cosmochim Acta 1989;53:2101–11.

Drake et al. (1989) published high temperature partitioning experimental results between solid iron metal, S-rich iron melt, and silicate basaltic melt for the moderate compatible elements of V, Cr, and Mn. Compatibility to metallic phases were Cr is greater than V is greater than Mn at high oxygen fugacity, and V is greater than Cr is greater than Mn at low oxygen fugacity. Solubilities into liquid metal were always larger than those in solid metal. These results suggest that the abundances of V, Cr, and Mn do not reflect core formation in the Earth. Instead, they are consistent with the relative volatilities of these elements (see Fig. 3.6). The similarity in the depletion patterns of V, Cr, and Mn inferred for the Earth's mantle and the Moon is NOT a sufficient condition for the Moon to have been derived from the Earth's mantle.

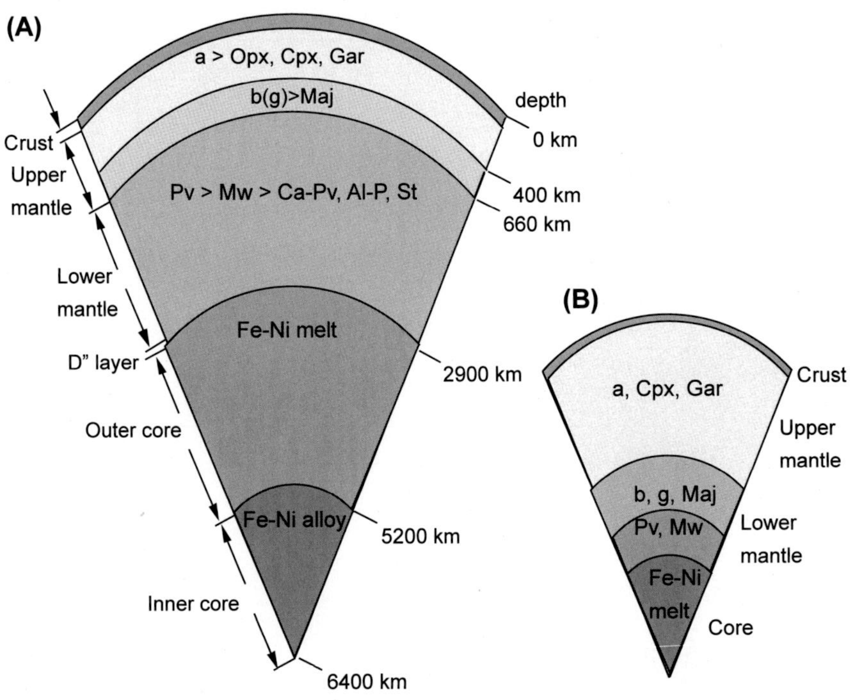

a: olivine
Opx:orthopyroxene
Cpx:clinopyroxene

Gar:garnet
b:modified spinel
g: spinel
Maj: majorite
(complicated garnet)

Pv: perovskite
Mw: magnesiowuestite
Ca-Pv: Ca-perovskite
Al-P: Al-rich phase
St: stishovite

FIGURE 3.7
Comparison of cross section of (A) the present Earth and (B) a planetesimal with 10–20% of Earth's mass (radius is 0.46–0.58 that of the Earth), which gives a Martian size to Theia as proposed by Ringwood et al. (1991). The size between (A) and (B) is proportional, but the thickness of each layer is not.

Ringwood et al. (1990) found that lithophile characters of Cr, V, and Mn are different at pressures, temperatures, and oxygen fugacities. Therefore they behave differently in the Earth's upper mantle and in the Moon condition. Their partitioning behaviors into the molten iron preferentially increase because of an increase of solubility of oxygen into the molten iron. Therefore, these elements' depletions in the Earth's mantle can be attributed to their siderophilic behavior during formation of the Earth's core. Thus a large proportion of the Moon was derived from the Earth's mantle after the Earth's core was segregated.

Ringwood et al. (1991) also conducted the partitioning experiments of Cr, V, and Mn at 1500–2000°C and at 3–25 GPa using a multi-anvil apparatus (see Box 3.4) between metallic iron and silicates, simulating the elemental partitioning in Theia (see Fig. 3.7). The partitionings of Cr, V, and Mn were always $D_{silicate/metal} > 1$, indicating they always remained lithophile. Thus there were no depletions of Cr, V, and Mn in Theia's mantle. Accordingly, depletions of Cr, V, and Mn in the Moon were not inherited from Theia's mantle. The similarity between the Earth's mantle and the Moon's mantle suggests that the Moon's mantle came from that of the Earth after core formation.

Gressmann and Rubie (2000) further investigated the partitioning of V, Cr, Mn, Ni, and Co between liquid metal and magnesiowüstite (MgO polymorph at high pressure), and observed that the partitionings into the metal increased very weakly with increasing pressure (2200 °C and 5–23 GPa), indicating a slight increase of siderophile behavior; temperature, however, had rather large effects on partitioning. If the deep magma ocean had a temperature of 3300 °C, pressure of 35 GPa, and oxygen fugacity with presence of FeO, the mantle abundance of V, Cr, Mn, Ni, and Co would be explained. Therefore, the Moon is likely to have formed largely from materials that were ejected either from the mantle of a large impacter or from the Earth's mantle (Box 3.6–3.7).

BOX 3.6 A.E. (TED) RINGWOOD (1930–93)

A.E. (Ted) Ringwood was a prominent Australian astroscientist specializing in high pressure–temperature (P–T) experiments. He received numerous awards: the W.B. Bowie Medal, Wollaston Medal, V.M. Goldschmidt Award, Clarke Medal, etc. For his achievement, one of the olivine high pressure polymorphs at the mantle transition zone (410–660 km deep) that contains hydrogen (water) is named ringwoodite.

BOX 3.7 HIGH PRESSURE EXPERIMENT APPARATUS

There are three types of high-pressure-high temperature experiment apparatus used by high pressure research groups at our institute, IPM (Institute for Planetary Materials, previously named ISEI, Institute for Study of Earth's Interior; Box 3.8). The piston-cylinder (Box 3.7.1) is used up to 1.5 GPa and 1500°C. The Kawai-type high pressure apparatus (Box 3.7.2) can produce 25 GPa and 2500°C by tungsten carbide (WC) anvils and more than 100 GPa by sintered diamond anvils. Although the high pressure area is extremely small, the diamond anvil cell apparatus can produce more than 100 GPa and 3000°C.

BOX 3.7.1 PISTON CYLINDER HIGH PRESSURE APPARATUS

Fig. Box 3.7.1 shows the piston cylinder apparatus. The red arrow in the figure shows the cylinder block. The cylinder block is water cooled. In the center of the block, there is a hole in the cylinder through which the sample is placed. The sample is sandwiched by pistons from the top and bottom. The pistons are pressed by oil pressure and the sample is compressed. For simulation experiments for relatively shallow geological phenomena, the piston cylinder apparatus is very effective because the operation is easy and the homogeneous pressure and temperature volume is large. The piston-cylinder can be used up to 1.5 GPa and 1500°C.

Continued...

BOX 3.7.1 PISTON CYLINDER HIGH PRESSURE APPARATUS continued

FIGURE BOX 3.7.1
A photograph of the piston cylinder high pressure apparatus at the Institute for Study of the Earth's Interior (ISEI) The *red arrow* indicates the cylinder, into which the pistons and the sample assembly are inserted and pressed from the bottom and the top.
Image credit: IPM, the Hacto group, Dr. D. Yamazaki.

3.5 CONSTRAINTS FROM HIGH FIELD STRENGTH ELEMENT

Zr, Hf, Nb, and Ta are called high field strength elements (HFSE). These elements easily hydrolyze in common acids such as hydrochloric acid or nitric acid, which interferes with precise concentration determination, when the simple calibration curve method is used. The only precise method is isotope dilution (ID; see Section 2.7.2); however, even when ID is used, an isotopic

BOX 3.7.2 KAWAI-TYPE HIGH PRESSURE APPARATUS

The Kawai-type high pressure apparatus, which is also called the multi-anvil apparatus, is shown in Fig. Box 3.7.2. This is a photograph of the one-axial press. The red arrow shows where the guide block exists. Fig. Box 3.7.3 shows a conceptual image of the guide block, which is designed from the six-split spheres shown in the figure (half of the split sphere is shown). The cubic assembly composed of eight WC cubic anvils is set in the guide block (see Fig. Box 3.7.3). Each of the WC anvils are truncated, and the octahedral sample assembly is placed in the center of the eight cubic anvils, as shown in Fig. Box 3.7.4A. From one face to another as shown in Fig. Box 3.7.4B, the octahedral sample assembly is made of $MgO+Cr_2O_3$. In this case, the sample shown as blue in the figure is stuffed in a Re foil, and covered with MgO and then $LaCrO_3$ in a ZrO_2 tube. The thermocouple for measuring temperature is placed at the Re foil bag. $LaCrO_3$ is used as a heater. This assemblage is placed in the press shown in Fig. Box 3.7.2. The sample is compressed and heated alternately to the target pressure and temperature

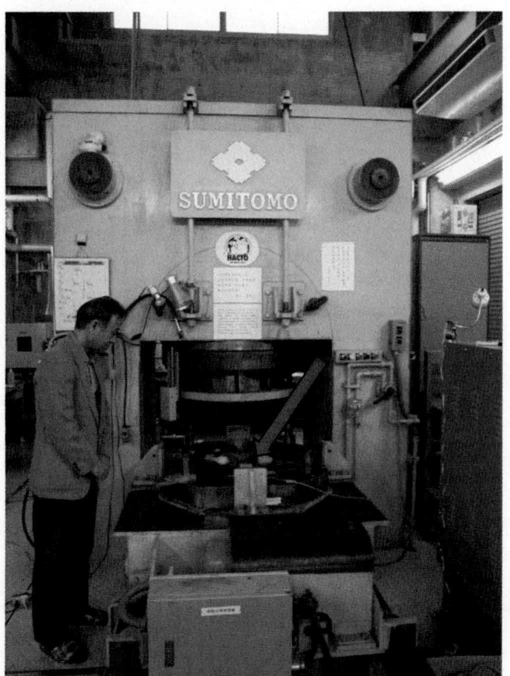

FIGURE BOX 3.7.2
A photograph of the Kawai-type high pressure apparatus in ISEI. The *red arrow* indicates the place of the guide block, where the cubic anvils with the sample assembly are placed and pressed.
Image credit: IPM, the Hacto group, Dr. D. Yamazaki.

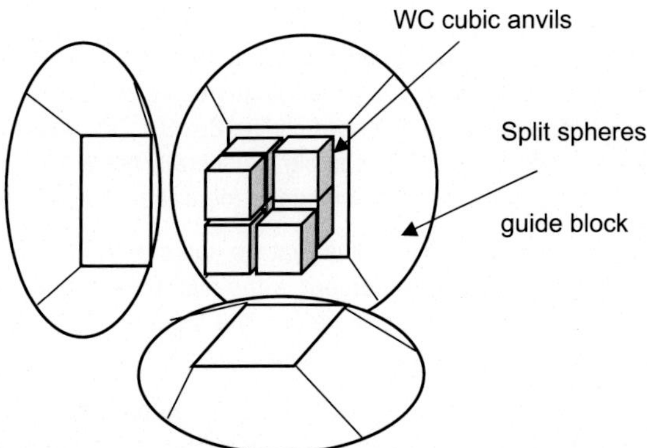

FIGURE BOX 3.7.3
A conceptual image of the guide block. The guide block is based on the split spheres in the figure, to press the assembly of the tungsten carbide (WC) cubic anvils. Only three split spheres are drawn. Actually, three faces pressing the cubic assembly are connected on each side of the guide block, and one-axial press is used to press the cubic anvils.

Continued...

BOX 3.7.2 KAWAI-TYPE HIGH PRESSURE APPARATUS continued

(A)

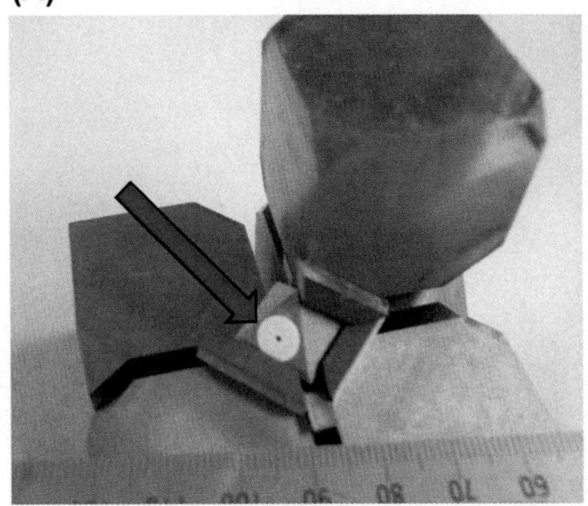

(B)

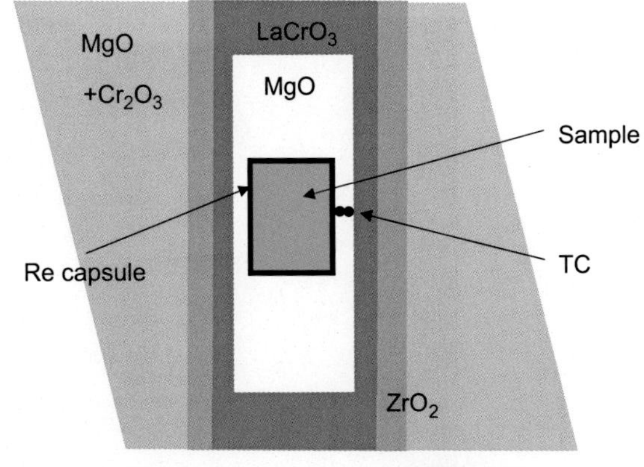

FIGURE BOX 3.7.4
(A) A photograph of the disassembled WC cubic anvils and the sample assembly indicated by the *red arrow*. The WC anvils are cut to put the octahedral sample assembly. (B) An example of the sample assembly. TC indicates the thermo-couple to measure temperature.
Image credit: IPM, the Hacto group, Dr. D. Yamazaki.

BOX 3.7.3 DIAMOND ANVIL CELL

In order to increase the achievable pressure, a harder material is more advantageous. In addition, the massive support theory requires the larger ratio of the bottom of the anvil against the top of the anvil. To satisfy the first criterion, diamond is the best material. To satisfy the second criterion, if a 3.5 mm ϕ diamond is pressed with the top of 0.04 mm ϕ, a high ratio of approximately 7700 can be obtained.

In Fig. Box 3.7.5A, the photograph of the two diamonds is shown. There is a sample at the location indicated by the red box. A right side cartoon is the enlarged cross-section. The sample is held by the Re gasket. The initial pressure is given by the cell assembly with four small screws (e.g., 6 mm ϕ) as shown in Fig. Box 3.7.5B, and the red arrow indicates the place of the diamonds. The sample is generally heated by the laser light from the top of the sample through the diamonds to more than 100 GPa and 3000°C.

The largest merit of the diamond anvil cell is that we can see the sample at high temperature and pressure conditions. The demerit is the high pressure area is too small to take out highly-pressurized samples. Therefore, the sample is studied in situ using lights (IRs, visible lights, X-rays) through diamonds.

Let us consider why such small screws in Fig. Box 3.7.5B can produce 100 GPa. When the tensile strength of the 6 mm ϕ screws is 2×10^3 N, the maximum strength from four screws is 8×10^3 N. When this strength is concentrated in the area of 0.04 mm ϕ, the pressure becomes

$$8\times10^3\text{N}/\left[\pi(0.02\times10^{-3}\text{m})^2\right]=6.4\times10^{12}\text{N/m}^2=6400\ \text{GPa}.$$

Therefore, such a small screw is enough to produce a high pressure of 100 GPa. Actually, 400 GPa and 6000K can be produced.

(A)

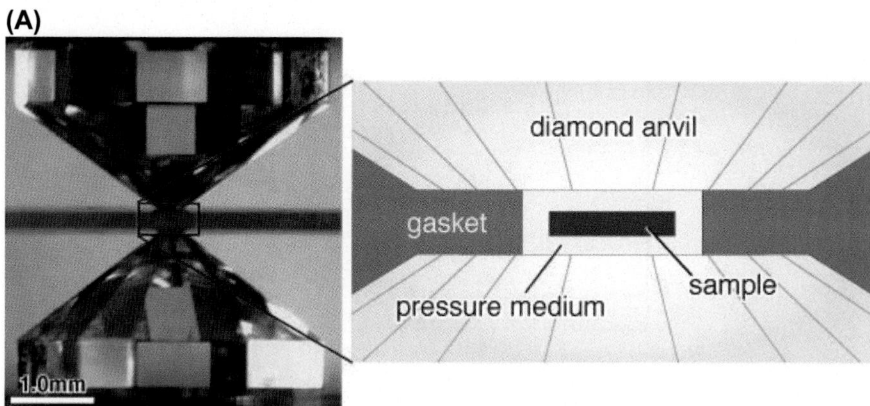

(B)

FIGURE BOX 3.7.5
The diamond anvil cell. (A) Two diamonds are facing with sandwiching a sample (*red arrow*). The right side figure is a cross-section of the sample, gasket, and diamond anvil. (B) The diamond anvil cell assembly. The shows the place of the diamond anvils.
Photographs and figure credits: IPM, S. Tateno.

BOX 3.8 SHUN'ITI AKIMOTO (1925–2004)

Schun'iti Akimoto was a Japanese high pressure physicist, who diffused the high pressure experimental techniques in Japan. After retirement from, Tokyo University, he was a director of the Institute for Study of the Earth's Interior (ISEI), Okayama University, Japan, which is now the Institute for the Planetary Materials (IPM). For his achievement, one of the high pressure silicate minerals, (Mg, Fe)SiO_3, was named akimotoite.

equilibrium must be achieved between sample and spike. As HFSE dissolves only in hydrofluoric acid (HF) forming fluoro–complex or oxo–fluoro–complex (Makishima et al., 1999), HFSE are called "fluorophile" elements (e.g., Makishima et al., 2009; Makishima, 2016), and HF is required to achieve isotopic equilibria even in ID (Box 3.7.3 and 3.8).

The Nb/Ta ratio is useful in determining the contributions of the proto-Earth and Theia (the impactor). At the high pressure, Nb partly becomes a siderophile element and enters the core (Wade and Wood, 2001). Therefore, the Nb/Ta ratio becomes lower than that of the chondrite. If the lunar materials are composed of the Earth's mantle, the lunar samples should also show a lower Nb/Ta ratio than the chondrite.

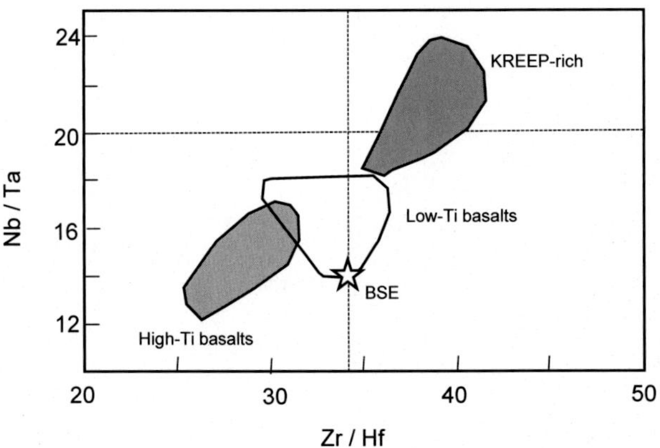

FIGURE 3.8
A schematic plot of high field strength element (HFSE) ratios of lunar samples: Zr:Hf versus Nb:Ta. KREEP-rich means KREEP-rich rocks. BSE means the bulk silicate Earth. The *horizontal* and *vertical dotted lines* indicate ratios of the CI chondrite.
The data are after Münker C. A high field strength element perspective on early lunar differentiation. Geochim Cosmochim Acta 2010;74:7340–61.

Münker (2010) developed a precise analytical measurement method of Zr/Hf and Nb/Ta ratios using isotope dilution-multicollector ICP mass spectrometry (ID-MC-ICP-MS; see Sections 2.7.2 and 2.7.4, respectively), and obtained precise Zr:Hf and Nb:Ta ratios. He used ^{183}W–^{180}Ta–^{180}Hf–^{176}Lu–^{94}Zr spikes combined with anion exchange chromatography with intermediate precision (2RSD) of less than ±0.6% and less than ±4% for Zr/Hf and Nb/Ta, respectively. He analyzed 22 lunar rocks, three lunar soils, and some meteorites. For those interested in anion exchange column chemistry, Makishima (2016) is a good textbook, although this is self-advertising. To understand errors (RSD, 2SD, 2SE, etc.) which appear in this book, see Box 4.4.

The analytical results are schematically shown in Fig. 3.8. The bulk silicate Earth (BSE) value is also plotted in Fig. 3.8 for comparison. The average Nb:Ta ratio of the lunar samples was 17.0 ± 0.8, significantly lower than that of the chondrite, which was 19.9 ± 0.6. The BSE value of 14 ± 0.3, which is far lower than that of the chondrite. Although Nebel et al. (2010) argues that the D″ layer as the source

BOX 3.9 THE D″ LAYER (D DOUBLE PRIME LAYER)

Between the core and mantle (core–mantle boundary or CMB), there is believed to be a liquid layer of approximately 200 km (see Fig. 3.7A). Oceanic crusts that deeply subducted and passed through the lower mantle could reach the core–mantle boundary and the source of the D″ layer materials. It is believed that some hot spots like the Hawaiian islands could start from this region, and could contain D″ layer materials.

of the Nb deposit (see Box 3.9), the lower Nb/Ta ratio of the lunar samples compared with the chondritic value is evidence of the relation between the Moon and Earth's mantle.

3.6 MASS FRACTIONATION OF OXYGEN ISOTOPIC RATIO OF $^{18}O/^{16}O$

Oxygen has three isotopes, and one isotopic mass fractionation using $^{18}O/^{16}O$ can be defined as:

$$\delta^{18}O = \left[\left(^{18}O/^{16}O\right)_{sample} / \left(^{18}O/^{16}O_{VSMOW}\right) - 1\right] \times 1000 \qquad (3.6.1)$$

where VSMOW means the Standard Mean Ocean Water of Vienna. Details on VSMOW are discussed in Tanaka and Nakamura (2012).

Fig. 3.9 shows a histogram for the $\delta^{18}O$ values of low Ti basalts including soil, regolith and apatite, and the Earth (an average and 2SD as a filled circle and a black bar, respectively; see Box 4.4 for 2SD). The figure is modified from Dauphas et al. (2014). The Moon data are from Wiechert et al. (2001), Spicuzza et al. (2007), Hallis et al. (2010), and Liu et al. (2010). The terrestrial data are from Eiler (2001). From Fig. 3.9, the $\delta^{18}O$ values of the Moon are identical to those of the terrestrial values. Both the oxygen isotope fractionations of the Moon and Earth are homogeneous. Combining $\delta^{18}O$ with another oxygen isotope ratio, more detailed discussion can be done by so-called three oxygen isotope ratios, which is performed in Section 3.7.

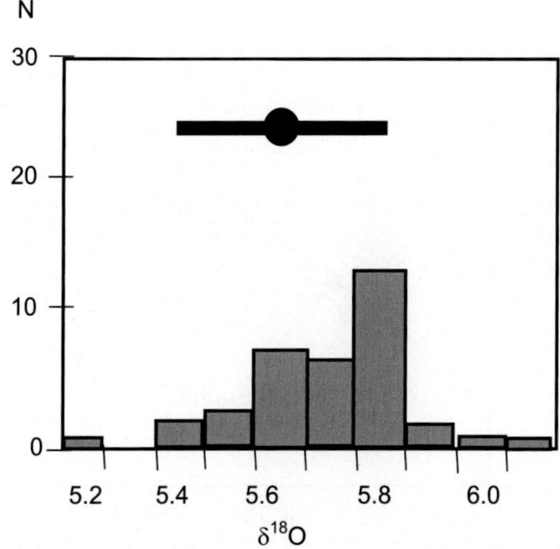

FIGURE 3.9
A histogram for the $\delta^{18}O$ values of low Ti basalts including soil, regolith and apatite, and Earth (a filled circle with 2SD as a bar).

The figure is modified from Dauphas N, Burkhardt C, Warren PH, Teng FZ. Geochemical arguments for and Earth-like Moon-forming impactor. Phil Trans R Soc A 2014;372:20130244.

3.7 MASS FRACTIONATION OF THREE OXYGEN ISOTOPE RATIOS

3.7.1 Definition of Three Oxygen Isotope Mass Fractionation

Oxygen has three isotopes, ^{16}O, ^{17}O and ^{18}O. The combination of two oxygen isotope ratios, $^{17}O/^{16}O$ and $^{18}O/^{16}O$, gives stronger constraints in the astrosciences, because the oxygen isotope discussion becomes not one but two dimensional. The oxygen isotope ratios are expressed by δ-notation:

$$\delta^{17}O = \left[\left(^{17}O/^{16}O\right)_{sample}\left(^{17}O/^{16}O_{VSMOW}\right) - 1\right] \times 1000 \quad (3.7.1)$$

$$\delta^{18}O = \left[\left(^{18}O/^{16}O\right)_{sample}/\left(^{18}O/^{16}O\right)_{VSMOW} - 1\right] \times 1000 \quad (3.7.2)$$

where VSMOW means Vienna Standard Mean Ocean Water (see Section 3.6). The data of the carbonaceous chondrites (CI, CM, and CV), ordinary chondrites (H and L), Earth, and the Moon are roughly plotted in Fig. 3.10 based on Clayton and Mayeda (1975).

3.7.2 Mass Fractionation Laws

Here we assume: ^{16}O, ^{18}O, and ^{17}O are denominators, target, and reference isotopes; R_m is an observed isotope ratio of $^{18}O{:}^{16}O$; R_c is a mass fractionation corrected ratio of $^{18}O{:}^{16}O$; R_r is a reference isotope ratio or a normalizing isotope ratio of $^{17}O{:}^{16}O$ (a constant); R_{rm} is an observed reference isotope ratio of

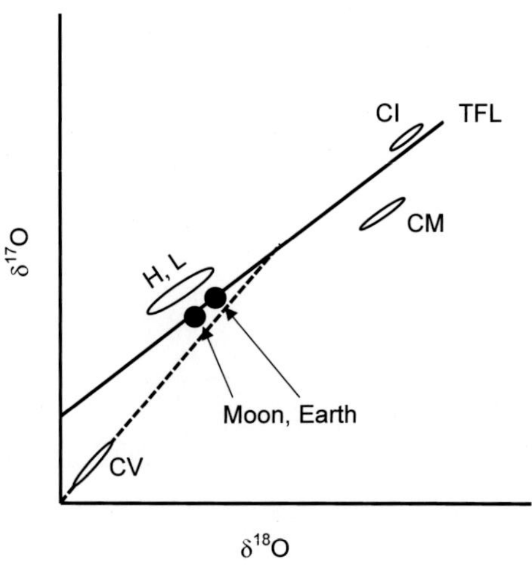

FIGURE 3.10
A plot of $\delta^{17}O$ against $\delta^{18}O$. TFL indicates a terrestrial fractionation line. The *dotted line* shows the line of CV chondrites. CI, CM, and CV are the carbonaceous chondrites. H and L indicate H and L ordinary chondrites, respectively. A slope of TFL is approximately 0.5, and that of a dotted line of CV is approximately 1. The $\delta^{18}O$ values of Earth are 5–7‰.

$^{17}O:^{16}O$; and α is a mass fractionation correction factor. Then, two isotope ratios of R_c and R_{rm} follow the mass fractionation laws as follows:

- The linear law

$$R_c = R_m \times [1 + \alpha \times (18 - 16)] \tag{3.7.3}$$

$$\alpha = (R_r/R_{rm} - 1)/(17 - 16) \tag{3.7.4}$$

- The power law

$$R_c = R_m \times (1 + \alpha)^{(18-16)} = R_m \times (R_r/R_{rm})^{[(18-16)/(17-16)]} \tag{3.7.5}$$

$$\alpha = (R_r/R_{rm})^{[1/(17-16)]} - 1 \tag{3.7.6}$$

- The exponential law

$$R_c = R_m \times (18/16)^{(\alpha \times 16)} \tag{3.7.7}$$

$$\alpha = [\log(R_r/R_{rm})/\log(17/16)]/16 \tag{3.7.8}$$

If the raw data of (R_m, R_{rm}) (=($^{18}O:^{16}O$, $^{17}O:^{16}O$) $_{observed}$) are mass-fractionated following the above three fractionation correction laws, the data of ($\delta^{18}O$, $\delta^{17}O$) should be plotted along a line with a slope of approximately 0.5 (Box 3.10).

BOX 3.10 MASS FRACTIONATION AND MASS DISCRIMINATION

"Mass fractionation" and "mass discrimination" can have similar meanings. In this book, mass fractionation indicates "the mass dependent fractionation that occurs in nature", whereas mass discrimination is "the mass dependent phenomena that occurs in the mass spectrometer or in artificial conditions such as evaporation or ion exchange column chemistry in sample preparation".

3.7.3 Application of Three Oxygen Isotope Fractionation to the Earth–Moon System

Clayton and Maeda (1975) showed that the oxygen isotope ratios of the Moon and Earth are plotted almost on the same point along the terrestrial fractionation line (TFL) as shown in Fig. 3.10. They also plotted data of extraterrestrial materials on Fig. 3.10; however, such oxygen isotope ratios are distinctive from all other Solar System bodies except enstatite chondrites and aubrites (not shown in Fig. 3.10). This strongly constrains the theory that the Moon and Earth were composed of the same source.

The TFL can be defined, which indicates all the terrestrial materials should be plotted on this line. The data of samples on the TFL are always on it irrespective of any possible fractionations, so far as the data follow the fractionation laws indicated above.

However, it should be noted that the reverse is false. Even if the oxygen isotope data of meteorites are on the TFL, the source of the meteorites may not necessarily be terrestrial. This idea sometimes appears in discussions about the origin of the Moon. Even if the Moon sample is on the TFL, the origin of the Moon is not necessarily Earth. Thus, the oxygen isotope ratio gives clear criteria that the sample is of extraterrestrial origin if the sample is not on the TFL.

This observation has been valid for over 40years (Clayton and Maeda, 1996; Wiechert et al., 2001; Spicuzza et al., 2007; Hallis et al., 2010). Because it is very unlikely that the two bodies, the proto-Earth and Theia have similar oxygen isotopic ratios on the TFL, it is required that two bodies mixed well and the evidence of two bodies were perfectly erased. This is consistent with other isotopes, such as Li, Mg, Si, K, Fe, Ti, Cr and Ni which are discussed in later sections (see Sections 3.8.1–3.8.7).

3.7.4 $\Delta^{17}O$—A New Oxygen Isotope Ratio Presentation

Pack and Herwatz (2014) invented a new data presentation for $\Delta^{17}O$, which is used for the Earth's mantle. They thought mass independent fractionation should occur even to a small degree in rocks and minerals, which was anticipated by Matsuhisa et al. (1978). The definition of the data presentation, is as follows:

$$\delta^{17}O = \left[\left({}^{17}O/{}^{16}O\right)_{sample}/\left({}^{17}O/{}^{16}O\right)_{VSMOW} - 1\right] \times 1000 \quad (3.7.9)$$

$$\delta^{18}O = \left[\left({}^{18}O/{}^{16}O\right)_{sample}/\left({}^{18}O/{}^{16}O_{VSMOW}\right) - 1\right] \times 1000 \quad (3.7.10)$$

Mass dependent fractionation between reservoirs (A, B) is defined as:

$$\alpha_{A-B}{}^{2/1} = \left(\alpha_{A-B}{}^{3/1}\right)^{\theta A-B} \quad (3.7.11)$$

The symbol α_{A-B} is the fractionation factor in $^{17}O/^{16}O$ ("2/1") and $^{18}O/^{16}O$ ("3/1") between reservoir A and B. In the case of equilibrium fractionation, A and B denote 2 phases. θ_{A-B} is termed the triple isotope fractionation exponent. For oxygen, θ_{A-B} can vary between 0.5 and 0.5305 (Matsuhisa et al., 1978; Young et al., 2002; Cao and Liu, 2011).

Small deviation of $^{17}O/^{16}O$ from a given reference line (RL) are expressed as:

$$\delta^{17}O = \delta'^{17}O - \beta \times \delta'^{18}O \quad (3.7.12)$$

where

$$\delta'^{17}O = 1000 \times \ln\left[\delta^{17}O_{VSMOW}/1000 + 1\right] \quad (3.7.13)$$

$$\delta'^{18}O = 1000 \times \ln\left[\delta^{18}O_{VSMOW}/1000 + 1\right] \quad (3.7.14)$$

and

$$\beta = 0.5305. \quad (3.7.15)$$

Note that $\Delta^{17}O$ is not a measured value but an artificially calculated value and a variable value, depending on the reference line.

3.7.5 Lunar Oxygen Isotopic Ratios

Wiechert et al. (2001) measured oxygen isotopic ratios of $^{17}O/^{16}O$, $^{18}O/^{16}O$ and $\Delta^{17}O$ of 31 lunar samples collected by the Apollo 11, 12, 15, 16, and 17 missions. All oxygen isotopic ratios were within ±0.016‰ (2SD) on a single mass-dependent fractionation line, which is identical to TFL within errors. This observation is consistent with the giant impact model. The proto-Earth and Theia, both of which had identical isotopic compositions, had mixed well. The similarity came from the identical heliocentric distances (distance from the sun) of both proto-Earth and Theia, because the isotopic composition is determined by the heliocentric distance. They used

$$\delta^{17}O = \delta^{17}O - 0.5245 \times \delta^{18}O. \quad (3.7.16)$$

They attributed the depletion of volatiles and higher FeO content of the Moon compared with that of Earth's mantle to secondary origin. Significant amounts of materials might be admixed to the Moon after the giant impact. However, they cannot change the oxygen isotopic ratios significantly. Their analytical precision can detect more than 3 and more than 5% of the primitive meteorites or differentiated planetesimals, respectively.

3.7.6 Development of Accuracy of Oxygen Isotopic Ratios

Herwartz et al. (2014) improved the accuracy of the oxygen isotope measurement. They carefully investigated the $\Delta^{17}O$ values of Wiechert et al. (2001), and concluded that the values of the lunar samples were not identical with that of the terrestrial values and should be elevated. They used $\Delta^{17}O$ to be

$$\delta'^{17}O - (0.532 \pm 0.006) \times \delta'^{18}O, \quad (3.7.17)$$

which is different from Eq. (3.7.16). This slope is identical to 0.5305 when the error is considered, which is high-temperature oxygen isotope fractionation (Young et al., 2002). They defined BSE values to be -0.101 ± 0.002‰ from their measurements of mantle minerals and mid-ocean ridge basalts (MORB). Three lunar basalt samples (high Ti basalt, low Ti gabbro and low Ti basalt) gave oxygen isotopic ratios in a small range of $\Delta^{17}O = -0.089 \pm 0.002$‰ ($1\sigma$, $n = 20$), which is higher than that of BSE (see Fig. 3.11).

Theia was estimated to be made not of carbonaceous chondrites, but of non-carbonaceous chondrites much the same as Earth, Mars, ordinary chondrites, enstatite chondrites, or achondrites with Ti, Cr, and Ni isotopic composition (Warren, 2011). They reevaluated lunar oxygen isotope data from three previous studies (Wiechert et al., 2001; Spicuzza et al., 2007; Hallis et al., 2010), and the slightly higher $\Delta^{17}O$ was consistent within uncertainties.

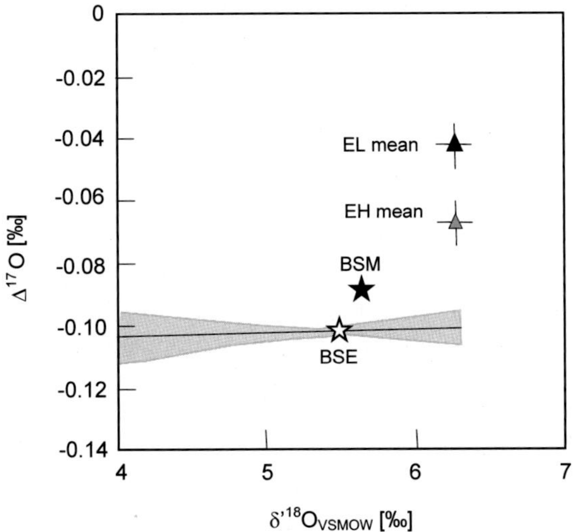

FIGURE 3.11
A plot of $\Delta^{17}O$ versus $\delta'^{18}O$ of averages of terrestrial, Moon and E chondrite samples as proposed by Herwartz et al. (2014). BSE means the bulk silicate Earth, and a horizontal line and a hatched area show the average and 1σ area of mantle minerals. BSM means the bulk silicate Moon. EL mean and EH mean indicate the averages of EL and EH chondrites, respectively. Error bars are written as 1σ.

It is suggested that the Moon received fractionally more impactor material than the Earth, and that the $\Delta^{17}O$ of Theia was most likely higher than that of the Earth and the Moon. Mixing of 4% of material isotopically resembling Mars would be sufficient to explain the observed 12 ppm difference between Earth and the Moon. It is against the general model (Cameron and Benz, 1991; Canup and Asphaug, 2001; Canup, 2012; Ćuk and Stewart, 2012; Reufer et al., 2012); therefore, larger fractions of Theia should be in the Moon. This implies that the $\Delta^{17}O$ composition of Theia was only slightly higher than that of the Earth.

Here the value of δf_T is introduced:

$$\delta f_T = (F_M/F_P - 1) \times 100 \tag{3.7.18}$$

where F is the mass fraction of proto-Earth material in the Moon (F_M) and in the final planet (F_P), respectively. The δf_T quantifies the percent compositional deviation of the Moon (or disk) from Earth. In most models, the impacter material in the Moon is larger than that in the Earth; therefore, δf_T is usually negative. The model-independent δf_T is obtained only from Nb:Ta ratios (see Section 3.5) to be 17.0 ± 0.8 and 14.0 ± 0.3 for the Moon and Earth's mantle, respectively. This translates δf_T to be −21 to −44. The model calculation indicates 70–90% of the Moon is made of Theia.

They obtained new enstatite chondrite data, which differed from Earth values by 59 ± 8 ppm (1σ, n = 14) and 35 ± 10 ppm (1σ, n = 10) for low iron enstatite chondrites (EL) and more metal-rich high iron enstatite chondrites (EH), respectively. If the oxygen isotopic composition of Theia is similar to that of

EL or EH chondrites, δf_T is −21 ± 9% and −36 ± 15% for EH, respectively. These estimates agree well with the more recent numerical models (Canup, 2012; Ćuk and Stewart, 2012; Reufer et al., 2012) and estimates from the Nb:Ta mass balance. If Theia resembled enstatite chondrites (ECs), the Moon will fall on a mixing line between ECs and Earth. The high-precision Ti isotope results suggests that the Earth and ECs are rather similar (Zhang et al., 2012). The simple average of $\varepsilon^{50}Ti$ of ECs are −0.12 (n = 5) and −0.29 (n = 2) for EH and EL, respectively, whereas $\varepsilon^{50}Ti$ of the lunar and terrestrial averages were −0.03 ± 0.04 and 0.01 ± 0.01, respectively. Thus ECs, and especially EHs, fit as the composition of Theia. High FeO content in the lunar mantle does not fit the low FeO content of ECs; however, Fe levels in ECs are high (EH and EL = 33 and 24 wt.%, respectively) (Javoy et al., 2010). Enstatite chondrites are sometimes considered to be the sole building blocks of Earth (Javoy et al., 2010). However; there is no need to have Earth created from one known type of a meteorite.

In addition, the late veneer (see Chapter 4) can change Earth's isotope ratios, and the present Moon shows the oxygen isotopes of the proto-Earth. From W isotope systematics, the late veneer is estimated to be 0.3–0.8% BSE (Walker, 2009). If 0.5% late veneer had low oxygen isotope composition of CV carbonaceous chondrites ($\Delta^{17}O$ = −4‰) (Clayton and Mayeda, 1999), $\Delta^{17}O$ of the bulk Earth would decrease approximately 20 ppm, which is consistent with the observed difference of $\Delta^{17}O$ between Earth and the Moon. The water in carbonaceous chondrites could be the source of the ocean water.

3.7.7 The Latest Oxygen Isotopic Evidence

Young et al. (2016) proposed the newer model based on the most recent measurements of the oxygen isotopic ratios of the Moon and Earth samples. They assumed that the San Carlos (SC) olivine is the standard of the Earth's mantle, because the oxygen isotope ratios of silicates are systematically different from those of the standard mean ocean water (SMOW). Thus, they obtained and expressed oxygen isotope ratios as $\delta'^{17}O$, $\delta'^{18}O$, and $\Delta'^{17}O$ based on the SC olivine. All the prime-δ and prime-Δ indicate the oxygen isotope ratios based on the SC olivine in this section. In this expression, $\Delta'^{17}O$ of the silicate Earth is approximately 0.1‰ lower than $\Delta^{17}O$ of that in SMOW. The following equation was the reference fractionation line for the SC olivine with a β value of 0.528:

$$\delta'^{17}O = \delta'^{17}O - 0.528 \times \delta'^{18}O \tag{3.7.19}$$

Based on this criterion, Young et al. (2016) found no discernible difference between the $\Delta'^{17}O$ values of the terrestrial mantle (the SC olivine; −1 ± 2 ppm, 1 SE) and those of lunar basalts powders (0 ± 2 ppm, 1 SE) or lunar fused powder beads (0 ± 3 ppm, 1 SE). When all uncertainty is taken into account, the $\Delta'^{17}O$ values of the terrestrial mantle is −1 ± 4.8 ppm (2 SE), which is indistinguishable from zero.

They compared the $\Delta'^{17}O$ value of Earth (SC olivine) with that of the Moon in references. In five studies: (Wiechert et al., 2001; Spicuzza et al., 2007; Hallis et al., 2010; Herwartz et al., 2014, and this study), two showed significant difference; one showed no difference; and the remaining two showed equivocal difference.

In particular, the results of Young et al. (2016) do not agree with the conclusions of those of Herwartz et al. (2014), which is discussed in Section 3.7.6. However, the same lunar sample agreed with each other. Thus Young et al. (2016) explained that the discrepancies are caused by unfortunate sample selection. (The author thinks this is an ad hoc, not scientific explanation.)

The lunar highland sample showed a significantly lower $\Delta'^{17}O$ value of −16 ± 3 ppm (1 SE), which is similar to Wiechert et al. (2001). Taking the lower values of the terrestrial anorthosite sample into account, the lower values seems to be related to a mass fractionation process in forming this rock type (terrestrial anorthosite and lunar anorthositic troctolite).

Differences in the $\Delta'^{17}O$ value between Theia and the proto-Earth were assumed to be 0.15 or 0.05‰ by Kaib and Cowan (2015) and Mastrobuono-Battisti et al. (2015), respectively, based on the recent N-body simulations of standard terrestrial planet formation scenarios with assumed gradients in $\Delta'^{17}O$ across the inner Solar System. Young et al. (2016) used a planetary accretion model (Rubie et al., 2015) which uses N-body accretion simulations based on the Grand Tack model (Walsh et al., 2011; see Section 2.10). They calculated the masses of Earth and Mars and the oxidation state of the Earth's mantle using a multi-reservoir model (silicate, oxidized iron and water) to describe the initial distribution of oxygen isotopes, including the effects of mass accretion after the giant impact. When the late veneer (see Chapter 4) considered to be less than 1% by mass was taken into account, $\Delta'^{17}O_{Theia}-\Delta'^{17}O_{proto\text{-}Earth}$ became nearly zero in all simulations. The $\Delta'^{17}O_{Moon}-\Delta'^{17}O_{Earth}$ values were +20% to approximately −60% for the Mars-sized impacter scenario, and +8% to approximately −12% in the proto-Earth-sized impacter scenarios.

As Young et al. (2016) observed that the Moon and Earth $\Delta'^{17}O$ values are similar, the oxygen isotopes of Theia and proto-Earth were too thoroughly mixed by the Moon-forming impact for the difference to be less than 5 ppm levels. This also constrains the late veneer theory, which is discussed in Chapter 4.

3.8 CONSTRAINTS FROM STABLE ISOTOPE RATIOS FOR THE ORIGIN OF THE PRESENT EARTH AND MOON

3.8.1 The δ^7Li Values

Lithium has only two isotopes, ^{6}Li and ^{7}Li. Precise isotope ratio determination for the two isotope elements has been very difficult by TIMS (see Section 2.7.1). The Li isotope analysis could become availabledue to the evolution of MC-ICP-MS. (see Section 2.7.4).

The Li isotope ratios are expressed as follows:

$$\delta^7\mathrm{Li}\,(‰) = \left[\left({}^7\mathrm{Li}/{}^6\mathrm{Li}\right)_{\mathrm{sample}}/\left({}^7\mathrm{Li}/{}^6\mathrm{Li}\right)_{\mathrm{standard}} - 1\right] \times 1000 \qquad (3.8.1)$$

The NIST L-SVEC Li is used as the standard material.

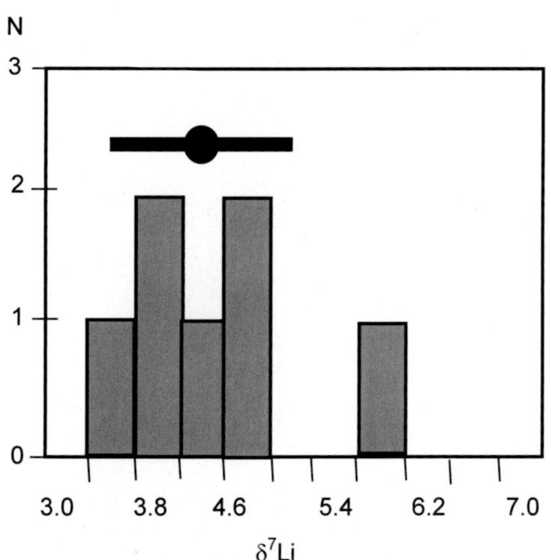

FIGURE 3.12
A histogram for the δ^7Li values of low Ti basalts including soil, regolith and apatite, and Earth (a filled circle with 2SD as a bar).
The figure is modified from Dauphas N, Burkhardt C, Warren PH, Teng FZ. Geochemical arguments for and Earth-like Moon-forming impactor. Phil Trans R Soc A 2014;372:20130244.

In Fig. 3.12, a histogram for the δ^7Li values of low Ti basalts including soil, regolith and apatite, and Earth (a filled circle with 2SD as a bar). The lunar data are from Magna et al. (2006) and Seitz et al. (2006). The Earth values are from Seitz et al. (2004) and Jeffcoate et al. (2007). The figure is modified from Dauphas et al. (2014).

As shown in Fig. 3.12, the Li isotopic signature of lunar basalts is similar to the isotopic signature of the Earth's mantle [MORB and ocean island basalts (OIBs)], or BSEs. This indicates core formation, volatile loss, and the presence of the crust and hydrosphere have not significantly influenced the Earth and Moon, which are already differentiated planetary bodies. It is suggested that if the giant impact had caused the formation of the Moon, all volatile elements such as Li could have accreted again without loss and formed the Moon.

3.8.2 The δ^{26}Mg Values

Mg isotope ratios are expressed as follows:

$$\delta^{25}\mathrm{Mg}\,(‰) = \left[\left({}^{25}\mathrm{Mg}/{}^{24}\mathrm{Mg}\right)_{\mathrm{sample}}/\left({}^{25}\mathrm{Mg}/{}^{24}\mathrm{Mg}\right)_{\mathrm{DSM3}} - 1\right] \times 1000 \qquad (3.8.2)$$

$$\delta^{26}\mathrm{Mg}\,(‰) = \left[\left({}^{26}\mathrm{Mg}/{}^{24}\mathrm{Mg}\right)_{\mathrm{sample}}/\left({}^{26}\mathrm{Mg}/{}^{24}\mathrm{Mg}\right)_{\mathrm{DSM3}} - 1\right] \times 1000 \qquad (3.8.3)$$

The DSM3 is the Mg standard solution made from pure Mg in Galy et al. (2003).

Fig. 3.13 shows a histogram for the δ^{26}Mg values of low Ti basalts, including soil, regolith and apatite, and Earth (a filled circle with 2SD as a bar). The figure is modified from Dauphas et al. (2014). The data are obtained by MC-ICP-MS

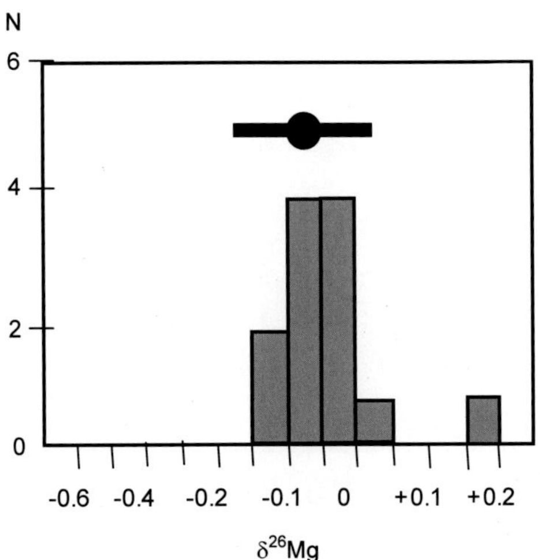

FIGURE 3.13
A histogram for the $\delta^{26}Mg$ values of low Ti basalts including soil, regolith and apatite, and Earth (a filled circle with 2SD as a bar).
The figure is modified from Dauphas N, Burkhardt C, Warren PH, Teng FZ. Geochemical arguments for and Earth-like Moon-forming impactor. Phil Trans R Soc A 2014;372:20130244.

(see Section 2.7.4). The Moon and Earth data are from Sedaghatpour et al. (2013) and Teng et al. (2010), respectively.

From Fig. 3.13, the Mg isotopic ratios of the Moon are indistinguishable from those of Earth. As Mg is a refractory element, the Mg isotopic results are consistent with the giant impact model.

When $\delta^{25}Mg$ is plotted against $\delta^{26}Mg$, all data are on the plotted on the line of a slope of approximately 2 (not shown), indicating the mass discrimination during measurement is appropriately corrected against the standard (Sedaghatpour et al., 2013). In addition, the data from Earth and the Moon are also plotted at a similar place on the same line, showing the mass-dependent fractionation is occurring along this line. In addition, the data of the chondrites are on the line, indicating the same origin of these three components.

3.8.3 The $\delta^{30}Si$ Values

The Si isotope ratios are expressed as follows:

$$\delta^{29}\mathrm{Si}\,(‰) = \left[\left({}^{29}\mathrm{Si}/{}^{28}\mathrm{Si}\right)_{\mathrm{sample}}/\left({}^{29}\mathrm{Si}/{}^{28}\mathrm{Si}\right)_{\mathrm{Standard}} - 1\right] \times 1000 \quad (3.8.4)$$

$$\delta^{30}\mathrm{Si}\,(‰) = \left[\left({}^{30}\mathrm{Si}/{}^{28}\mathrm{Si}\right)_{\mathrm{sample}}/\left({}^{30}\mathrm{Si}/{}^{28}\mathrm{Si}\right)_{\mathrm{Standard}} - 1\right] \times 1000 \quad (3.8.5)$$

The standard is the NBS-28 standard.

Fig. 3.14 shows a histogram for the $\delta^{30}Si$ values of low Ti basalts including soil, regolith and apatite, and Earth (a filled circle with 2SD as a bar). The figure

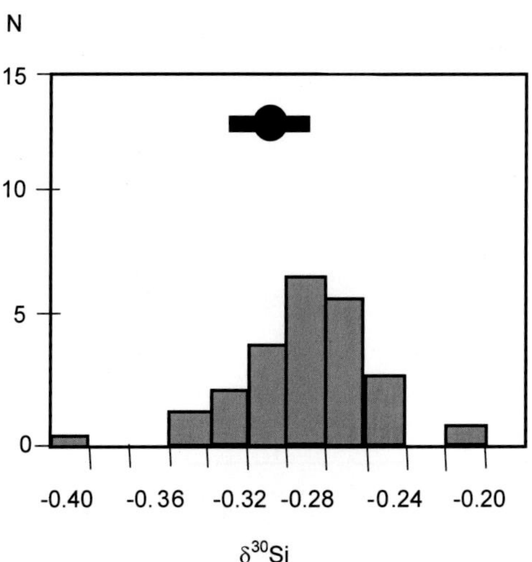

FIGURE 3.14

A histogram for the δ^{30}Si values of low Ti basalts including soil, regolith and apatite, and Earth (a filled circle with 2SDs as a bar).

The figure is modified from Dauphas N, Burkhardt C, Warren PH, Teng FZ. Geochemical arguments for and Earth-like Moon-forming impactor. Phil Trans R Soc A 2014;372:20130244.

is modified from Dauphas et al. (2014). The Si isotope data are obtained by MC-ICP-MS (see Section 2.7.4). The Moon data are from Armytage et al. (2012), Fitoussi and Bourdon (2012), and Zambardi et al. (2013). The Earth data are from Fitoussi and Bourdon (2012), Zambardi et al. (2013), Fitoussi et al. (2009), Armytage et al. (2011), and Savage et al. (2010, 2011).

In Fig. 3.14, the δ^{30}Si values of the Moon samples are indistinguishable from those of the Earth. However, the Si isotope ratios of the chondrite is systematically lower than those of Earth (and the Moon) (not shown). The difference between the BSE and the chondrite is considered to be the equilibrium fractionation in silicon in metal and silicate during core formation (Georg et al., 2007; Fitoussi et al., 2009; Armytage et al., 2011; Zambardi et al., 2013). In this model, the heavier silicon is in the mantle, and the lighter silicon is in the core. In addition, 3–16 wt.% of silicon must be in the core. This value is consistent with the density deficit of the core (Hirose et al., 2013).

Fig. 3.15 is a plot of δ^{30}Si versus Mg/Si of chondrites, Earth and the Moon. The line and the dotted curves show the regression line and 95% confidence interval. From this figure, the BSE is on the regression line of the chondrites. The Moon is not in the range of the chondrites, indicating the Moon cannot be made by mixing between enstatite chondrite and other chondrites. In this figure, the Mg:Si ratios of enstatite chondrites are much lower than BSE (Dauphas et al., 2014). Fitoussi and Bourdon (2012) concluded that the Earth cannot be made of the enstatite chondrites. This contradicts against generally accepted observations that the ordinary and carbonaceous chondrites do not match as the building block of the silicate bulk Earth, but the enstatite chondrites fit most isotope budgets (e.g., oxygen).

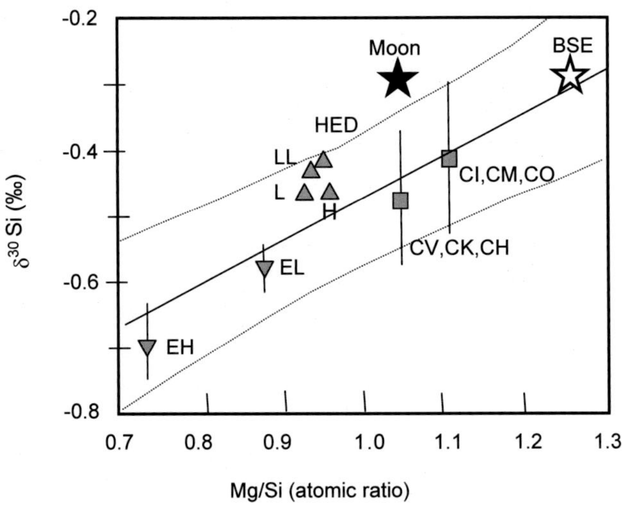

FIGURE 3.15
A plot of δ^{30}Si versus Mg/Si of chondrites, the Earth, and the Moon. The *line* is a regression line of all chondrites. The *dotted curves* indicate 95% confidence interval.
The figure is modified from Dauphas N, Burkhardt C, Warren PH, Teng FZ. Geochemical arguments for and Earth-like Moon-forming impactor. Phil Trans R Soc A 2014;372:20130244.

To create the Moon from the proto-Earth mantle, which corresponds to the bulk silicate Earth, and the enstatite chondritic Theia, an increase of δ^{30}Si is required. (The mixture of BSE and EH in Fig. 3.15 is around the square point of the carbonaceous chondrite CV, CK, and CH. So, a simple increase of δ^{30}Si can be the Moon point.) Such Si isotopic fractionation requires a counterpart with very light Si isotope ratios. However, such materials have not been found so far. The only process that explains the removal of such light Si materials is that the light-isotopic Si gases made by the giant impact went into the sun when clearing-up the super-Earths as proposed by Batgin and Laughlin (2015) (see Section 2.12). However, this idea seems too ad hoc.

3.8.4 The δ^{41}K Values

Humayn and Clayton (1995a) developed the K isotopic determination method. The sample was digested with HF + HNO_3, K was separated using cation exchange columns, and ^{41}K:^{39}K was determined using SIMS (see Box 2.7) with repeatability of 0.5‰ (2SE). The δ^{41}K values and errors are defined as:

$$\delta^{41}\mathrm{K}\,(\text{‰}) = 1000 \times \begin{array}{l} \left[\left(R_{sample}/R_{std}\right) - 1\right] \\ \pm \left(R_{sample}/R_{std}\right)\left[\left(2\sigma_{sample}/R_{sample}\right)^2 + \left(\left(2\sigma_{std}/R_{std}\right)^2\right)\right]^{1/2} \end{array} \tag{3.8.6}$$

where R_{sample} and R_{std} are ^{41}K:^{39}K ratios of the sample, and the standard, σ_{sample} and σ_{std} are standard errors (SE) of R_{sample} and R_{std}, respectively. The standard is NIST SRM 985.

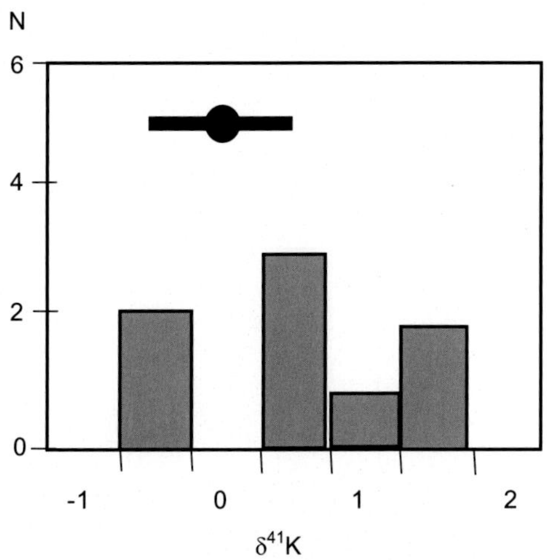

FIGURE 3.16
A histogram for the $\delta^{41}K$ values of low Ti basalts including soil, regolith and apatite, and Earth (a filled circle with 2SD as a bar).
The figure is modified from Dauphas N, Burkhardt C, Warren PH, Teng FZ. Geochemical arguments for and Earth-like Moon-forming impactor. Phil Trans R Soc A 2014;372:20130244.

In Fig. 3.16, a histogram for the $\delta^{41}K$ values of low Ti basalts including soil, regolith and apatite, and Earth (a filled circle with 2SD as a bar) is shown. The lunar data and the Earth values are by Humayn and Clayton (1995a,b), respectively. The figure is modified from Dauphas et al. (2014). The lunar basalts in Fig. 3.16 show similar values to the terrestrial values. With the exception of far heavier (0–20‰) lunar soils, $\delta^{41}K$ values of carbonaceous chondrites of CI, CM, and CV, the ordinary chondrites H, L, LL, and the enstatite chondrites show no $\delta^{41}K$ anomaly, and are identical to terrestrial values (Humayn and Clayton, 1995b). It is very interesting but the problem is not yet solved.

3.8.5 The δ^{56}Fe Values

The $\delta^{56}Fe$ value is defined as:

$$\delta^{56}\mathrm{Fe}\,(‰) = \left[\left({}^{56}\mathrm{Fe}/{}^{54}\mathrm{Fe}\right)_{\mathrm{sample}}/\left({}^{56}\mathrm{Fe}/{}^{54}\mathrm{Fe}\right)_{\mathrm{Standard}} - 1\right] \times 1000 \qquad (3.8.7)$$

where the standard IRMM-014 is used. The Fe isotope ratios are determined by TIMS using the double spike method (Johnson and Beard, 1999) and N-TIMS (Walczyk, 1997) (see Section 2.7.1; for the double spike method, see Makishima, 2016). But the most popular method is MC-ICP-MS using the double spike method (see Section 2.7.4; Millet et al., 2012; Makishima, 2013, 2016). The intermediate precision of $\delta^{56}Fe$ is generally greater than 0.05‰.

Fig. 3.17 is a histogram for the $\delta^{56}Fe$ values of low Ti basalts including soil, regolith and apatite, and Earth (a filled circle with 2SD as a bar). The figure is modified from

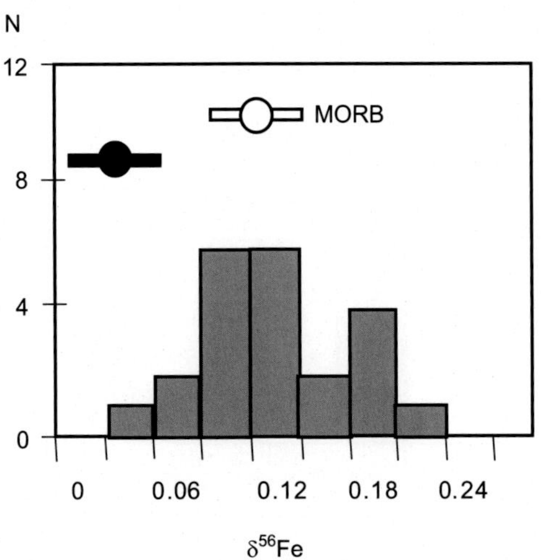

FIGURE 3.17
A histogram for the δ^{56}Fe values of low Ti basalts including soil, regolith, and apatite. The δ^{56}Fe of the Earth is shown as the *black circle* with 2SD as a bar. The *white circle* and *bar* indicate those of MORB. *The figure is modified from Dauphas N, Burkhardt C, Warren PH, Teng FZ. Geochemical arguments for and Earth-like Moon-forming impactor. Phil Trans R Soc A 2014;372:20130244.*

Dauphas et al. (2014). The Moon data are from Li et al. (2010), Wiesli et al. (2003), Poitrasson et al. (2004), and Weyer et al. (2005). The Earth data are from Weyer and Ionov (2007) and Craddock et al. (2013). In particular, Craddock et al. (2013) estimated the δ^{56}Fe value of the abyssal peridotites to be is 0.010 ± 0.010 (2SD)‰, resulting in the terrestrial mantle to be 0.025 ± 0.025‰, which is indistinguishable from thatof chondrites ($\delta^{56}Fe = +0.005 \pm 0.008$‰, $n = 42$). In contrast, as shown in Fig. 3.17, the δ^{56}Fe of MORBs were $+0.110 \pm 0.003$‰, which are apparently higher than those of the terrestrial mantle and consistent with the lunar basalts.

However, Craddock et al. (2013) do not accept the graphical difference of δ^{56}Fe between residual abyssal peridotites and MORBs shown in Fig. 3.17, and concluded that there is no difference of δ^{56}Fe between abyssal peridotites and MORBs. They also conclude that there are no mechanisms to fractionate the Fe isotopic ratios by melting. Thus, they concluded that there are no differences in the Fe isotopic ratios between the BSE and the Moon.

In conclusion, Li, O, Mg, Si, K, and Fe isotopes of Moon samples are indistinguishable from those of the BSE.

3.8.6 The ε^{54}Cr Values

Cr has four isotopes, and generally the ε^{54}Cr value is used, which is defined as:

$$\varepsilon^{54}Cr = \left[\left({}^{54}Cr/{}^{52}Cr\right)_{sample} / \left({}^{54}Cr/{}^{52}Cr\right)_{Standard} - 1\right] \times 10000 \quad (3.8.8)$$

A plot of Δ^{17}O versus ε^{54}Cr is shown in Fig. 3.18. Black circles show CI, CM, CV, CO, CK, and CR carbonaceous chondrites. A gray circle indicates CB

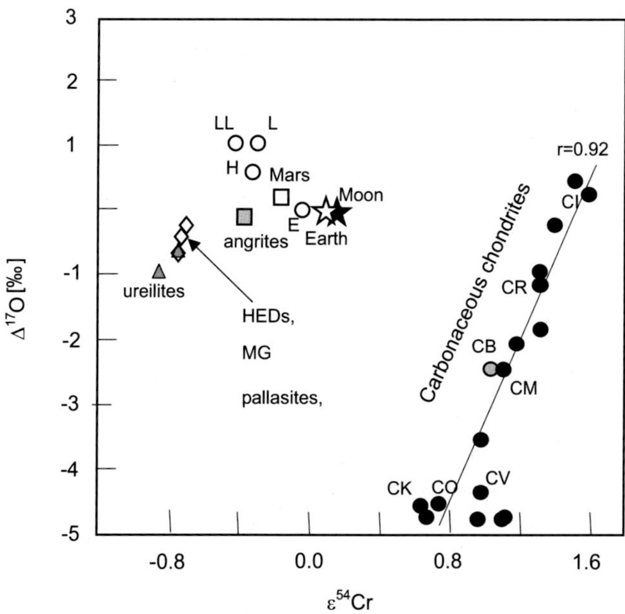

FIGURE 3.18
A plot of $\Delta^{17}O$ versus $\varepsilon^{54}Cr$. *Black circles* show CI, CM, CV, CO, CK, and CR carbonaceous chondrites. A *gray circle* indicates CB carbonaceous chondrite, for which the point is based on different analyses for $\Delta^{17}O$ and $\varepsilon^{54}Cr$. *White and black stars* and a *white square* show Earth, Moon, and Mars values, respectively. *White circles* indicate L, LL, H ordinary chondrites, and enstatite (E) chondrites. A *gray square* and *gray triangles* show angrites and ureilites, respectively. *Diamonds* indicate HEDs, MG pallasites, and mesosiderites. The *line* in carbonaceous chondrites is a regression line for all carbonaceous chondrites including CB, which gave r = 0.92.
This plot is modified after Warren PH. Stable-isotopic anomalies and the accretionary assemblage of the Earth and Mars: a subordinate role for carbonaceous chondrites. Earth Planet Sci Lett 2011;311:93–100.

carbonaceous chondrite, for which the point is based on different analyses for $\Delta^{17}O$ and $\varepsilon^{54}Cr$. White and black stars and a white square show Earth, Moon and Mars values, respectively. White circles indicate L, LL, H ordinary chondrites, and enstatite (E) chondrites. A gray square and gray triangles show angrites and ureilites, respectively. Diamonds indicate HEDs, MG pallasites, and mesosiderites. In this figure, Earth and the Moon are plotted at almost the same points.

Carbonaceous chondrites shows a line. The line is a regression line for all carbonaceous chondrites including CB, which gave r = 0.92. This indicates that the carbonaceous chondrites are made by the mixing of a CK-CO-CV reservoir and a CI reservoir. In contrast, in the plot of $\Delta^{17}O$ versus $\varepsilon^{54}Cr$, the ordinary chondrites (L, LL, H), ureilites, angrites, HEDs, MG pallasites, and mesosiderites are formed from the individual reservoir.

Compared with other isotope ratios, the Cr isotope ratio is less effective isotope ratio, because the difference of the Cr isotope is in ε-unit, showing the variation is 10 times smaller than other isotopes shown in δ or ‰ units. However, when the extinct isotope pair of ^{53}Mn–^{53}Cr is combined, two isotope pairs of Cr becomes useful.

3.8.7 The ε^{50}Ti Values

Zhang et al. (2012) determined ^{50}Ti:^{47}Ti ratios in lunar samples measured by MC-ICP-MS (see Section 2.7.4). They developed a new data correcting method for secondary effects caused by cosmic-ray exposure on the lunar surface using ^{150}Sm:^{152}Sm and ^{158}Gd:^{160}Gd isotope systematics. The Ti isotope ratios are presented as:

$$\varepsilon^{50}\text{Ti} = \left[\left(^{50}\text{Ti}/^{47}\text{Ti}\right)_{\text{sample}} / \left(^{50}\text{Ti}/^{47}\text{Ti}\right)_{\text{rutile}} - 1\right] \times 10^4 \tag{3.8.9}$$

where the terrestrial samples had an ε^{50}Ti value of 0.01 ± 0.01 in ε-units (weighted average; $n = 19$).

The largest difference of Ti from other transition elements is that Ti is either very refractory or is in rutile, a very refractory mineral (meaning that it has high evaporation temperature).

They plotted ε^{50}Ti versus ε^{54}Cr, which is shown in Fig. 3.19. Black circles show CI, CM, CV, CO, CK, CR carbonaceous chondrites, and Tagish Lake (TL). A gray

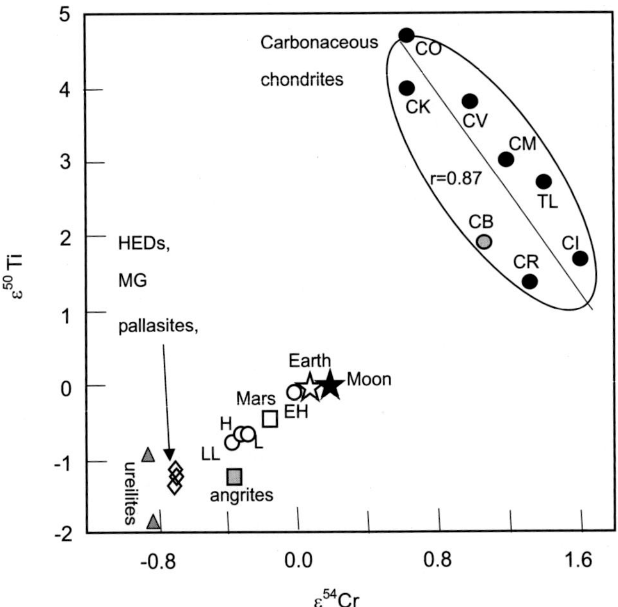

FIGURE 3.19
A plot of ε^{50}Ti versus ε^{54}Cr. *Black circles* show CI, CM, CV, CO, CK, CR carbonaceous chondrites, and Tagish Lake (TL). A *gray circle* indicates CB carbonaceous chondrite, for which the point is based on different analyses of samples for ε^{50}Ti and ε^{54}Cr. *White and black stars* and a *white square* show Earth, Moon, and Mars values. *White circles* indicate L, LL, H ordinary chondrites, and EH enstatite chondrites. A *gray square* and *gray triangles* show angrites and ureilites, respectively. *Diamonds* indicate HEDs, MG pallasites, and mesosiderites. The *line* in carbonaceous chondrites is a regression line for all carbonaceous chondrites including CB, which gave $r = 0.87$.
This plot is modified after Warren PH. Stable-isotopic anomalies and the accretionary assemblage of the Earth and Mars: a subordinate role for carbonaceous chondrites. Earth Planet Sci Lett 2011;311:93–100.

circle indicates CB carbonaceous chondrite, for which the point is based on different analyses of samples for $\varepsilon^{50}Ti$ and $\varepsilon^{54}Cr$. White and black stars and a white square show values for Earth, the Moon and Mars. White circles indicate L, LL, H ordinary chondrites, and EH enstatite chondrites. A gray square and gray triangles show angrites and ureilites, respectively. Diamonds indicate HEDs, MG pallasites, and mesosiderites. The line is a regression line for all carbonaceous chondrites including CB, which gave $r=0.87$.

A majority of the lunar rocks were in this range. However, a few were lower than this ratio. When corrected with the cosmic-ray exposure, all lunar data are identical with the data from Earth to within ±0.04 ε-units. In contrast, bulk meteorites spanned six ε-units.

The isotopic homogeneity of this highly refractory element suggests that lunar material was derived from the proto-Earth mantle, an origin that could be explained by efficient impact ejection. Or Earth and the Moon exchanged material between the Earth's magma ocean and the protolunar disk (Pahlevan and Stevenson, 2007). Or by fission from a rapidly rotating post-impact Earth (Ćuk and Stewart, 2012).

3.9 ASTROPHYSICAL MODELS FOR THE GIANT IMPACT

3.9.1 Basic Simulations of the Giant Impacts

Fig. 3.20 shows the basic calculation of the giant impact performed by Agnor and Asphaug (2004) and Asphaug (2010). Some of the giant impact scenarios for the accretion efficiency (ξ) are indicated as functions of v_{rel}/v_{esc}, the colliding masses $M_2{:}M_1$, and the impact angle θ. ξ is expressed as $\xi=(M_F-M_1)/M_2$, where M_1 and M_2 correspond to the masses of the proto-Earth and the impacter, respectively. M_F is the mass of the largest final body. The v_{rel} and v_{esc} values are the relative velocity of the initial two bodies and the escape velocity from the final body. M_1 and M_2 are masses of Earth and the impacter, respectively.

It is easy to understand Fig. 3.20 by assuming extreme cases at $M_2{:}M_1=1{:}10$ (the case of Mars and Earth). When there is almost no difference in velocity of the two bodies ($v_{rel}/v_{esc}=\sim 0$), the two bodies will merge into the one body at the impact angles $\theta=0$, 30, 45, and 60. Thus, the accretion efficiency (ξ) will be approximately 1, which is shown as "efficient accretion" in Fig. 3.20. In contrast, if the relative velocity (v_{rel}/v_{esc}) is greater than 2.5, two bodies do not accrete irrespective of the impact angles (θ), and the proto-Earth is eroded or even disrupted ($\xi<0$). These cases are indicated as "erosion, disruption" in Fig. 3.20.

When $M_2{:}M_1$ is 1:10 (the case of Mars and Earth), as the hitting angle increases, the accretion efficiency (ξ) decreases approximately 0.9 to 0 in the relative velocity (v_{rel}/v_{esc}) range of 0.6 to 1.5. This range is referred to as "hit-and-run collisions." In this range, two bodies hit but separate into two similar bodies. When $M_2{:}M_1=1{:}2$ or 1:1, the accretion efficiencies (ξ) drastically drop from approximately 1 to 0 at the relative velocity (v_{rel}/v_{esc}) of approximately 0.6 at $\theta=45$.

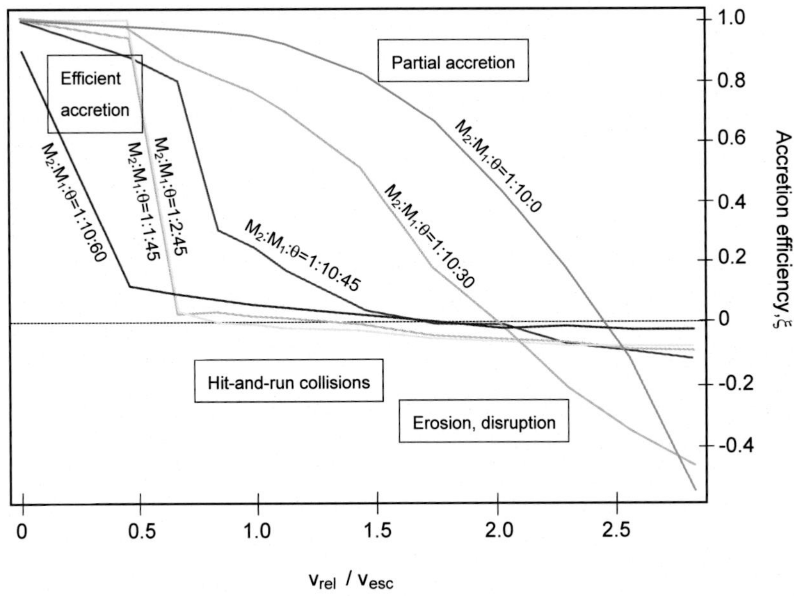

FIGURE 3.20
Model calculation results for accretion efficiency (ξ) in giant impacts as a function of v_{rel} for colliding masses M_2:M_1 and for varying impact angles θ. M_1 and M_2 are masses of Earth and the impactor, respectively. The accretion efficiency (ξ) is expressed as $\xi = (M_F - M_1)/M_2$ where MF is the largest final body. The v_{rel} value is the relative velocity of initial two bodies. The v_{esc} value is the escape velocity from the final body. M_2:M_1 = 10:1 corresponds to the relation between Earth and Mars. Relations of M_2:M_1:θ = 10:2:45 (yellow) and M_2:M_1:θ = 10:1:45 (orange) overlap, so that it is difficult to discern. These calculations are based on Agnor and Asphaug (2004) and Asphaug (2010).

3.9.2 Cosmochemical Constraints of the Giant Impact

As discussed in Sections 3.4–3.8, the giant impact models must satisfy the following astrochemical constraints:

1. The mass of the core of the Moon is approximately 4% of the total mass of the Moon. The mass of the core of the Earth is 34% of the mass of the Earth.
2. Highly siderophile element (HSE) concentrations of the BSE are depleted 100 times less than those of the CI chondrites. These concentrations are approximately 100 times higher than those of calculated values using the partition coefficients. The chondritic material should have added to BSE as the late veneer (>1000 km diameter and at 4.4–4.2 Ga; Bottke et al., 2010; Schlichting et al., 2012). This could have also brought water to the Earth.
3. HSEs in the Moon are 10^{-4} times less than those of CI chondrite, perhaps depleted by the core formation (discussed in Chapter 4).
4. Isotopic compositions for most elements (Li, O, Mg, Si, K, Fe, Cr, and Ti) on the Moon are similar to those of BSE.

3.9.3 The Standard Model

The giant impact models have been unified to the standard model over a quarter of a century. Ida et al. (1997) calculated hundreds of particles after the giant impact. Canup and Asphaug (2001) estimated $M_2=0.1–0.15\ M_1$, $v_{rel}=\sim v_{esc}$, and $\theta=\sim 45$ degrees (indicated by the red line in Fig. 3.20). In such a Mars-size impact, v_{esc} is supersonic; therefore, materials are vaporized. The protolunar disk which is made by melts and vapors are formed after the giant impact, then accrete to form the Moon. Canup (2004) estimated the temperature of the disk to be approximately 3000K. Massive diffusion occurred between the Earth and protolunar isotopic system (Pahlevan and Stevenson, 2007). The melt-vapor protolunar disk could explain the efficient differentiation of the approximately 4% iron fraction and magma ocean formation on the Moon. From the magma ocean, the rafting crust was formed by plagioclase-feldspar, which became lunar highlands of predominantly anorthosite composition (Warren, 1985; Shearer et al., 2006).

Belbruno and Gott (2005) proposed that Theia was formed at one of the Trojan points (see Box 2.10 for the Trojan point) near one AU. The iron cores of the two planets agglomerate without much mixing of silicate with iron (Nimmo and Agnor, 2006). Dahl and Stevenson (2010) found only a fraction of the Earth's core equilibrates with silicate during a giant impact. In the impact-triggered fission (Ćuk and Stewart, 2012) and hit-and-run collision (Reufer et al., 2012), Theia's mantle either escapes or becomes silicate atmosphere of the Earth. Canup (2012) makes the Moon and the Earth by colliding of proto-Earths with similar isotopic composition to solve the isotopic problem.

3.9.4 The Latest Model Proving the Similarity Between the Proto-Earth and Theia (The Impacter)

Mastrobuono-Battisti et al. (2015) performed extensive N-body simulations of terrestrial planet formation. To study the compositions of planets and their impacters, they analyzed 40 dynamic simulations of the late stages. Each simulation started from a disk of 85–90 planetary embryos and 1000–2000 planetesimals extending from 0.5 AU to 4.5AU. Within 100–200 Myrs, each simulation typically produced three to four rocky planets formed from collisions between embryos and planetesimals. They also calculated the oxygen isotope ratios and $\Delta^{17}O$ of Earth and the Moon. Using the large quantity of simulation data, they compared the composition of each surviving planet with that of its last giant impacter, which is the last planetary embryo that impacted the planet (the final impact).

Unlike the simulation of Pahlevan and Stevenson (2007), they analyzed a single statistically limited simulation, which included only 150 particles, and compared the compositions of any impacters on any planets during the simulation (not only giant impacts, due to small number statistics) to the composition of the planets.

As a result, they found that a significant fraction of all planetary impacters have compositions similar to the planets they struck, in contrast with the distinct compositions of different planets existing in the same planetary system. This means that the planet–impacter pairs are robustly more similar in composition than are pairs of surviving planets in the same system. The impacter and planet having a similar composition is also applicable to the origin of the $\Delta^{17}O$ similarity between Earth and the Moon.

This conclusion—that the proto-Earth and the impacter have a similar composition from the beginning—solves a lot of problems in standard models.

References

Agnor C, Asphaug E. Accretion efficiency during planetary collisions. Astrophys J 2004;613:L157–60.

Albarede F. The recovery of spatial isotopic distributions from stepwise degassing data. Earth Planet Sci Lett 1978;39:387–97.

Armytage R, Georg R, Savage P, Williams H, Halliday A. Silicon isotopes in meteorites and planetary core formation. Geochim Cosmochim Acta 2011;75:3662–76.

Armytage R, Georg R, Williams H, Halliday A. Silicon isotopes in lunar rocks: implications for the Moon's formation and the early history of the Earth. Geochim Cosmochim Acta 2012;77:504–14.

Asphaug E. Similar-sized collisions and the diversity of planets. Chem Erde Geochem 2010;70:199–219.

Batygin K, Laughlin G. Jupiter's decisive role in the inner Solar System's early evolution. Proc Natl Acad Sci USA 2015;112:4214–7.

Belbruno E, Gott Jr III . Where did the Moon come from? Astron J 2005;129:1724–45.

Bottke WF, Walker RJ, Day JMD, Nesvorny D, Elkins-Tanton L. Stochastic late accretion to Earth, the Moon, and Mars. Science 2010;330:1527–30.

Cameron AGL. Formation of the prelunar accretion disk. Icarus 1985;62:319–27.

Cameron AGL, Ward WR. The origin of the Moon. Lunar Planet Inst Sci Conf Abs 1976;7:120–2.

Cameron AGL, Benz W. The origin of the Moon and the single impact hypothesis IV. Icarus 1991;92:204–16.

Cao X, Liu Y. Equilibriumm mass-dependent fractionation relationships for triple oxygen isotopes. Geochim Cosmochim Acta 2011;75:7435–45.

Canup M. Dynamics of lunar formation. Annu Rev Astron Astrophys 2004;42:441–75.

Canup RM. Forming a Moon with an Earth-like composition via a giant impact. Science 2012;338:1052–5.

Canup RM, Asphaug E. Origin of the Moon in a giant impact near the end of the Earth's formation. Nature 2001;412:708–12.

Clayton RN, Mayeda TK. Genetic relations between the Moon and meteorites. Proc Lunar Sci Conf 6th 1975:1761–9.

Clayton RN, Maeda TK. Oxygen isotope studies of achondrites. Geochim Cosmochim Acta 1996;60:1999–2017.

Clayton RN, Mayeda TK. Oxygen isotope studies of carbonaceous chondrites. Geochim Cosmochim Acta 1999;63:2089–104.

Craddock PR, Warren JM, Dauphas N. Abyssal peridotites reveal the near-chondritic Fe isotopic composition of the Earth. Earth Planet Sci Lett 2013;365:63–76.

Ćuk M, Stewart ST. Making the Moon from a fast-spinning Earth: a giant impact followed by resonant despinning. Science 2012;338:1047–52.

Dahl TW, Stevenson DJ. Turbulent mixing of metal and silicate during planet accretion-and interpretation of the Hf-W chronometer. Earth Planet Sci Lett 2010;295:177–86.

Daly RA. Origin of the Moon and its topography. Proc Am Phil 1946;90:104–19.

Darwin GH. On the precession of a viscous spheroid, and on the remote history of the Earth. Phil Trans R Soc Lond 1879;170:447–538.

Dauphas N, Burkhardt C, Warren PH, Teng FZ. Geochemical arguments for and Earth-like Moon-forming impactor. Phil Trans R Soc A 2014;372:20130244.

Drake MJ. Geochemical constraints on the origin of the Moon. Geochim Cosmochim Acta 1983;47:1759–67.

Drake MJ, Newsom CJ, Capobianco CJ. V, Cr and Mn in the Earth, Moon, EPB and SPB and the origin of the Moon: experimental studies. Geochim Cosmochim Acta 1989;53:2101–11.

Dreibus G, Wänke H. On the chemical composition of the moon and the eucrites parent body and comparison with composition of the Earth: the case of Mn, Cr and V. Lunar Planet Sci Abst 1979;10:315–7.

Eiler JM. Oxygen isotope variations of basaltic lavas and upper mantle rocks. Rev Mineral Geochem 2001;43:319–64.

Fitoussi C, Bourdon B. Silicon isotope evidence against an enstatite chondrite Earth. Science 2012;335:1477–80.

Fitoussi C, Bourdon B, Kleine T, Oberli F, Reynolds BC. Si isotope systematics of meteorites and terrestrial peridotites: implications for Mg/Si fractionation in the solar nebula and for Si in the Earth's core. Earth Planet Sci Lett 2009;287:77–85.

Galy A, Yoffe O, Janney PE, et al. Magnesium isotope heterogeneity of the isotopic standard SRM980 and new reference materials for magnesium-isotope-ratio measurements. J Anal At Spectrom 2003;18:1352–6.

Georg RB, Halliday AN, Schauble EA, Reynolds BC. Silicon in the Earth's core. Nature 2007;447:1102–6.

Gressmann CK, Rubie DC. The origin of the depletions of V, Cr and Mn in the mantles of the Earth and Moon. Earth Planet Sci Lett 2000;184:95–107.

Halliday AN. Terrestrial accretion rates and the origin of the Moon. Earth Planet Sci Lett 2000;176:17–30.

Hallis L, Anand M, Greenwood R, Miller MF, Franchi I, Russell S. The oxygen isotope composition, petrology and geochemistry of mare basalts: evidence for large-scale compositional variation in the lunar mantle. Geochim Cosmochim Acta 2010;74:6885–99.

Hartmann WK, Davis D. Satellite-sized planetesimals and lunar origin. Icarus 1975;24:504–15.

Herwartz D, Pack A, Friedrichs B, Bischoff A. Identification of the giant impactor Theia in lunar rocks. Science 2014;344:1146–50.

Hirose K, Labrosse S, Hernlund J. Composition and state of the core. Annu Rev Earth Planet Sci 2013;41:657–91.

Humayn M, Clayton RN. Precise determination of the isotopic composition of potassium: application to terrestrial rocks and lunar soils. Geochim Cosmochim Acta 1995a;59:2115–30.

Humayn M, Clayton RN. Potassium isotope cosmochistry; genetic implications of volatile element depletion. Geochim Cosmochim Acta 1995b;59:2131–48.

Ida S, Canup RM, Stewart GR. Lunar accretion from an impact-generated disk. Nature 1997;389:353–7.

Javoy M, Kaminski E, Guyot F, et al. The chemical composition of the Earth: enstatite chondrite models. Earth Planet Sci Lett 2010;293:259–68.

Jeffcoate A, Elliott T, Kasemann S, Ionov D, Cooper K, Brooker R. Li isotope fractionation in peridotites and mafic melts. Geochim Cosmochim Acta 2007;71:202–18.

Johnson CM, Beard BL. Correction of instrumentally produced mass fractionation during isotopic analysis of Fe by thermal ionization mass spectrometry. Int J Mass Spectrom 1999;193:87–99.

Kaib NA, Cowan NB. The feeding zones of terrestrial planets and insights into Moon formation. Icarus 2015;252:161–74.

Liu Y, Spicuzza MJ, Craddock PR, Day J, Valley JW, Dauphas N, Taylor LA. Oxygen and iron isotope constraints on near-surface fractionation effects and the compoisition of lunar mare basalt source regions. Geochim Cosmochim Acta 2010;74:6259–62.

Lock SJ, Stewart ST, Petaev MI, et al. A new model for lunar origin: equilibration with Earth beyond the hot spin stability limit. Proc 47th Lunar Planet Sci Conf (Lunar and Planetary Science Institute) 2016:2881. pdf.

Lu YH, Makishima A, Nakamura E. Coprecipitation of Ti, Mo, Sn and Sb with fluorides and application to determination of B, Ti, Zr, Nb, Mo, Sn, Sb, Hf and Ta by ICP-MS. Chem Geol 2007;236:13–26.

Magna T, Wiechert U, Halliday AN. New constraints on the lithium isotope compositions of the Moon and terrestrial planets. Earth Planet Sci Lett 2006;243:3336–53.

Makishima A. Thermal ionization mass spectrometry (TIMS). Silicate digestion, separation, measurement. Weinheim: Wiley-VCH; 2016.

Makishima A, Masuda A. Primordial Ce isotopic composition of the solar system. Chem Geol 1993;106:197–205.

Makishima A, Nakamura E, Nakano T. Determination of zirconium, niobium, hafnium and tantalum at $ng\,g^{-1}$ levels in geological materials by direct nebulization of sample HF solutions into FI-ICP-MS. Geostand Newsl 1999;23:7–20.

Makishima A, Nakamura E. Determination of major, minor and trace elements in silicate samples by ICP-QMS and ICP-SFMS applying isotope dilution-internal standardization (ID-IS) and multi-stage internal standardization. Geostand Geoanal Res 2006;30:245–71.

Makishima A, Tanaka R, Nakamura E. Precise elemental and isotopic analyses in silicate samples employing ICP-MS: application of HF solution and analytical techniques. Anal Sci 2009;25:1181–7.

Makishima A. A simple and fast separation method of Fe employing extraction resin for isotope ratio determination by multicollector ICP-MS. Int J Anal Mass Spectrom Chromatogr 2013;1:95–102.

Mastrobuono-Battisti A, Perets HB, Raymond SN. A primordial origin for the compositional similarity between the Earth and the Moon. Nature 2015;520:212–5.

Matsuhisa Y, Goldsmith JR, Clayton RN. Mechanisms of hydrothermal crystallization of quartz at 250°C and 15 kbar. Geochim Cosmochim Acta 1978;42:173–82.

Meyer C. Vitrophyric pigeonite basalt. Lunar Sample Compendium 2010:15597.

Meyer C. Ilmenite basalt (high K). Lunar Sample Compendium 2011a:10024.

Meyer C. Ferroan anorthosite. Lunar Sample Compendium 2011b:15415.

Millet MA, Baker JA, Payne C. Ultra-precise stable Fe isotope measurements by high resolution multiple-collector inductively coupled plasma mass spectrometry with a ^{57}Fe-^{58}Fe double spike. Chem Geol 2012;304–305:18–25.

Münker C. A high field strength element perspective on early lunar differentiation. Geochim Cosmochim Acta 2010;74:7340–61.

Neal CR, Taylor LA. Petrogenesis of mare basalts – a record of lunar volcanism. Geochim Cosmochim Acta 1992;56:2177–211.

Nebel O, Van Westrenen W, Vroon P, Wille M, Raith M. Deep mantle storage of the Earth's missing niobium in late-stage residual melts from a magma ocean. Geochim Cosmochim Acta 2010;74:4392–404.

Nimmo F, Agnor CB. Isotopic outcomes of N-body accretion simulations: constraints on equilibration processes during large impacts from Hf/W observations. Earth Planet Sci Lett 2006;243:26–43.

Pack A, Herwartz D. The triple oxygen isotope composition of the Earth mantle and understanding $\Delta^{17}O$ variations in terrestrial rock and minerals. Earth Planet Sci Lett 2014;390:138–45.

Pahlevan K, Stevenson DJ. Equilibration in the aftermath of the lunar-forming giant impact. Earth Planet Sci Lett 2007;262:438–49.

Papanastassiou DA, Wasserburg GJ. Lunar chronology and evolution from Rb-Sr studies of Apollo 11 and 12 samples. Earth Planet Sci Lett 1971;11:37–62.

Poitrasson F, Halliday AN, Lee DC, Levasseur S, Teutsch N. Iron isotope differences between Earth, Moon, Mars and Vesta as possible records of contrasted accretion mechanisms. Earth Planet Sci Lett 2004;223:253–66.

Reufer A, Meier MMM, Bentz W, Wieler R. A hit-and-run giant impact scenario. Icarus 2012;221:296–9.

Ringwood AE. On the chemical evolution and densities of the planets. Geochim Cosmochim Acta 1959;15:257–83.

Ringwood AE. The chemical composition and origin of the Earth. In: Hurley PM, editor. Advances in earth sciences. Cambridge: MIT Press; 1966. p. 287–356.

Ringwood AE, Kesson SE. Basaltic magmatism and the bulk composition of the Moon, II. Siderophile and volatile elements in Moon, Earth and chondrites: implications for lunar origin. Moon 1977;16:425–64.

Ringwood AE. Origin of the Earth and Moon. New York: Springer-Verlag; 1979.

Ringwood AE. Terrestrial origin of the Moon. Nature 1986;322:323–8.

Ringwood AE, Kato T, Hibberson W, Ware N. High pressure geochemistry of Cr, V and Mn and implications for the origin of the Moon. Nature 1990;347:174–6.

Ringwood AE, Kato T, Hibberson W, Ware N. Partitioning of Cr, V, and Mn between mantles and cores of differentiated planetesimals: implications for giant impact bypotheses of lunar origin. Icarus 1991;89:122–8.

Rubie DC, Jacobson SA, Morbidelli A, et al. Accretion and differentiation of the terrestrial planets with implications for the compositions of early-formed Solar System bodies and accretion of water. Icarus 2015;248:89–108.

Savage P, Georg R, Armytage R, Williams H, Halliday A. Silicon isotope homogeneity in the mantle. Earth Planet Sci Lett 2010;295:139–46.

Savage PS, Georg RB, Williams HM, Burton KW, Halliday AN. Silicon isotope fractionation during magmatic differentiation. Geochim Cosmochim Acta 2011;75:6124–39.

Schlichting HE, Warren PH, Yin Q-Z. The last stages of terrestrial planet formation: dynamical friction and the late veneer. Astrophys J 2012;752:8.

Sedaghatpour F, Teng F-Z, Liu Y, Sears DW, Taylor LA. Magnesium isotopic composition of the Moon. Geochim Cosmochim Acta 2013;120:1–16.

Seitz H-M, Brey GP, Lahaye T, Durali S, Weyer S. Lithium isotopic signatures of peridotite xenoliths and isotopic fractionation at high temperature between olivine and pyroxenes. Chem Geol 2004;212:163–77.

Seitz H-M, Brey GP, Weyer S, et al. Lithium isotope compositions of Martian and lunar reservoirs. Earth Planet Sci Lett 2006;245:6–18.

Shearer CK, Hess PC, Wieczorek MA, et al. Thermal and magmatic evolution of the Moon. Rev Mineral Geochem 2006;60:365–518.

Spicuzza MJ, Day J, Taylor LA, Valley JW. Oxygen isotope constraints on the origin and differentiation of the Moon. Earth Planet Sci Lett 2007;253:254–65.

Tanaka R, Nakamura E. Determination of ^{17}O-excess of terrestrial silicate/oxide minerals with respect to Vienna Standard Mean Ocean Water (VSMOW). Rapid Commun Mass Spectrom 2013;27:285–97.

Teng F-Z, Li W-Y, Ke S, et al. Magnesium isotopic composition of the Earth and chondrites. Geochim Cosmochim Acta 2010;74:4150–66.

Touboul M, Puchtel IS, Walker RJ. Tungsten isotopic evidence for disproportional late accretion to the Earth and Moon. Nature 2014;520:530–3.

Wade J, Wood B. The Earth's 'missing' niobium may be in the core. Nature 2001;409:75–8.

Walczyk T. Iron isotope ratio measurements by negative thermal ionization mass spectrometry using FeF_4^- molecular ions. Int J Mass Spectrom Ion Process 1997;161:217–27.

Walker RJ. Highly siderophile elements in the Earth, Moon and Mars: update and implications for planetary accretion and differentiation. Chem Erde 2009;69:101–25.

Walsh KJ, Morbidelli A, Raymond SN, O'Brien DP, Mandell AM. A low mass for Mars from Jupiter's early gas-driven migration. Nature 2011;475:206–9.

Wang K, Jacobsen SB. Potassium isotopic evidence for a high-energy giant impact origin of the Moon. Nature 2016;538:487–90.

Wänke H, Baddenhausen, Blum K, et al. On the chemistry of lunar samples and achondrites. Primary matter in the lunar highlands: a re-evaluation. In: Proc Lunar Sci Conf 8th. 1977. p. 2191–213.

Warren PH. The magma ocean concept and lunar evolution. Annu Rev Astron Astrophys 1985;18:201–40.

Warren PH. A concise compilation of petrologic information on possibly pristine nonmare Moon rocks. Am Mineral 1993;78:360–76.

Warren PH. Stable-isotopic anomalies and the accretionary assemblage of the Earth and Mars: a subordinate role for carbonaceous chondrites. Earth Planet Sci Lett 2011;311:93–100.

Weyer S, Anbar AD, Brey GP, Münker C, Mezger K, Woodland AB. Iron isotope fractionation during planetary differentiation. Earth Planet Sci Lett 2005;240:251–64.

Weyer S, Ionov DA. Partial melting and melt percolation in the mantle: the message from Fe isotopes. Earth Planet Sci Lett 2007;259:119–33.

Wiechert U, Halliday A, Lee D-C, Snyder G, Taylor L, Rumble D. Oxygen isotopes and the Moon-forming giant impact. Science 2001;294:345–8.

Wiesli RA, Beard BL, Taylor LA, Johnson CM. Space weathering processes on airless bodies: Fe isotope fractionation in the lunar regolilth. Earth Planet Sci Lett 2003;216:457–65.

Young ED, Galy A, Nagahara H. Kinetic and equilibrium mass-dependent isotope fractionation laws in nature and their geochemical and cosmochemical significance. Geochim Cosmochim Acta 2002;66:1095–104.

Young ED, Kohl IE, Warren PH, Rubie DC, Jacobsen SA, Morbidelli A. Oxygen isotopic evidence for vigorous mixing during the Moon-forming giant impact. Science 2016;351:493–6.

Zambardi T, Poitrasson F, Corgne A, Meheut M, Quitte G, Anand M. Silicon isotope variations in the inner Solar System: implications for planetary formation, differentiation and composition. Geochim Cosmochim Acta 2013;121:67–83.

Zhang J, Dauphas N, Davis A, Leya I, Dedkin A. The proto-Earth as a significant source of lunar material. Nat Geosci 2012;5:251–5.

CHAPTER 4

What Is the Late Veneer, and Why Is It Necessary?

Chapter Outline

4.1 INTRODUCTION

The late veneer—the addition of materials after the giant impact—is required to reconcile the abundance of highly siderophile elements (HSEs) and platinum group elements (PGEs). Isotopic data from tungstein (W) clearly reveals the necessity of the late veneer. See Fig. 3.1, a cartoon that shows the relationship between the proto-Earth, Theia (the impacter), the giant impact, and the late veneer in time sequence, to help you understand various events. The late veneer is required not only in cosmochemistry, but also in astrophysics (e.g., Schlichtling et al., 2012). In this chapter, the late veneer is discussed in detail.

Origins of the Earth, Moon, and Life. http://dx.doi.org/10.1016/B978-0-12-812058-3.00004-1

4.2 CONSTRAINTS FROM HIGHLY SIDEROPHILE ELEMENTS AND PLATINUM GROUP ELEMENTS

In Section 4.2, HSEs are discussed, HSEs contains PGEs (Ru, Rh, Pd, Os, Ir and Pt), rhenium (Re) and gold (Au). Generally, they have very high evaporation temperatures; however, sometimes sulfur (S), selenium (Se) and tellurium (Te) are included, which have very low evaporation temperatures. When oxygen fugacity (fO_2) is high, Re and Os are oxidized into rhenium (VII) oxide (Re_2O_7) and osmium (VIII) tetroxide (OsO_4), which have low boiling temperatures and are very volatile.

4.2.1 Primitive Upper Mantle

For the representation of the Earth's mantle, the primitive upper mantle (PUM) is used, because we have no idea about the HSE composition of the lower mantle. However, the upper mantle concentration of HSE is estimated from the mantle peridotites. In Fig. 4.1, the iridium (Ir) concentration versus aluminum oxide (Al_2O_3) (wt.%) is plotted. The Ir concentrations of peridotites are in a concentrated range of $3.5 \pm 0.4\,ng\,g^{-1}$ (2SD). McDonough and Sun (1995) estimated the most fertile (least depleted) mantle, which is the PUM, has Al_2O_3 (wt.%) = 4.5%.

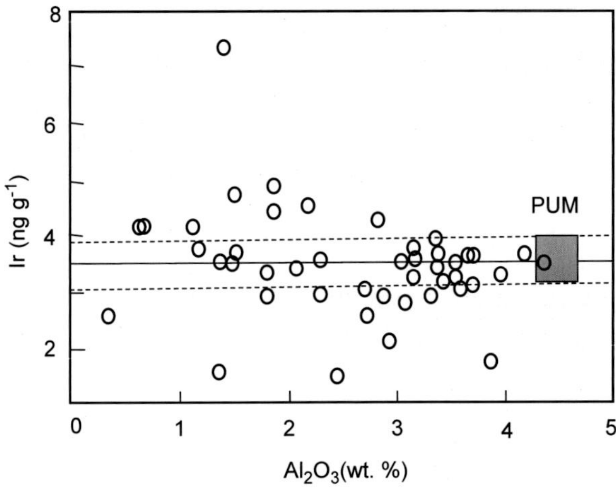

FIGURE 4.1
A plot of Ir content ($ng\,g^{-1}$) versus Al_2O_3 (wt.%) in peridotites from all over the world. PUM means Primitive Upper Mantle. The *horizontal line* indicates the average, and the *dotted line* indicates the range of 2SD. *The alumina data are from McDonough WF, Sun S-s. The composition of the Earth. Chem Geol 1995;120:223–53. The figure is modified from Becker H, Horan MF, Walker RJ, Gao S, Lorand J-P, Rudnick RL. Highly siderophile element composition of the Earth's primitive upper mantle constraints from new data on peridotite massifs and xenoliths. Geochim Cosmochim Acta 2006;70:4528–50 and Walker RJ. Highly siderophile elements in the Earth, Moon and Mars: update and implications forplanetary accretion and differentiation. Chem Erde 2009;69:101–25.*

Palladium, when plotted Pd:Ir ratios versus Al_2O_3 (wt.%), showed positive correlation from the origin (Pattou et al., 1996; Becker et al., 2006; not shown). When Al_2O_3 (wt.%) is 4.5%, Pd concentration was obtained to be $7.1 \pm 1.3\,ng\,g^{-1}$ (2 SDs) by Becker et al. (2006).

Table 4.1 **The HSE Concentrations of PUM, the Averages of Carbonaceous Chondrites, Ordinary Chondrites, and Enstatite Chondrites, and $(X/Ir)_A/(X/Ir)_{Carbonaceous}$ Where X Are HSEs and A is the Target Material (e.g., PUM)**

	PUM (ng g^{-1})	2SD (ng g^{-1})	Carbonaceous (ng g^{-1})	Ordinary (ng g^{-1})	Enstatite (ng g^{-1})
Re	0.35	0.06	51.55	53.38	55.94
Os	3.9	0.5	642.6	678.6	637.4
Ir	3.5	0.4	608.5	584.6	582.7
Ru	7.0	0.9	862.2	880.4	873.3
Pt	7.6	1.3	1150	1185	1186
Pd	7.1	1.3	688.5	657.9	850.9
Normalized to carbonaceous chondrite					
	PUM		Carbonaceous	Ordinary	Enstatitie
Re/Ir	1.18		1	1.08	1.13
Os/Ir	1.06		1	1.10	1.04
Ru/Ir	1.41		1	1.06	1.06
Pt/Ir	1.15		1	1.07	1.08
Pd/Ir	1.79		1	0.99	1.29

Becker et al. (2006) plotted Re/Ir, Os/Ir and Pt/Ir ratios and obtained PUM values similar to the chondritic values. However, Ru/Ir and Pd/Ir (already discussed above), were suprachondritic values. In conclusion, Re, Os, Ru, and Pt concentrations of PUM were estimated to be 0.35 ± 0.06, 3.9 ± 0.5, 7.0 ± 0.9, and 7.6 ± 1.3 ngg^{-1}, respectively.

These PUM values as well as the chondritic vales (Horan et al., 2003; Walker et al., 2002) are summarized in Table 4.1. In addition, $(X/Ir)_A/(X/Ir)_{Carbonaceous}$ where X is HSEs and A is the target materials (e.g., PUM) are also shown. From the Re and Pd enrichments to Ir normalized to those of the carbonaceous chondrite, the PUM ratios resemble the enstatite chondrite.

4.2.2 Comparison of PUM With the Moon

As the Moon could be stratified and our sample collection from the Moon is very limited, it is very difficult to estimate HSE abundance on the Moon. One solution is to compare the HSE abundances of similar major element abundances of PUM and lunar basalts, especially with similar magnesium oxide (MgO) contents (Warren et al., 1999). In Fig. 4.2, a plot of osmium (Os) contents versus MgO in terrestrial and lunar volcanic rocks (low-Ti and high-Ti basalts) is shown. Although the lunar samples are biased, the Os concentration is about 1/100 lower from the terrestrial basalt trend.

4.2.3 HSE Depletion of PUM

For PUM, HSE is very depleted, about 1/100 of that of CI chondrite. As the partitioning of HSE into the metallic melts is extremely high, it is easy to explain the depletion of HSE in the silicate mantle (PUM). In Fig. 4.3, using metal-silicate equilibrium constants at high pressure-temperature conditions, the equilibrated mantle

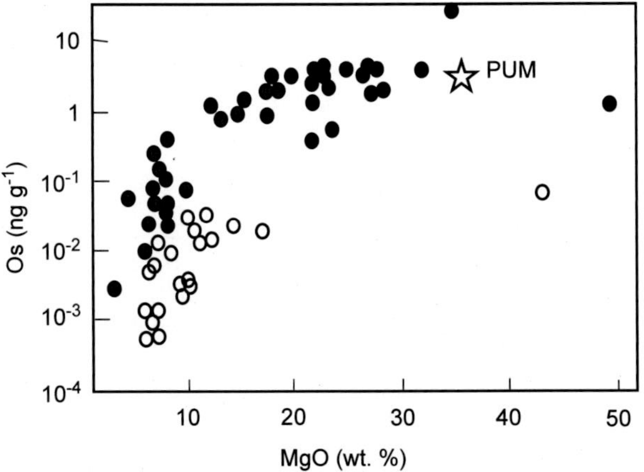

FIGURE 4.2

A plot of Os (ng g^{-1}) content versus MgO (wt.%). The vertical axis is logarithmic. *Solid and open circles* indicate terrestrial and lunar volcanic rocks, respectively. A *star* shows the Primitive Upper Mantle (PUM). The terrestrial data are mostly from the Caribbean Large Igneous Province (Walker et al., 1999).

The lunar data are from Walker RJ, Horan MF, Shearer CK, Papike JJ. Low abundances of highly siderophile elements in the lunar mantle: evidence for prolonged late accretion. Earth Planet Sci Lett 2004;224:399–413 and Day JMD, Pearson DG, Taylor LA. Highly siderophile element constraints on accretion and differentiation of the Earth-Moon system. Science 2007;315:217–9.

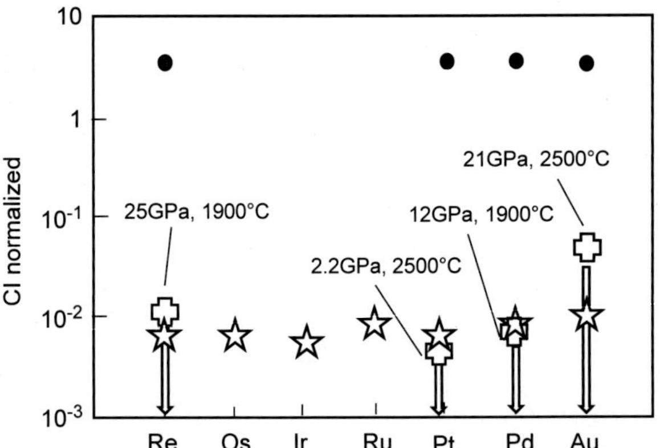

FIGURE 4.3

Concentration of Re and PGE elements (HSEs) in the upper mantle (PUM) normalized to those of CI chondrites, which is shown as *stars*. D values are used, concentration of the equilibrated mantle becomes lower as indicated by the *arrows*. D values of Re, Pt, Pd, and Au of Righter and Drake (1997), Cottrell and Walker (2006), Righter et al. (2008), and Danielson et al. (2005) are used, respectively. Note that the vertical axis is logarithmic normalized to CI concentrations of Anders and Grevisse (1989).

The figure is modified from Walker RJ. Highly siderophile elements in the Earth, Moon and Mars: update and implications for planetary accretion and differentiation. Chem Erde 2009;69:101–25.

are shown as open crosses. It is very difficult to obtain accurate equilibrium constants, because precise analysis of trace HSE in silicates of run products is required. The HSEs in the run products are analyzed by spot analytical methods (see Box 4.1). Electron probe detection limits are too poor (~200 μg g^{-1}), and laser ablation inductively coupled plasma mass spectroscopy (LA-ICP-MS) (see Box 4.2) is generally

BOX 4.1 SPOT ANALYTICAL METHODS

In cosmochemistry, the elemental and isotopic analyses of samples are divided into bulk analysis and spot analysis (Fig Box 4.1). In bulk analysis, significant amounts of the sample are digested. "Significant amounts" means recognizable or visible amounts that can be measured accurately by a balance. Usually, 1 mg to 1 g is used in the bulk analysis. It is difficult to digest more than 1 g perfectly by a general acid digestion method. It is difficult to handle a sample smaller than 1 mg or to measure its weight precisely by static electricity. After the sample is perfectly digested and prepared in the sample solution, we determine the amounts of elements or isotope ratios of the target element in the sample solution. In the measurement, we excite elements by flame [flame atomic absorption spectrophotometry (FAAS)], heat [thermal ionization mass spectrometry (TIMS)], or flameless atomic absorption spectrophotometry (FLAAS) or inductively coupled plasma (ICP). Then we determine elements by lights or nuclei. The light comes from electron transfer from one energy level to the other by the excitation. Therefore, we use "electrons," and the method is called photospectrometry. When we count the number of "nuclei," it is called mass spectrometry.

	Analyzed amount	Exciting method	Measurement
Bulk analysis	$1\sim10^{-3}$ g	Flame (FAAS) Heat (FLAAS/TIMS) Plasma (ICP-AES/ICP-MS)	Light Light/Nucleus Light/Nucleus
Spot analysis	$3\text{x}10^{-6}\sim3\text{x}10^{-9}$ g	Electron (EPMA) Nucleus (SIMS) Light (Laser) (LA-ICP-MS)	Light (X-ray) Nucleus Nucleus

FIGURE BOX 4.1
Bulk analysis and spot analysis.

There are three ways to perform a spot analysis: either an electron beam, ion beam, or laser are bombarded onto a spot area with a diameter from 10 to 100 μm ($3\times10^{-6}\sim3\times10^{-9}$ g if the density is ~3). The bombarded elements emit light (characteristic X-ray), secondary ions, or ablated particles that are determined by inductively coupled plasma mass spectrometry. The methods used are secondary electron microanalysis by energy dispersive spectrometry (SEM-EDS), electron probe microanalysis (EPMA) using wave length dispersive spectrometry (WDS), secondary ion mass spectrometry (SIMS), or LA-ICP-MS. In laser ablation, infrared, visible, and ultraviolet laser lights are used. For ICP-MS, Q-pole type ICP-MS (ICP-QMS), sector magnet type ICP-MS (ICP-SFMS), and multicollector ICP-MS (MC-ICP-MS) are used depending on analytical purposes. The summary of spot elemental analyses is shown in Fig. Box 4.2.

All analytical methods except LA-ICP-MS (see Box 4.2) are explained elsewhere (e.g., Makishima, 2016).

BOX 4.1 SPOT ANALYTICAL METHODS—CONT'D

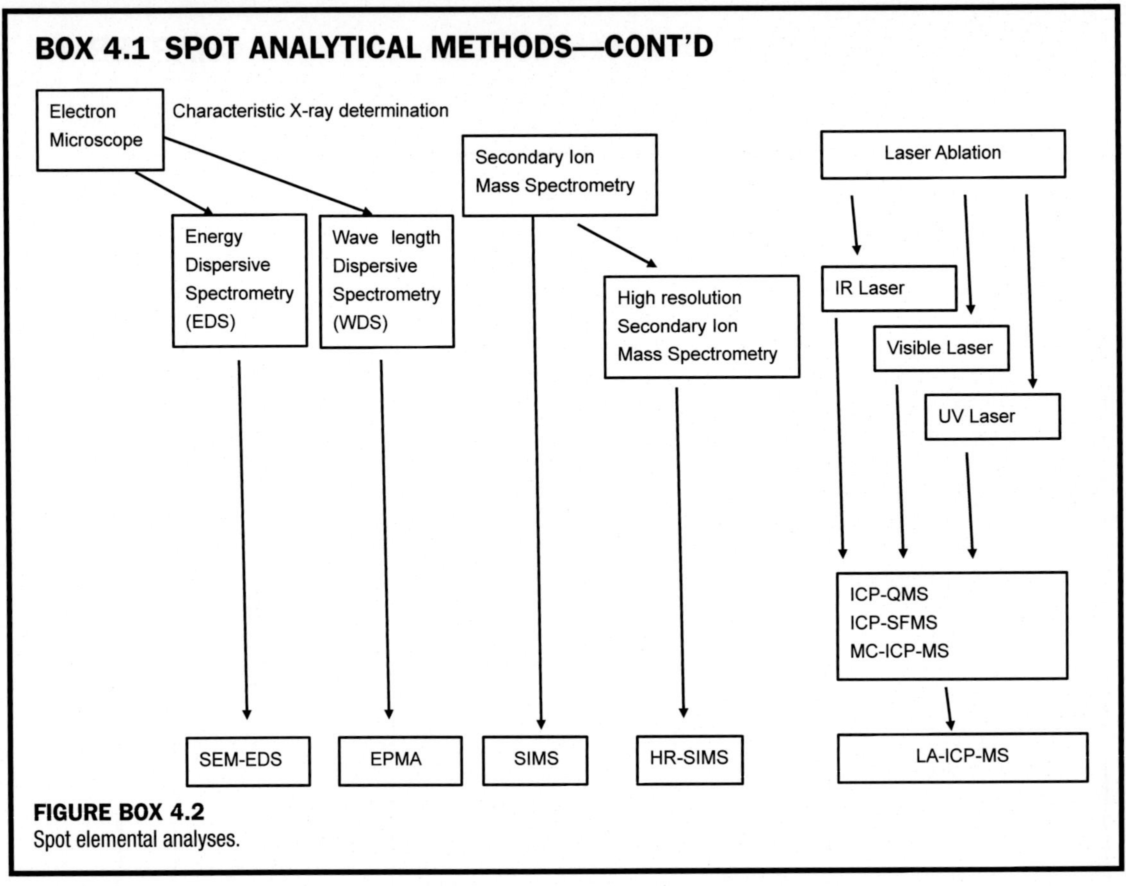

FIGURE BOX 4.2
Spot elemental analyses.

BOX 4.2 LASER ABLATION-INDUCTIVELY COUPLED PLASMA MASS SPECTROMETRY (LA-ICP-MS)

LA-ICP-MS is a hybrid analytical system of ICP-MS with laser ablation system (see Fig. Box 4.3). As ICP-MS is already explained in this book (see Section 2.7.3), in this Box section, laser ablation is explained.

In LA-ICP-MS, laser light with visible light wavelength is not short enough to obtain good ablation. Good laser ablation means less elemental fractionation. The key to less elemental fractionation is sublimation of the entire sample in a short time and formation of small uniform particles, all of which go to ICP-MS. High energy in a small area by a visible to infrared wavelength causes melting only of the small area, and evaporation from the low evaporation elements (or molecules) occurs. Therefore, a shorter wavelength laser, such as an ultraviolet (UV) laser, is required.

To get UV laser light, the argon–fluorine (Ar–F) excimer laser has become popular. In the excimer laser, not crystal but Ar and F_2 gases are used as laser media. When Ar is excited, it becomes reactive, and ArF is formed. This ArF causes an inverted population of electrons and laser light is emitted. The wavelength is 193 nm. This laser is strong and stable.

A schematic diagram of LA-ICP-MS is shown in Fig. Box 4.3. The laser beam is focused on the sample. In a commercial Ar–F excimer laser system, the bottom of the laser pit becomes flat from approximately 10 to 100 μm. As oxygen gas absorbs the UV light, the laser beam line as well as the atmosphere around the sample chamber is filled with nitrogen gas. The sample chamber can be controlled to x-y-z direction by a computer.

BOX 4.2 LASER ABLATION-INDUCTIVELY COUPLED PLASMA MASS SPECTROMETRY (LA-ICP-MS)—CONT'D

In LA-ICP-MS, how to make and carry the particles of the sample effectively into the plasma without elemental fractionation is very important. Helium gas was found to be the best gas in the sample chamber. When He is put into the plasma, the plasma is cooled. The plasma becomes lower in temperature, and therefore unstable.

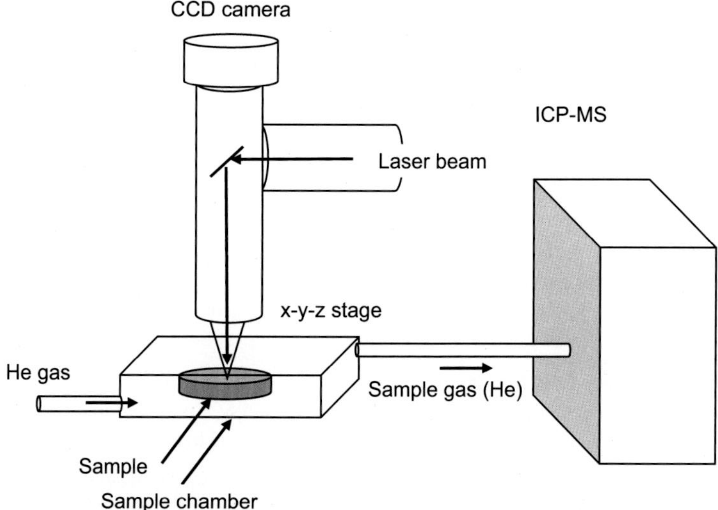

FIGURE BOX 4.3
Laser ablation–inductively coupled plasma mass spectrometry system (LA-ICP-MS).

BOX 4.3 EMISSION AND ABSORPTION OF LIGHT

The atom has electron shells of K, L, M, N, O, ..., K shell has two electrons of 1s orbit (see Fig. Box 4.4A). One orbit can have $2n^2$ electrons. L shell has 2s and three 2p

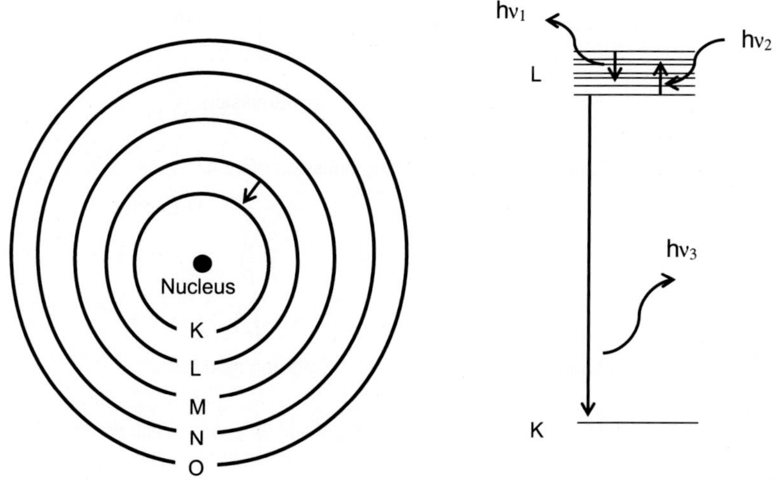

FIGURE BOX 4.4
Model of the electron and energy levels. (A) A simplified shell model of the atom. Energy levels of electrons are shown as K, L, M, N, and O. The *arrow* in (A) shows the electron transition of the L energy level to the K energy levels. (B) Energy levels corresponding to the shell model.

BOX 4.3 EMISSION AND ABSORPTION OF LIGHT—CONT'D

orbits, totaling eight electrons. M shell has 3s, three 3p, and five 3d orbits, totaling 18 electrons. N shell has 4s, three 4p, five 4d, and seven 4f orbits, totaling 32 electrons. O shell has 5s, three 5p, five 5d, seven 5f, and nine 5g orbits, totaling 50 electrons. Energy levels corresponding to the shell model are shown in Fig. Box 4.4B. The lights of $h\nu_1$ and $h\nu_2$ in the figure are lights emitted or absorbed by electron transitions within the L shell, which are UV/visible lights. The light of $h\nu_3$ is X-ray emitted by the electron transition of the L shell to the K shell.

When there is an electron in the ground state (see Fig. Box 4.5A, left) and light with energy of hν comes to the electron, the electron absorbs the energy and enters the excited state. This is "absorption" of light. In contrast, when there is an electron in the exited state, the electron goes down to the ground state with emitting light with the energy of hν (see Fig. Box 4.5A, right). This is the "emission" of light. In Fig. Box 4.5B, energy spectra of absorption and emission are presented.

Photospectroscopy is the method that determines elements by the emitted or absorbed lights from the photospectra (Fig. Box 4.5B). Unfortunately, there are many energy levels in the photospectra because there are similar energy levels in many electrons. Therefore, assignment and overlaps of the spectra become more complicated than mass spectrometry. The mass spectra are only one sheet (Table 1.1), but the light spectra become one book or more. This is one of the disadvantages of photospectrometry.

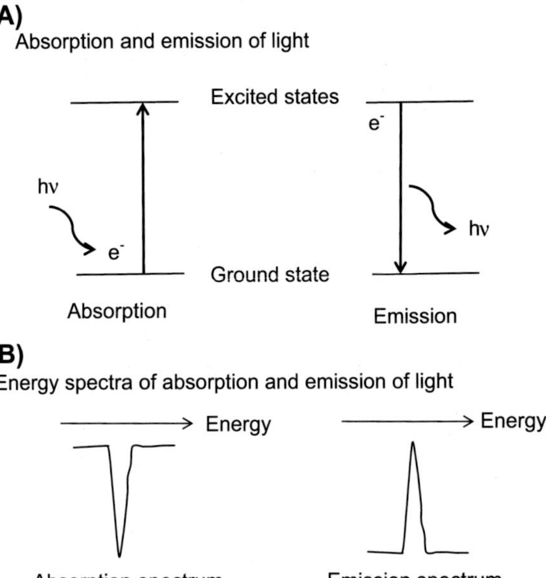

FIGURE BOX 4.5
Schematic diagrams of (A) absorption and emission of light and (B) energy spectra of absorption and emission of light. We assume that there are ground and excited state electron energy levels (e-). The energy difference between the two states is hν.

used. Although the silicate spots (~50 μm ϕ) free from metals are chosen, HSE often forms micronuggets, which hinder precise analysis of silicates. As a result, the partitioning constants can be lower than the true values. If the measurements or experiments are free from these nugget effects, HSE concentration of the calculated mantle could be lower, which is shown as arrows in Fig. 4.3. As a result, the HSE abundance of PUM seems to be more overabundant than expected.

To explain composition of PUM, terms and models of "late accretion," "heterogeneous accretion," or "late veneer" appeared. These terms generally are used to indicate small mass addition (~0.1%) after almost all of the Earth had formed, to add desirable amounts of HSE or volatiles (water or carbons), etc. The late veneer often appears later in the discussion, because it is easy to add or change elemental abundances or isotopic ratios. As a conclusion, without the late veneer, the abundance of HSE cannot be explained. The late veneer is further discussed in Sections 6.4 and 8.2, etc.

4.3 CONSTRAINTS FOR THE LATE VENEER FROM W ISOTOPES

4.3.1 Problems in Early Works, and Solutions

Although the ^{182}Hf–^{182}W decay system is an ideal isotopic system for discussion of the giant impact and Moon formation in the early solar system (Jacobsen, 2005; Halliday, 2008), early research on the Moon cannot be used by the effects of neutron capture. The samples were interfered with the effects of cosmic rays transforming ^{181}Ta into ^{182}W by neutron capture, because the Moon samples were collected from the surface. In addition, the burn out process of ^{182}W could occur (Leya et al., 2000).

To avoid such effects, using Ta-free metals separated from KREEP basalts (see Section 3.3.2 for KREEP) or melts by the impacts was one strategy (Kleine et al., 2005; Touboul et al., 2007). These studies using crystallization products of the lunar magma ocean revealed that the W isotopic composition was uniform and similar to the terrestrial mantle. These results suggested that the formation of the Moon and crystallization of the lunar magma ocean occurred after most ^{182}Hf decayed out, which is after 60 Myr from the formation of the Solar System (Touboul et al., 2007, 2009).

Touboul et al. (2015) determined the Os isotope ratios, HSE concentration, and W isotope ratios of these metals from the impact melt rocks, to examine the early history of the Moon. The HSEs in the metals are considered to be equilibrated with the melt or vapor during impact (Tera et al., 1974).

Kruijer et al. (2015) developed a new strategy. They measured HSE contents and investigated the contribution of meteorites. In addition, they measured $\varepsilon^{180}Hf$ and $\varepsilon^{182}W$ of the KREEP samples. The KREEP samples were chosen because they are residual liquids of the lunar magma ocean, and it is considered that the initial values of $\varepsilon^{182}W$ (before irradiation values) were similar and similarly irradiated for Hf–Ta–W isotopes. Actually, when $\varepsilon^{182}W$ is plotted against $\varepsilon^{180}Hf$, the samples formed a clean correlation line (Slope = −0.549 ± 0.019, MSWD = 0.36)

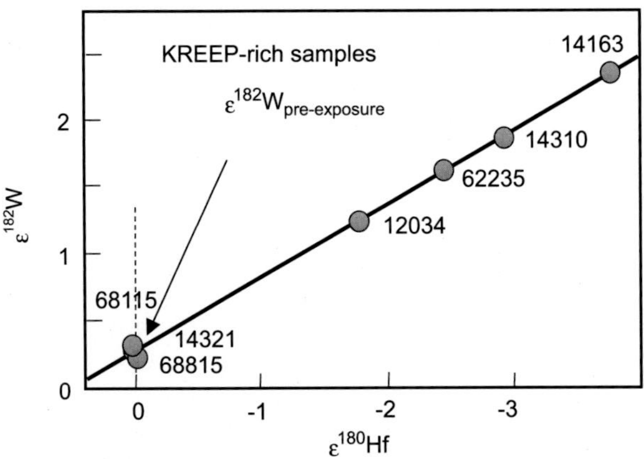

FIGURE 4.4

A plot of $\varepsilon^{182}W$ versus $\varepsilon^{180}Hf$ determined for KREEP-rich samples. The intercept at $\varepsilon^{180}Hf = 0$ indicates the pre-exposure of $\varepsilon^{182}W$ (= +0.27 ± 0.04 at the 95% confidence level).

The figure is modified after Kruijer TS, Kleine T, Fischer-Gödde M, Sprung P. Lunar tungsten isotopic evidence for the lateveneer. Nature 2015;520:534–7.

(see Fig. 4.4). Thus only the less irradiated samples were used in discussion. The two articles by Touboul et al. (2015) and Kruijer et al. (2015) are continuing articles in the same issue of *Nature*, and they arrived at similar conclusions.

4.3.2 Lunar W Isotopic Ratios

Touboul et al. (2015) discussed the incorporation of both stony or iron meteorites, and from two values of 68115, 114, 68815, 394, and 68815, 396 metals determined the average lunar W isotopic ratio to be $\varepsilon^{182}W = +0.206 \pm 0.051$ (2SD) relative to the present Earth mantle, which was normalized to $^{186}W/^{184}W = 0.92767$ (for normalizing and negative thermal ionization mass spectrometry (N-TIMS), see Makishima, 2016).

Kruijer et al. (2015) determined the lunar W isotopic ratio to be $\varepsilon^{182}W = +0.27 \pm 0.04$ (2 SDs) relative to the present Earth mantle, which was also normalized to $^{186}W/^{184}W = 0.92767$ (therefore, both data can be compared directly).

It is amazing that two data overlap within a small range of error bars. The grand average is $\varepsilon^{182}W = 0.24$.

4.3.3 Percentage of the Late Veneer From the Lunar W Isotopic Ratio

In Fig. 4.5, the $\varepsilon^{182}W$ values of the Moon determined by Touboul et al. (2015) and Kruijier et al. (2015) are plotted as white and black circles, respectively.

A gray triangle in Fig. 4.5 indicates the relation between the $\varepsilon^{182}W$ values and the mass fraction of the late accreted material (the late veneer) by Touboul et al. (2015). These values are estimated by mass balance assuming W contents in BSE and the impacted Theia (the chondrite) to be 13 and 200 ppb, respectively. The

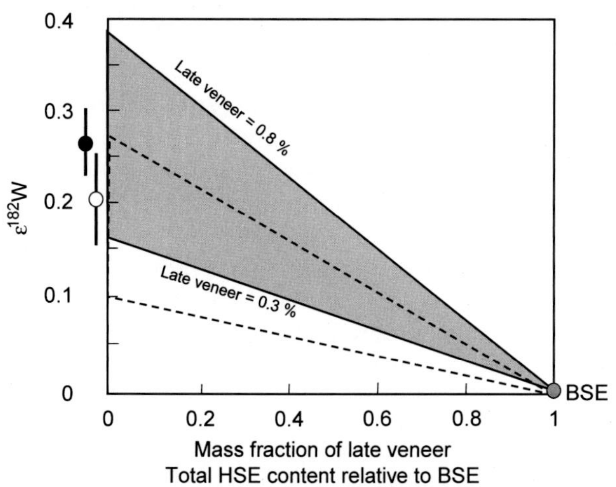

FIGURE 4.5
A plot of $\varepsilon^{182}W$ versus Mass fraction of the late accreted material (late veneer) to the present bulk silicate Earth (a *gray circle*). *White and black circles* are $\varepsilon^{182}W$ of the Moon by Touboul et al. (2015) and Kruijier et al. (2015), respectively (the horizontal positions of these points have no meaning). A *gray triangle* and a *triangle with dotted lines* are the estimated relation between the $\varepsilon^{182}W$ values and the mass fractions of the late accreted material by Touboul et al. (2015) or of total HSE content relative to BSE by Kruijier et al. (2015), respectively. (See text for details Fig. 5.1. Relative frequency of zircons on the Moon).
The figure is after Grange ML, Nemchin AA, Pidgeon RT, Merle RE, Timms NE. What lunar zircon ages can tell? Lunar Planet Sci 2013;44:1884.

upper and lower lines from the bulk silicate Earth (BSE) indicate when the mass fractions of the late veneer were 0.8% and 0.3%, respectively.

Kruijier et al. (2015) estimated the late veneer component to be 80% of (CI + CV + CM) material and 20% of IVA-like iron meteorite material to explain HSE abundances including Se, Te, and S ratios in the Earth's primitive upper mantle. The late veneer value of $\varepsilon^{182}W$ of the mixture is −2.60, and its mass is 0.35% of the Earth mass. The dotted line is the range of the 95% confidence level. The error mainly comes from the uncertainty of the W concentration of BSE of 13 ± 5 ppb.

As both estimations are highly dependent on many assumptions, the mass fraction of the late veneer cannot be defined with high accuracy. As far as the HSE concentrations of BSE cannot be determined to definite values and chemical compositions of added meteorites cannot be fixed, the precision of the estimation cannot be improved.

4.3.4 After the Giant Impact

Bottke et al. (2015) paid attention to the ejecta of the giant impact. They estimated that numerous kilometer-sized ejecta fragments by the giant impact should have struck main-belt asteroids at $>10\,km\,s^{-1}$ velocities, which caused heating and degassing of the target rocks. The impacts produced more than 1000 times the heat than a typical impact at $5\,km\,s^{-1}$. By fitting the heating model of stony meteorites, the Moon was estimated to be formed at approximately 4.47 Ga.

4.4 CONSTRAINTS ON THE LATE VENEER FROM OXYGEN ISOTOPE RATIOS

Using the oxygen isotope ratios, Young et al. (2016) tried to constrain the composition and size of the late veneer of primitive bodies that impacted the silicate Earth. A large flux of the late veneer planetesimals is implied by a higher average W isotope ratios for the Moon than for the Earth and by differences in HSE concentrations between the mantles of Earth and the Moon (Walker et al., 2015). The interpretation of these data is that Earth and the Moon began with the same W isotopic ratios, but that the Earth inherited a greater fraction of low tungsten isotopic materials as chondritic planetesimals after the Moon-forming giant impact (Kruijer et al., 2015; Touboul et al., 2014; see Section 4.3.3 and Fig. 4.5).

If the Earth–Moon system was mixed well by the giant impact, which is a constraint of the W isotope ratios, then the nearly identical $\Delta'^{17}O$ values of Earth and the Moon can constrain the $\Delta'^{17}O$ values of the late veneer impacter. The mass flux ratio of the Earth/Moon is estimated to be in a range from 200 to 1200 (Walker et al., 2015; Schlichting et al., 2012; Bottke et al., 2015). Assuming a mass of the late veneer to be 2×10^{22} kg (Walker et al., 2015), and a maximum Earth/Moon flux ratio of 1200 (Bottke et al., 2015), the late veneer fraction is 0.00477.

Combining this value with the measured value for $\Delta'^{17}O_{Moon} - \Delta'^{17}O_{Earth}$ of zero (Young et al., 2016; see Section 3.7.7) requires that the late veneer impacter had average $\Delta'^{17}O$ values within approximately 0.2‰ or less of Earth, which is similar to enstatite chondrites (Newton et al., 2000). Alternatively, with the maximum value of $\Delta'^{17}O_{Moon} - \Delta'^{17}O_{Earth}$ of ± approximately 5 ppm (Young et al., 2016; when the largest analytical uncertainty is assumed), the $\Delta'^{17}O$ value of the late veneer is ±1.1‰, which includes aqueously altered carbonaceous chondrites or some ordinary chondrites.

If the value of $\Delta'^{17}O_{Moon} - \Delta'^{17}O_{Earth}$ is 12 ppm (Herwartz et al., 2014 which Young et al., 2016 concluded was impossible), the late veneer of $\Delta'^{17}O$ becomes −2.7‰, suggesting that the late veneer impactor was mainly composed of relatively unaltered and dry carbonaceous chondrites (Clayton et al., 1976).

Thus, the result of Young et al. (2016) suggests that if the late veneer impacter was mainly composed of carbonaceous chondrites, the parent bodies must have included substantial fractions of high-$\Delta'^{17}O$ water either in the form of aqueous alteration minerals or as water ice (Sections 4.5.1–4.5.7).

4.5 STATISTICS

As more than 100 isotopic ratios are measured for one sample in the isotope ratio measurement, statistics are important. Sometimes 2SD is used, and sometimes 2SE is used. Standard deviation is also used. To understand the errors of measurements or error bars, understanding the statistics is important. However, using statistics is very simple. This Section 4.5 (Sections 4.5.1–4.5.7) are based on Makishima (2016).

4.5.1 Average and Standard Deviation

When there is one element solution, and the isotopic ratio or the concentration of the element is measured based on the same method, we get the first analytical value (A_1), the second value (A_2), ..., n-th value (A_n), by the same method with the same analytical condition in short time. Then we obtain average (B) as

$$B = (A_1 + A_2 + \ldots + A_n)/n$$

The standard deviation (sometimes denoted as SD or σ) is obtained as

$$SD = \left(\sum (Ai - B)^2\right)/(n-1)^{\frac{1}{2}}$$

which is easily obtained by a "STDEV" function in a Microsoft EXCEL worksheet software. We can calculate the relative standard deviation in percentage (RSD %) as

$$RSD\,\% = SD/B \times 100$$

Two RSD % means twofold of RSD %. We call SD of the same sample in short time (e.g., the repeated measurements of the same sample) as "repeatability,"

4.5.2 The Normal Population

The normal population is one of ideal variations of data. The data scatters as a bell shape. The horizontal axis is the data value, and the vertical axis is the probability. When this curve is integrated from infinite menus to infinite plus, the integrated probability becomes 1. From−σ to +σ, 68.3% of the data are included. From −2σ to +2σ, 95.4% of the data are included. From −3σ to +3σ, 99.7% of the data are included.

4.5.3 The Standard Error

We assume a normal population with an average of B and a standard deviation of σ. Here, what we want to do is to estimate the average and standard deviation of the parent population from the subpopulation. In many cases we cannot take all data into the statistics. Then we randomly take the number of n samples and make a subpopulation.

The average and the standard deviation of the subpopulation are B′ and SE. From the statistics, B is equal to B′. The standard deviation of the subpopulation is called a standard error, SE. The standard error (SE) should be:

$$SE = SD/n^{1/2}$$

In other words, the probability that the estimated B is within the range of B ± SE is 0.68. B ± 1.96SE is 0.95 and B ± 2.85SE is 0.99.

In the isotope ratio measurement, more than 100 ratios are measured as subpopulation, and the standard error is used as the error. Thus the standard deviation of obtained ratios are divided by $n^{1/2}$, and the error of one measurement is presented as SE.

4.5.4 ISO Suggestion

In this book, two types of errors are used, repeatability and intermediate precision, because ISO (International Standard Organization) redefined these words. We used "precision" ("" indicates the older usage), but this should now be called repeatability. Now, ISO's precision is almost similar to what was previously called "accuracy," which includes the difference from the true value.

When the sample is homogeneous powder, intermediate precision is the error of repeated measurements starting from digesting the sample. "Repeatability or reproducibility" was used for this meaning. Now, reproducibility has a different meaning of difference among laboratories. As we do not know the true value, intermediate precision is used to evaluate accuracy. Both definitions are randomly used even in analytical chemistry papers. Therefore, care should be taken when you read or write papers.

"Error" and "uncertainty" are vague words that sometimes express repeatability, and sometimes mean intermediate precision or accuracy. They sometimes indicate RSD, sometimes 2 RSD %. Therefore, we must be careful and clarify the definition when we use "error" or "uncertainty".

4.5.5 Variance, Covariance, and Correlation Coefficient

We assume there are two variables, (x, y), and a series of paired values (x1, y1), (x2, y2), …, (xN, yN). For example, these data are two analyses of two isotopic or chemical ratios. We can calculate the means for both variables:

$$Y = \left(\sum y_i\right)/N$$

$$X = \left(\sum x_i\right)/N$$

We can define the variances of x and y:

$$V_y = \sum (y_i - Y)^2/N$$

$$V_x = \sum (x_i - X)^2/N$$

The standard deviations, which are the square roots of the variances, are obtained as:

$$\sigma y = \sqrt{V_y} = \sqrt{\left[\sum (y_i - Y)^2/N\right]}$$

$$\sigma x = \sqrt{V_x} = \sqrt{\left[\sum (x_i - X)^2/N\right]}$$

The covariance is defined as:

$$V_{yx} = V_{xy} = \sum (y_i - Y)(x_i - X)/N$$

The covariance is an index that shows correlation between y and x. If there is no correlation, the covariance is 0. If positive or negative correlations exist, the covariance becomes positive or negative. To make the covariance free from the units of y and x, the correlation coefficient r is introduced:

$$r = V_{yx}/\sqrt{(V_{xx}\,V_{yy})}$$
$$= \sum(y_i - Y)(x_i - X)/\sqrt{\sum(y_i - Y)^2}\,\sqrt{\sum(x_i - X)^2}$$

When there is no correlation between y and x, r = 0; when y and x are fully positively correlated, r = 1; when fully negatively correlated, r = −1. Usually, we regard that when |r| is greater than 0.90 there is a correlation between y and x. If |r| is less than 0.90, there is a poor correlation between y and x.

4.5.6 Weighted Average, Variance, and Covariances

When we take the uncertainty of each data point into account, the weighted average is calculated as:

$$X_\sigma = \left[\sum (x_i/\sigma_i^2)\right] / \left[\sum (1/\sigma_i^2)\right]$$

The variance and covariance is analogously calculated as:

$$V_x = \left[\sum \left((x_i - X_\sigma)^2/\sigma_i^2/(N-1)\right)\right] / \left[\sum (1/\sigma_i^2)/N\right]$$

$$V_{yx} = \left[\sum ((y_i - Y_\sigma)(x_i - X_\sigma)/\sigma_i^2/(N-1))\right] / \left[\sum (1/\sigma_i^2)/N\right]$$

4.5.7 The Least Square Method

The least square method is when there are (x,y) data sets that are fitted by the straight line, y = ax + b. The usual criterion is to minimize the distance between the points and the fitted line.

The solutions are:

$$a = Y - r\,X\,\sigma_y/\sigma_x$$

$$b = r\,\sigma_y/\sigma_x$$

where r is the correlation coefficient given by section 4.5.5. Then the equation of the best straight line fit is;

$$y = Yx - r\,\sigma_y/\sigma_x\,(Xx - 1)$$

The uncertainty on the slope, (p) is:

$$(p) = \sigma_y\sqrt{(1 - r^2)}/\left[\sigma_x\sqrt{N}\right]$$

The uncertainty on the ordinate at the origin, (I) is:

$$(I) = \sigma_x\sqrt{(1 - r^2)}/\left[\sigma_y\sqrt{N}\right]$$

However, these solutions do not include uncertainty of each datum. To include them, sophisticated mathematics is required. They are described in York (1969) and included in the program package ISOPLOT by Ludwig (1999) and later versions. These statistics are used in drawing isochrons for dating as indicated in Sections 2.3, 2.5, 4.5.6, and 4.5.7.

References

Anders E, Grevesse N. Abundances of the elements meteoritic and solar. Geochim Cosmochim Acta 1989;53:197–214.

Becker H, Horan MF, Walker RJ, Gao S, Lorand J-P, Rudnick RL. Highly siderophile element composition of the Earth's primitive upper mantle constraints from new data on peridotite massifs and xenoliths. Geochim Cosmochim Acta 2006;70:4528–50.

Bottke WF, Vokrouhlichy D, Marchi S, et al. Dating the Moon-forming impact event with asteroidal meteorites. Science 2015;348:321–3.

Clayton RN, Onuma N, Mayeda TK. A classification of meteorites based on oxygen isotopes. Earth Planet Sci Lett 1976;30:10–8.

Cottrell E, Walker RJ. Constraints of core formation from Pt partitioning in mafic silicate liquids at high temperature. Geochim Cosmochim Acta 2006;70:1565–80.

Danielson LR, Sharp TG, Hervig RL. Implications for core formation of the Earth from high pressure temperature Au partitioning experiments. In: Lunar and Planetary Science Conference XXXVI, Abstract #1955 (CD-ROM). Houston: Lunar and Planetary Institute; 2005.

Day JMD, Pearson DG, Taylor LA. Highly siderophile element constraints on accretion and differentiation of the Earth-Moon system. Science 2007;315:217–9.

Grange ML, Nemchin AA, Pidgeon RT, Merle RE, Timms NE. What lunar zircon ages can tell?, Lunar Planet Sci 2013;44:1884.

Halliday AN. A young Moon-forming giant impact at 70–110 million years accompanied by late-stage mixing, core formation and degassing of the Earth. Philos Trans R Soc A 2008;366:4163–81.

Herwartz D, Pack A, Friedrichs B, Bischoff A. Identification of the giant impactor Theia in lunar rocks, Science 2014;344:1146–50

Horan MF, Walker RJ, Morgan JW, Grossman JN, Rubin A. Highly siderophile elements in chondrites. Chem Geol 2003;196:5–20.

Jacobsen SB. The Hf-W isotopic system and the origin of the Earth and Moon. Annu Rev Earth Planet Sci 2005;33:531–70.

Kleine T, Palme H, Metzger K, Halliday AN. Hf-W chronometry of lunar metals and the age and early differentiation of the Moon. Science 2005;320:1671–4.

Kruijer TS, Kleine T, Fischer-Gödde M, Sprung P. Lunar tungsten isotopic evidence for the late veneer. Nature 2015;520:534–7.

Leya I, Wieler R, Halliday AN. Cosmic-ray production at tungsten isotopes in lunar samples and meteorites and its implications for Hf-W cosmochemistry. Earth Planet Sci Lett 2000;175:1–12.

Ludwig KR. ISOPLOT/version 2.01. A geochronological tool kit for Microsoft Excel. Berkeley (CA): Geochronological Center; 1999.

Makishima A. Thermal ionization mass spectrometry (TIMS). Silicate digestion, separation, measurement. Weinheim: Wiley-VCH; 2016.

McDonough WF, Sun S-s. The composition of the Earth. Chem Geol 1995;120:223–53.

Newton J, Franchi IA, Pillinger CT. The oxygen-isotopic record in enstatite chondrite. Meteorit Planet Sci 2000;35:689–98.

Pattou L, Lorand JP, Gros M. Non-chondritic platinum-group element ratios in the Earth's mantle. Nature 1996;379:712–5.

Righter K, Drake MJ. Metal-silicate equilibrium in a homogeneously accreting Earth: new results for Re. Earth Planet Sci Lett 1997;146:541–53.

Righter K, Huayun M, Danielson L. Partitioning of palladium at high pressures and temperatures during core formation. Nat Geosci 2008;1:321–3.

Schlichting HE, Warren PH, Yin Q-Z. The last stages of terrestrial planet formation: dynamical friction and the late veneer. Astrophys J 2012;752:8.

Tera F, Papanastassiou DA, Wasserburg GJ. Isotopic evidence for a terminal lunar cataclysm. Earth Planet Sci Lett 1974;22:1–21.

Touboul M, Kleine T, Bourdon B, Palme H, Wieler R. Late formation and prolonged differentiation of the Moon inferred from W isotopes in Lunar metals. Nature 2007;450:1206–9.

Touboul M, Kleine T, Bourdon B, Palme H, Wieler R. Tungsten isotopes in ferroan anorthosites: implications for the age of the Moon and the lifetime of its magma ocean. Icarus 2009;199:245–9.

Touboul M, Walker RJ. High precision tungsten isotope measurement by thermal ionization mass spectrometry. Int J Mass Spectrom 2012;309:109–17.

Touboul M, Puchtel IS, Walker RJ. Tungsten isotopic evidence for disproportional late accretion to the Earth and Moon. Nature 2015;520:530–3.

Warren PH, Kallemeyn GW, Kyte FT. Origin of planetary cores: evidence from highly siderophile elements in martian meteorites. Geochim Cosmochim Acta 1999;63:2105–22.

Walker RJ, Storey M, Kerr A, Tarne J, Arndt NT. Implications of ^{187}Os heterogeneities in mantle plume: evidence from Gorgona Island and Curacao. Geochim Cosmochim Acta 1999;66(63):713–28.

Walker RJ, Horan MF, Morgan JW, Becker H, Grossman JN, Rubin A. Comparative ^{187}Re-^{187}Os systematics of chondrites: implications regarding early solar system processes. Geochim Cosmochim Acta 2002;66:4187–201.

Walker RJ, Horan MF, Shearer CK, Papike JJ. Low abundances of highly siderophile elements in the lunar mantle: evidence for prolonged late accretion. Earth Planet Sci Lett 2004;224:399–413.

Walker RJ. Highly siderophile elements in the Earth, Moon and Mars: update and implications for planetary accretion and differentiation. Chem Erde 2009;69:101–25.

Walker RJ, Berminghan K, Liu JG, Puchtel IS, Touboul M, Worsham EA. In search of late-stage planetary building blocks. Chem Geol 2015;411:125–42.

York D. Least squares fitting of a straight line with correlated errors. Earth Planet Sci Lett 1969;5:320–4.

Young ED, Kohl IE, Warren PH, Rubie DC, Jacobsen SA, Morbidelli A. Oxygen isotopic evidence for vigorous mixing during the Moon-forming giant impact. Science 2016;351:493–6.

CHAPTER 5

The Age of the Moon

Chapter Outline

5.1 INTRODUCTION

Through the use of astrophysics, Jacobsen et al. (2014) determined the Moon-forming age to be 95±32 Myr after condensation by an N-body simulation (see Box 2.8). If we assume the condensation age to be 4.56 Ga, the Moon was formed in 4.46 Ga. They showed that earlier formation is ruled out at a 99.9% confidence level.

Carlson et al. (2014) divided lunar rocks into five types by cosmochemistry and astrogeology based on time (see Table 5.1).

Table 5.1 Age Estimates for Early Lunar Differentiation Events

Observation	Age (Ga)	References
Giant impact by heating model	4.47	Bottke et al. (2015)
FAN ages	4.360 ± 0.003	Borg et al. (2011)
	4.31 ± 0.07	Nyquist et al. (2010)
Peak in lunar zircon age distribution	4.320	Grange et al. (2013)
The second peak	4.200	Grange et al. (2013)
Oldest point on lunar zircon	4.417 ± 0.006	Nemchin et al. (2009)
Zircon Hf model ages	4.38 – 4.48	Taylor et al. (2009)
Mare basalt ^{146}Sm–^{142}Nd source age	4.32	Nyquist et al. (1995)
	4.35	Rankenburg et al. (2006)
	4.45	Boyet and Carlson (2007)
	4.33	Brandon et al. (2009)
Pb model age for lunar highlands	4.42	Terra and Wasserburg (1974)
KREEP Sm–Nd and Lu–Hf model ages	4.36 ± 0.04	Gaffney and Borg (2014)
	4.36 ± 0.04	Sprung et al. (2013)
	4.47 ± 0.07	Nyquist et al. (2010)
	~4.26	Lugmair and Carlson (1978)
^{182}Hf-^{182}W lunar model age	<4.50	Touboul et al. (2007)
Mg-suites	4.283, 4.421	Carlson et al. (2014)
urKREEP	4.368 ± 0.029	Gaffney and Borg (2013)

The table is modified from Carlson RW, Borg LE, Gaffney AM, Boyet M. Rb-Sr, Sm-Nd and Lu-Hf isotope systematics of the lunar Mg-suite: the age of the lunar crust and its relation to the time of Moon formation. Phil Trans R Soc A 2014;372:20130246.

Origins of the Earth, Moon, and Life. http://dx.doi.org/10.1016/B978-0-12-812058-3.00005-3

1. Initial crystallization: In the magma ocean model of the Moon, an initially extensively molten Moon first crystallized mafic silicates that sank into the mantle to form the source regions of much later mare basalt magmatism (Walker et al., 1975; Warren, 1985).
2. Ferroan anorthosite and floating plagioclase: After 70–80% crystallization of the magma ocean, plagioclase began to crystallize from a dense iron-rich differentiated magma, causing the plagioclase to float to form the ferroan anorthosite (FAN) series of lunar highland rocks (Dowty et al., 1974).
   From the model calculation (Elkins-Tanton et al., 2011), the floatation of the crust may have occurred over a 1000 years; then solidification of the magma ocean ended in a few tens of millions of years. It is strange that the magma ocean continued until approximately 4300 Ma. To keep the Moon molten, some heat source is required. Tidal heating by the Earth is one possibility.
3. Mafic cumulates with Eu anomaly. The extraction of plagioclase from the magma ocean left the mafic cumulates of the lunar interior with a deficiency in Eu relative to neighboring rare earth elements (REEs). This is reflected as a negative Eu anomaly (see Box 3.2) in some mare basalts (Taylor and Jakes, 1974).
4. Crystallization of KREEP: Further crystallization made residual liquid strongly enriched in incompatible elements named KREEP, due to its enrichment in potassium (K), REE, and phosphorus (P), as well as other incompatible elements (see Section 3.3.2; Warren and Wasson, 1979).
5. Mg-suite cumulate rocks: The final liquid component became the lunar highland crusts, which are plagioclase rich rocks but are distinguished from FANs by their higher Mg:Fe ratios and the presence of abundant mafic phases. The Mg-suite cumulate rocks are usually partial melts of cumulates in the lunar interior (Shearer and Papike, 2005), or parental magmas originating from large impacts (Hess, 1994).

There are at least five ages of the Moon. Thus, those who use the Moon age must clarify which age is used in the discussion.

5.2 AGE OF FANS

As discussed in Section 5.1, the FAN suite of lunar crustal rocks is considered to be the primary lunar floated-cumulated crust that crystallized in the second stage of magma ocean solidification. (It should be remembered that the first solidified silicates in the first stage of magma ocean solidification sank into the deep mantle.)

According to this model, FANs represent the oldest age of lunar crustal rock types. Attempts to date this rock suite precisely have failed because individual isochron measurements are not typically matched to the cosmochemical history of the samples and have not been confirmed by each isotopic system (Hanan and Tilton, 1987; Carlson and Lugmair, 1988; Alibert et al., 1994; Borg et al., 1999; Norman et al., 2003).

Nyquist et al. (2010) obtained a Sm–Nd mineral isochron age of 4.47 ± 0.07 Ga for FAN-67075 and also obtained the Sm–Nd isochron age of 4.31 ± 0.07 Ga for anorthositic clasts of lunar meteorites Y86032 and Dho 908.

By making improvements to standard isotopic techniques, Borg et al. (2011) determined the age of crystallization of FAN 60025 to be 4.360 ± 0.003 Ga using the ^{207}Pb–^{206}Pb isotopic system of 4.3592 ± 0.0024 Ga, the ^{147}Sm–^{143}Nd isotopic system of 4.367 ± 0.011 Ga, and the ^{146}Sm–^{142}Nd isotopic system of $4.318^{+0.030}_{-0.038}$ *Ga*, which is model dependent. These extraordinarily young ages require that either the Moon solidified significantly later than most previous estimates, or the standard model that FANs are floated cumulates of a primordial magma ocean is incorrect. This problem has not been solved yet; thus, further studies are required.

5.3 ZIRCON AGES

Zircon crystallizes from the melt that saturates in zirconium. Zircon is formed in felsic samples, but is sometimes found in mafic cumulates. Most lunar zircons are found in KREEP, but some zircons are found in impact melt breccias. Hafnium, which is present in a concentration of approximately 1% in zircon, has a high neutron cross-section with thermal neutrons; therefore, elements in zircon are protected against neutron cosmic rays, which change isotope ratios or cause fission of uranium (Box 5.1).

BOX 5.1 NEUTRON CROSS-SECTION

The neutron cross-section is likely a reaction between an incident neutron and a target nucleus. It is expressed as a "barn" unit, which is the dimension of $10^{-28}\,m^2$. When the cross-section is larger, the reaction occurs more often.

We can assume a situation in which an incident neutron hits a target nucleus. On Earth, this takes place in an atomic reactor. On the Moon or an asteroid, neutrons from the sun or cosmic neutrons in space hit the Moon and asteroid surfaces.

The probability of the reaction between the neutron and the target nucleus is dependent not only on the characteristics of the nucleus (for example, zirconium has very low neutron cross-section, which is why it is used in the tube to put uranium in the nuclear reactor), but also on the energy (velocity) of the neutron. Usually, a slow neutron has higher probability, and a fast neutron has lower probability of reaction. There is an energy threshold in the neutron–nucleus reaction. A low-energy neutron corresponds to the temperature of approximately 290K, and is called a thermal neutron. A faster neutron causes scattering, not the reaction (Walker et al., 1989).

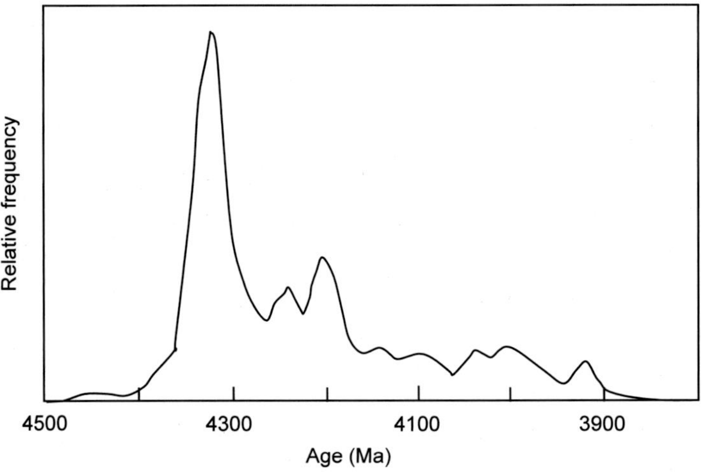

FIGURE 5.1
Relative frequency of zircons on the Moon.
The figure is based on Grange ML, Nemchin AA, Pidgeon RT, Merle RE, Timms NE. What lunar zircon ages can tell? Lunar Planet Sci 2013;44:1884.

Grange et al. (2013) devised an age distribution of zircon (see Fig. 5.1). They found the peaks at 4.320, 4.240, 4.200, and 3.920 Ga between 4.4 and 3.9 Ga. The peak of 4.320 Ga was the most prominent, followed by the 4.200 Ga. They attributed the periodicity to the radioactive decay heat of the KREEP reservoir, which is enriched in incompatible elements such as K, U, and Th.

Taylor et al. (2009) obtained 4.38–4.48 Ga from the Hf model ages of zircons. Nemchin et al. (2009) found the oldest zircon of 4.417 ± 0.006 Ga, which is considered to be the oldest magmatic activity or magma ocean on the Moon.

5.4 AGE OF KREEP ROCKS

Lugmair and Carlson (1978) observed that lunar KREEP samples 12034, 14307, 15382, 65015, 15426, and 75075 are aligned on a line of approximately 4.26 Ga of the Sm–Nd isochron. However, the variations of Sm/Nd were not large; all data did not result in a clear age (To obtain precise age data in the isochron method, large variation in the parent/daughter ratio (horizontal axis) is required.).

Gaffney and Borg (2013) obtained Lu–Hf data for KREEP samples of 15386, 72275,383, 77215, and 78238, and the model age of 4.36 ± 0.04 Ga. Sprung et al. (2013) obtained Lu–Hf data of KREEP-rich rocks 12034, 14310, 65015, 62235, 68115, and 68815, and the model age of 4.36 ± 0.04 Ga.

Terra and Wasserburg (1974) obtained the Pb model age of 4.42 Ga for lunar highlands using U–Th–Pb systematics. Touboul et al. (2007) obtained the ^{182}Hf–^{182}W lunar model age of less than 4.50 Ga.

5.5 AGE OF MG-SUITE ROCKS

Carlson et al. (2014) determined Rb–Sr, ^{146,147}Sm–^{142,143}Nd, and Lu–Hf isotopic analyses of Mg-suite lunar crustal rocks 67667, 76335, 77215, and 78238, including an internal isochron for norite 77215. Isochron ages for 77215 determined by their study were Rb–Sr = 4.450 ± 0.270 Ga, ^{147}Sm–^{143}Nd = 4.283 ± 0.023 Ga, and Lu–Hf = 4.421 ± 0.068 Ga. The initial Nd and Hf isotopic compositions of all samples indicate that a source region was slightly enriched for incompatible elements, which previously suggested that the Mg-suite crustal rocks contain a component of KREEP. The Sm/Nd–^{142}Nd/^{144}Nd correlation shown by both a FAN and Mg-suite rocks is consistent with the trend defined by mare and KREEP basalts, the slope of which corresponds to ages between 4.35 and 4.45 Ga. And the ages are in good agreement with the model of lunar formation by the giant impact into Earth in c. 4.4 Ga.

Gaffney and Borg (2014) named the final formed product of magma ocean crystallization urKREEP. They obtained 4.368 ± 0.029 Ma from the model ages of Hf and Nd isotope systematics.

5.6 AGE OF THE MOON MANTLE DIFFERENTIATION BY THE ^{146}SM-^{142}ND METHOD

The ^{146}Sm–^{142}Nd method, which uses an α-decay of ($T_{1/2}$ = 103 Ma; see Section 2.3), is very useful for the estimation of the formation age of the mantle. The method requires neutron correction, and the obtained age is the model age; however, it provides important information. Furthermore, it can combine with the ^{147}Sm–^{143}Nd method.

Nyquist et al. (1995) determined 4.32 Ga to be the lunar initial differentiation age, which corresponds to the initial crystallization age in Section 5.1. Rankenburg et al. (2006) determined that lunar mantle formation was 215^{+23}_{-21} *Myr* from the formation of the solar system, which corresponds to 4.35 Ga. Boyet and Carlson (2007) calculated the early lunar differentiation to be 4.45 Ga by the ^{146}Sm–^{142}Nd and ^{147}Sm–^{143}Nd methods.

Brandon et al. (2009) remeasured six lunar basalts samples, from Hi–Ti basalts, Lo–Ti basalts, and KREEP, for Sm–Nd and Lu–Hf isotope compositions. They also re-evaluated the Moon evolution models. The model that the bulk Moon has a superchondritic ^{147}Sm/^{144}Nd ratio of 6–8.8% (which means Sm/Nd ratios are higher than that of chondrite) best explains the Nd–Hf isotope compositions of the measured lunar mare basalts. The ^{142}Nd–^{143}Nd isotope systematics are best interpreted with an age of 229^{+24}_{-20} *Ma* after nebular condensation. The mare basalt sources were closed approximately 150–190 Ma after accretion of the Moon. A long cooling history of a magma ocean is consistent with U–Pb age distributions in zircons (Nemchin et al., 2009; Grange et al., 2013).

5.7 IMPLICATION OF THE LUNAR AGE

As the Moon surface has been intact since its solidification, the activity of the solar system is recorded on the surface as the size frequency of craters in a given surface of the Moon. However, if the activity of the solar system is assumed, the crater size frequency or the crater size distribution is a function of the age of the Moon surface.

This means that age dating on the surface of the Moon is possible (Michael and Neukum, 2009). The absolute age of the Moon surface determined by cosmochronology provides a good calibration on the crater-size-frequency age determination method.

References

Alibert C, Norman MD, McCulloch MT. An ancient age for a ferroan anorthosite clast from lunar breccia 67016. Geochim Cosmochim Acta 1994;58:2921–6.

Borg LE, Norman M, Nyquist L, et al. Isotopic studies of ferroan anorthosite 62236: a young lunar crustal rock from a light rare-earth element-depleted source. Geochim Cosmochim Acta 1999;58:2921–6.

Borg LE, Connelly JN, Boyet M, Carlson RW. Chronological evidence that the Moon is either young or did not have a global magma ocean. Nature 2011;477:70–2.

Bottke WF, Vokrouhlický D, Marchi S, Swindle T, Scott ERD, Weirich JR, Levison H. Dating the Moon-forming impact event with asteroidal meteorites. Science 2015;348:321–323.

Boyet M, Carlson RW. A highly depleted moon or a non-magma ocean origin for the lunar crust? Earth Planet Sci Lett 2007;262:505–16.

Brandon AD, Lapen TJ, Debaille V, Beard BL, Rankenburg K, Neal C. Re-evaluating $^{142}Nd/^{144}Nd$ in lunar mare basalts with implications for the early evolution and bulk Sm/Nd of the Moon. Geochim Cosmochim Acta 2009;73:6421–45.

Carlson RW, Lugmair GW. The age of ferroan anorthosite 60025: oldest crust on a young Moon? Earth Planet Sci Lett 1988;90:119–30.

Carlson RW, Borg LE, Gaffney AM, Boyet M. Rb-Sr, Sm-Nd and Lu-Hf isotope systematics of the lunar Mg-suite: the age of the lunar crust and its relation to the time of Moon formation. Phil Trans R Soc A 2014;372:20130246.

Dowty E, Prinz M, Keil K. Ferroan anorthosite: a widespread and distinctive lunar rock type. Earth Planet Sci Lett 1974;24:15–25.

Elkins-Tanton LT, Burgess S, Yin Q-Z. The lunar magma ocean: reconciling the solidification process with lunar petrology and geochronology. Earth Planet Sci Lett 2011;304:326–36.

Gaffney AM, Borg LE. A young age for KREEP formation determined from Lu-Hf isotope systematics of KREEP basalts and Mg-suite samples. Lunar Planet Sci 2013;44:1714.

Gaffney AM, Borg LE. A young solidification age for the lunar magma ocean. Geochim Cosmochim Acta 2014;140:227–40.

Grange ML, Nemchin AA, Pidgeon RT, Merle RE, Timms NE. What lunar zircon ages can tell? Lunar Planet Sci 2013;44:1884.

Hanan BB, Tilton GR. 60025-relict of primitive lunar crust. Earth Planet Sci Lett 1987;84:15–21.

Hess PC. Petrogenesis of lunar troctolites. J Geophys Res 1994;99:19083–93.

Jacobsen SA, Morbidelli A, Raymond SN, O'Brien DP, Walsh KJ, Rubie DC. Highly siderophile elements in Earth's mantle as a clock for the Moon-forming impact. Nature 2014;508:84–7.

Lugmair GW, Carlson RW. Sm-Nd constraints on early lunar differentiation and the evolution of KREEP. Lunar Planet Sci Conf, vol. 9. 1978. p. 689–704.

Michael GG, Neukum G. Planetary surface dating from crater size-frequency distribution measurements: partial resurfacing events and statistical age uncertainty. Earth Planet Sci Lett 2009;294:223–9.

Nemchin A, Timms N, Pidgeon R, Geisler T, Reddy S, Meyer C. Timing of crystallization of the lunar magma ocean continued by the oldest zircon. Nat Geosci 2009;2:133–6.

Norman MD, Borg LE, Nyquist LE, Bogard DD. Chronology, geochemistry and petrology of a ferroan noritic anorthosite clast from Descartes breccia 67215: clues to the age, origin, structure, and history of the lunar crust. Meteorit Planet Sci 2003;38:645–61.

Nyquist LE, Wiesmann H, Bansal B, Shih C-Y, Keith JE, Harper CL. ^{146}Sm-^{142}Nd formation interval f or the lunar mantle. Geochim Cosmochim Acta 1995;59:2817–37.

Nyquist LE, Shih C-Y, Reese YD, et al. Lunar crustal history recorded in lunar anorthosites. Lunar Planet Sci 2010;41:1383.

Rankenburg K, Brandon AD, Neal CR. Neodymium isotope evidence for a chondritic composition of the Moon. Science 2006;312:1359–72.

Shearer CK, Papike JJ. Early crustal building processes on the Moon: models for the petrogenesis of the magnesian suite. Geochim Cosmochim Acta 2005;69:3445–61.

Sprung P, Kleine T, Scherer EE. Isotopic evidence for chondritic Lu/Hf and Sm/Nd of the Moon. Earth Planet Sci Lett 2013;380:77–87.

Taylor SR, Jakes P. The geochemical evolution of the Moon. In: Proc. 5th Lunar Sci Conf. New York: Pergamon Press; 1974. p. 1287–305.

Taylor DJ, McKeegan KD, Harrison TM. Lu-Hf zircon evidence for rapid lunar differentiation. Earth Planet Sci Lett 2009;279:157–64.

Terra F, Wasserburg GJ. U-Th-Pb systematics on lunar rocks and inferences about lunar evolution and the age of the moon. In: Proc. 5th Lunar Sci Conf. New York: Pergamon Press; 1974. p. 1500–71.

Touboul M, Kleine T, Bourdon R, Palme H, Wider R. Late formation and prolonged differentiation of the Moon inferred from W isotopes in lunar metals. Nature 2007;450:1206–9.

Walker D, Longhi J, Hays JF. Differentiation of a very thick magma body and implications for the source regions of mare basalts. In: Proc. 6th Lunar Sci. Conf. New York: Pergamon Press; 1975. p. 1103–1120.

Walker FW, Parrington JR, Feiner F. Nuclides and isotopes. 14th ed. USA: General Electric Company; 1989.

Warren PH, Wasson JT. The origin of KREEP. Rev Geophys Space Phys 1979;17:73–88.

Warren PH. The magma ocean concept and lunar evolution. Annu Rev Earth Planet Sci 1985;13:201–40.

CHAPTER 6

Age of the Earth From Geological Records Remaining on the Earth Surface

Chapter Outline

6.1 INTRODUCTION

After the late veneer, the Earth seemed to be in a stabilized condition. Within the Earth, the core was formed. On the surface, the atmosphere was formed and cooled, and the sea is believed to have appeared. Therefore, during 4.6 Ga, most evidence of large geological events on the surface had been washed away. In Table 6.1, the early important differential events that can be discussed even now are shown (Carlson et al., 2014). In this chapter, each event summarized in Fig. 6.1 is explained. The pink boxes are based on observation of chondrites. The orange boxes indicate events related to

Origins of the Earth, Moon, and Life. http://dx.doi.org/10.1016/B978-0-12-812058-3.00006-5

Table 6.1 Age Estimates for Early Earth's Differentiation Events

Observation	Age (Ga)	References
Hf–W age of core formation	4526 4538-4468	Kleine et al. (2009) Rudge et al. (2010)
U–Pb age of the Earth	4.55 ~4450	Patterson (1956) Allegre et al. (2008)
I–Pu–Xe age of the Earth's atmosphere	~4450	Staudacher and Allegre (1982) Pepin and Porcelli (2006) Mukhopadhyay (2012)
^{146}Sm–^{142}Nd model age for Isua	4350–4470	Caro et al. (2006) Rizo et al. (2011)
Peak in oldest Hadean zircons	4350	Holden et al. (2009)
^{146}Sm–^{142}Nd model age for Nuvvuagittuq crust	4340–4400	O'Neil et al. (2012)

The Table is modified from Carlson RW, Borg LE, Gaffney AM, Boyet M. Rb-Sr, Sm-Nd and Lu-Hf isotope systematics of the lunar Mg-suite: the age of the lunar crust and its relation to the time of Moon formation. Phil Trans R Soc A 2014;372:20130246.

the Moon. The yellow boxes are related to topics about the early Earth. The green and blue boxes show constraints from extinct nuclei and astrophysics, respectively.

6.2 CORE FORMATION AGE FROM Hf–W SYSTEMATICS

From the combination of ^{182}W–^{142}Nd evidence, Kleine et al. (2009) proposed that the bulk Earth may have superchondritic samarium/neodymium (Sm/Nd) and halfnium/tungstein (Hf/W) ratios and that the formation of the core must have terminated more than approximately 42 Myr (4526 Ma) after the formation of calcium aluminum-rich inclusions (CAIs) (4568.3 ± 0.7 Ma; Burkhardt et al., 2008), which is consistent with the Hf–W age for the formation of the Moon.

Using the Hf–W and U–Pb isotopic systems, Rudge et al. (2010) suggested the rapid accretion of Earth's main mass within approximately 10 Myr from the formation of the Solar System. The Earth's accretion terminated 30–100 Myr after the formation of the Solar System. They also proposed the disequilibrium model, in which some fraction of the embryos' metallic cores (descending small metal droplets) was allowed to directly enter the Earth's core, without equilibrating with the Earth's mantle. Their results indicate that only 36% of the Earth's core must have formed in equilibrium with Earth's mantle.

Unfortunately, this model does not include the discussion of highly siderophile elements (HSEs), because there are no appropriate radiogenic isotope systems for HSEs. However, if there were appropriate isotope systems (e.g., the Re–Os isotope system could be applicable), the model could be extended to HSEs. In addition, if the calculations could explain the present concentration of HSEs in the mantle, the late veneer would not be required. This is speculation of the author, and of course, future studies are required!

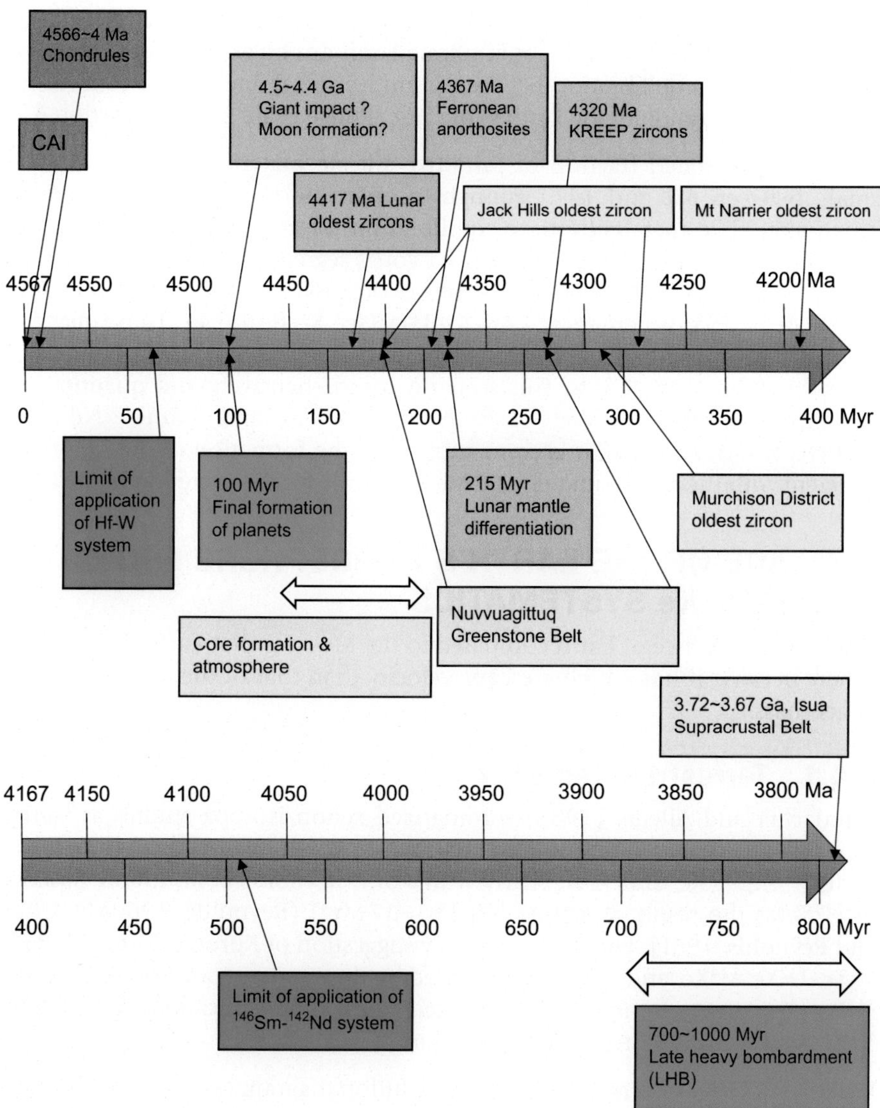

FIGURE 6.1
Summary of important events including hypothetical ones in the early Earth's history. The *pink boxes* are related to the chondrites. The *orange boxes* indicate events related to the Moon. The *yellow boxes* are related to the early Earth. The *green and blue boxes* show the limits of extinct nuclei and geophysical requirements.

6.3 U–Pb AGE OF THE EARTH

By using the U–Pb method, the pioneer Patterson (1956) used the age of the stony meteorites to calculate Earth's age to be 4.55 ± 0.07 Ga.

Allegre et al. (2008) reject the rapid accretion and early differentiation of the Earth (30–40 Ma) after the birth of the Solar System at 4.567 Ga (Amelin et al., 2002,

2006) (see Box 2.2). They used Hf–W, U–Pb, and I–Xe systematics on the Earth. The W isotopic composition of the bulk silicate Earth can be explained by an incomplete isotopic re-equilibration between primitive metal and silicate components during the segregation of the Earth's core (e.g., Rudge et al., 2010; see Section 6.2).

The nonequilibrated fraction of primitive silicate material is estimated to be small, between 6% and 14%, enough to open the ^{182}Hf–^{182}W chronometer. This incomplete metal/silicate re-equilibration affects the U–Pb chronometer only slightly. The mean age of the Earth's core's segregation is between 4.46 and 4.38 Ga. This evaluation overlaps the time of outgassing of the atmosphere based on the ^{129}I–^{129}Xe systematics, 4.46–4.43 Ga (see Section 6.4). Thus, they concluded that the period of approximately 4.45 Ga relates to the major primitive differentiation of the Earth. This scenario comprehensively and quantitatively explains the ^{182}Hf–^{182}W, $^{235, 238}$U–$^{207, 206}$Pb, ^{129}I–^{129}Xe, and ^{146}Sm–^{142}Nd terrestrial records. In addition, it is compatible with the formation of the Moon and coherent with the approximately 102 Ma time scale for the accretion of the Earth.

6.4 AGE OF THE EARTH'S ATMOSPHERE FROM I–Pu–Xe SYSTEMATICS

One large difference of Earth compared to the Moon is that Earth has an atmosphere because it has a higher escape velocity than that of the Moon due to its larger mass.

6.4.1 Terrestrial Xenology

Staudacher and Allegre (1982) summarized xenon isotope studies as "terrestrial xenology." Xenon has nine stable isotopes: ^{124}Xe, ^{125}Xe, ^{128}Xe, ^{129}Xe, ^{130}Xe, ^{131}Xe, ^{132}Xe, ^{134}Xe, and ^{136}Xe. Early works of J. Reynolds et al. found variation of ^{129}Xe as the result of extinct ^{129}I ($T_{1/2}$ = 17 Myr) (Reynolds, 1960a,b; Jeffery and Reynolds, 1961). Following the early suggestion of Kuroda (Kuroda, 1960), ^{131}Xe, ^{132}Xe, ^{134}Xe, and ^{136}Xe are increased by the spontaneous fission of extinct ^{244}Pu ($T_{1/2}$ = 82 Myr) and of ^{238}U over geological time (Alexander et al., 1971). ^{130}Xe can be used as the reference isotope.

When a ^{129}Xe/^{130}Xe versus ^{134}Xe/^{130}Xe evolution diagram (see Fig. 6.2) is made, the horizontal and vertical axes indicate the decay of ^{129}I and the fissions of ^{244}Pu and ^{238}U, respectively. The atmosphere is plotted in the center, and mid-ocean ridge basalts (MORBs) are plotted along the mixing line between the atmosphere and the hypothetical lower mantle. Other terrestrial samples can be made by the mixing of vertically evolved materials by the fission of ^{238}U and MORB samples. As result, the terrestrial samples are created by the triangle formed by the vertical line and the MORB line.

Staudacher and Allegre (1982) made a box model consisting of seven components (atmosphere, upper mantle, lower mantle, ocean, continent, sediments, and oceanic crust). Using this model, the degassing of the terrestrial mantle, the time interval between the formation of meteorites, and the formation of the

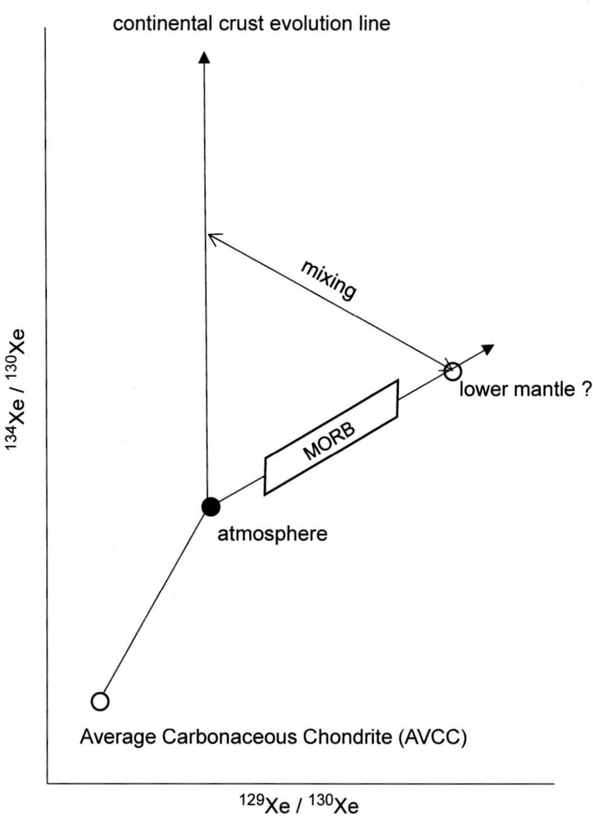

FIGURE 6.2
Theoretical $^{129}Xe/^{130}Xe$ versus $^{134}Xe/^{130}Xe$ evolution diagram.

Earth were estimated. The mean age of the atmosphere was determined to be 25 Myr (99.9% degassing of the mantle) or 10 Myr (60% degassing). The age of the Earth was estimated to be 4.48–4.47 Ga, which is 50–70 Myr younger than the formation of the meteorites.

6.4.2 Discovery of New Components of Xe

Pepin and Porcelli (2006) measured CO_2 well gases and MORBs. The nonradiogenic Xe in the atmosphere could not be the primordial base composition in the mantle. Solar-like components, for example, U–Xe, solar wind Xe, or both were required. If the present estimate for U/I in the bulk silicate Earth (BSE) could be applied to all interior volatile reservoirs, the differing $^{129}Xe_{rad}/^{136}Xe_{244}$ ratios in MORB and the well gases point to two episodes of major mantle degassing, presumably driven by giant impacts, respectively, approximately 20–50 and 95–100 Ma after Solar System origin, assuming current values for initial $^{129}I/^{127}I$ and $^{244}Pu/^{238}U$. The earlier time range for degassing of the well gas source spans Hf–W calculations for the timing of a moon-forming impact. The second, later impact further outgassed the upper mantle and MORB source. (The author thinks this could be the late veneer!?)

6.4.3 Early Heterogeneity Kept in the Icelandic OIB Source

The isotope ^{129}Xe is produced from the radioactive decay of extinct ^{129}I. The isotope ^{136}Xe is produced from extinct ^{244}Pu and extant ^{238}U. The lower ratios of $^{129}Xe/^{130}Xe$ in ocean island basalts (OIBs) than those of MORBs are the evidence for the existence of a relatively undergassed primitive deep-mantle reservoir (Marty, 1989; Trieloff and Kunz, 2005; Holland and Ballentine, 2006). The low $^{129}Xe/^{130}Xe$ in OIBs can be explained by mixing between subducted atmospheric Xe and MORB Xe; therefore, a less degassed deep-mantle reservoir is not required.

Mukhopadhyay (2012) measured noble gases (He, Ne, Ar, Xe) from an Icelandic OIB, which showed differences in elemental abundances and $^{20}Ne/^{22}Ne$ ratios between the Iceland mantle plume and the MORB source. These observations indicated that the lower $^{129}Xe/^{130}Xe$ ratios in OIBs are due to a lower I/Xe ratio in the OIB mantle source and cannot be explained only by mixing the atmospheric Xe with the MORB-type Xe. The Iceland plume source has a higher proportion of Pu-derived fissionic Xe to U-derived fissionic Xe; therefore, the plume source is required to be less degassed than MORBs. These results showed that the Earth's mantle accreted volatiles from at least two separate sources and that neither the Moon-forming giant impact nor the 4.45 Gyr mantle convection erased the signature of the Earth's heterogeneous accretion and early differentiation.

At the end of Section 6.4.2, the author joked that there could be evidence of the late veneer from the study using the Xe isotopes. At Section 6.4.3, a new old reservoir was also found. The Xe isotopic study seems to give clear evidence for the late veneer in near future.

6.5 APPLICATION OF ^{142}Nd ISOTOPE SYSTEMATICS TO THE OLDEST CRUSTS ON EARTH

6.5.1 Application of $^{146,147}Sm$–$^{142,143}Nd$ Systematics to West Greenland Samples

Caro et al. (2006) developed a new ultra-high–precision $^{142}Nd/^{144}Nd$ measurement method and applied it to early Archaean rocks. The $^{142}Nd/^{144}Nd$ ratio of the Nd standard solution can be determined with intermediate precision of 2 ppm (2σ), allowing resolution of 5 ppm. The 3.6–3.8 Ga West Greenland metasediments, metabasalts, and orthogneisses, which are considered to be the oldest rocks on the Earth, displayed positive ^{142}Nd anomalies ranging from 8 to 15 ppm. Using a simple two-stage model with an initial $\varepsilon^{143}Nd$ value of 1.9 ± 0.6 ε units, both ^{147}Sm–^{143}Nd and ^{146}Sm–^{142}Nd systematics constrained mantle differentiation to 50–200 Ma after the formation of the solar system. This chronological constraint fits the differentiation of the Earth's mantle in the late stage of crystallization of a magma ocean.

They developed a two-box model describing ^{142}Nd and ^{143}Nd isotopic evolution of the depleted mantle in the crust–mantle system. The early terrestrial protocrust had a lifetime of c. 0.7–1 Ga to produce the observed Nd isotope signature of Archaean rocks. In this two-box mantle–crust system, the evolution of isotopic and chemical heterogeneity of the depleted mantle was modeled as a function of the mantle stirring time. Using the dispersion of $^{142}Nd/^{144}Nd$ and

$^{143}Nd/^{144}Nd$ ratios observed in the early Archean rocks, the stirring time of the early Earth's mantle was constrained to be 100–250 Ma, faster than that of modern oceanic basalts by a factor of 5.

6.5.2 Application of the $^{146,147}Sm$–$^{142,143}Nd$ and ^{176}Lu–^{176}Hf Systematics to West Greenland Samples

Rizo et al. (2011) first applied a combined $^{146,147}Sm$–$^{142,143}Nd$ and ^{176}Lu–^{176}Hf study to mafic rocks (amphibolites) from the western part of the Isua Supracrustal Belt (ISB, SW Greenland). The whole-rock isochrons of Sm–Nd and Lu–Hf gave identical ages within a margin of error, 3.72 ± 0.08 and 3.67 ± 0.07 Ga, respectively. The excess of ^{142}Nd in Isua samples was observed to be 7–16 ppm relative to the terrestrial Nd standard. This indicates that early differentiated reservoirs escaped complete homogenization by mantle convection until the Archean eon. The intercept of the Sm–Nd whole-rock isochron is consistent with ^{142}Nd results and with a superchondritic initial $^{143}Nd/^{144}Nd$ ratio ($\varepsilon^{143}Nd_{3.7\,Ga} = +1.41 \pm 0.98$). In contrast, the corresponding initial $\varepsilon^{176}Hf_{3.7\,Ga} = -1.41 \pm 0.57$ is subchondritic. Since Lu/Hf and Sm/Nd fractionate similarly during mantle processes, the Sm–Nd and Lu–Hf isotope systems display inconsistent parent–daughter behavior in the source of Isua amphibolites.

Based on high-pressure and -temperature phase partition coefficients, a model was proposed that satisfies ^{147}Sm–^{143}Nd, ^{176}Lu–^{176}Hf, ^{142}Nd results, and trace element characters. A deep-seated source composed largely of magnesium perovskite (98% MgPv) containing 2% calcium perovskite (CaPv; see Box 6.1) satisfactorily explains the Nd and Hf isotopic discordance observed for Isua amphibolites. The negative high field strength element anomalies characterizing Isua basalts could have been inherited from such an early (4.53–4.32 Ga) deep mantle cumulate. A deep-seated source was involved in the formation of ISB lavas.

6.1 PEROVSKITES

Generally, when perovskite is used, it indicates $CaTiO_3$. However, other perovskites are indicated in cosmochemistry. Please see Fig. 3.7A, which is a cross-section of the present Earth. The lower mantle deeper than 660 km (~24 GPa) is composed of magnesium or calcium perovskites, which are indicated as Pv and Ca–Pv in the figure, respectively. Their chemical formulas are (Mg, Fe)SiO_3 and $CaSiO_3$, respectively.

In cosmochemistry, chemical compositions of the mantle and basalt are considered to be reference compositions. For the former, the pyrolitic mantle composition is used, and for the latter, the MORB composition is used (McDonough and Sun, 1995). Both model compositions of the mantle and basalt form Pv and Ca–Pv at a pressure higher than 24 GPa, which decomposes into post-perovskites at higher pressures than approximately 110 GPa.
Many materials have the perovskite composition and structure, $A^{2+}B^{4+}X^{2-}$ 3 with useful industrial applications. MgCNi3 is a metallic perovskite that attracts attention because it shows superconductivity. Yttrium barium copper oxide is a ceramic perovskite that also has superconductivity. Methylammonium lead triiodide (CH3NH3PbI3) solar cells are dye-sensitized with a high-power conversion efficiency of 20%.

6.5.3 Application of the $^{146,147}Sm$–$^{142,143}Nd$ Systematics to the Ujaraaluk Unit in the Nuvvuagittuq Greenstone Belt, Canada

O'Neil et al. (2012) studied the Ujaraaluk unit in the Nuvvuagittuq Greenstone Belt (NGB) in Northern Quebec, Canada. NGB is dominated by mafic and ultramafic rocks metamorphosed to upper amphibolite facies. Rare felsic intrusive rocks provide zircon ages of up to ~3.8 Ga (David et al., 2009; Cates and Mojzsis, 2007) establishing the minimum formation age of the NGB as the Eoarchean. Primary U-rich minerals that provide reliable formation ages for the dominant mafic lithology called the Ujaraaluk unit have been found.

Metamorphic zircons, retiles, and monazites were present in the unit and gave variably discordant results, with $^{207}Pb/^{206}Pb$ ages ranging from 2.8 to 2.5 Ga. The younger ages overlapped 2686 ± 4 Ma zircon ages for intruding pegmatites (David et al., 2009) and Sm–Nd ages for garnet formation in the Ujaraaluk rocks, suggesting this era as the time of peak metamorphism and metasomatism in the NGB, simultaneous with regional metamorphism of the Superior craton.

The slope of Sm–Nd data for Ujaraaluk was 3.6 ± 0.2 Ga with scattering on the isochron (MSWD = 134). This "isochron" seems to consist of a series of younger (approximately 3.2–2.5 Ga) slopes for the different geochemical groups within the Ujaraaluk, which is possibly older than 4 Ga. The ^{146}Sm–^{142}Nd systematics are less affected by metamorphism at 2.7 Ga because of ^{146}Sm extinction prior to approximately 4 Ga. The ^{142}Nd dataset for the Ujaraaluk and associated ultramafic rocks showed a good correlation between Sm/Nd ratio and $^{142}Nd/^{144}Nd$ that corresponds to an age of $4388^{+15}_{-17}\ Ma$. The dataset included samples with superchondritic Sm:Nd ratios.

The upper Sm/Nd ratio end of the Ujaraaluk correlation is defined by rocks that are interpreted to be cumulates to compositionally related extrusive rocks, indicating that this crystal fractionation had to occur while ^{146}Sm decay was active, well before 4 Ga.

Intruding gabbros gave ^{143}Nd and ^{142}Nd isochron ages of 4115 ± 100 Ma and $4313^{+41}_{-69}\ Ma$, respectively, also supporting a Hadean age for the gabbros and providing a minimum age for the intruded Ujaraaluk unit. 3.6 Ga tonalites surrounding the NGB, 3.8 Ga trondhjemitic intrusive veins, and a 2.7 Ga pegmatite showed a lack of ^{142}Nd compared to the terrestrial standard.

A subset of least disturbed Ujaraaluk samples had coherent isotopic compositions for both short-lived and long-lived Nd isotopic systems giving ^{143}Nd and ^{142}Nd isochron ages overlapping within error of 4321 ± 160 Ma (MSWD = 6.3) and $4406^{+14}_{-17}\ Ma$ (MSWD = 1.0), respectively (see Fig. 6.3). The NGB thus preserved over 1.6 Gyr of early Earth history in mafic crust formed in the Hadean.

6.6 OLDEST ZIRCON ON EARTH

High resolution secondary ion mass spectrometry (HR-SIMS; see Fig. Box 2.6) was a great success in zircon analysis. Using the HR-SIMS, SHRIMP, zircons from Mt. Narryer in Western Australia having U–Pb ages of 3800–4500 Myr were

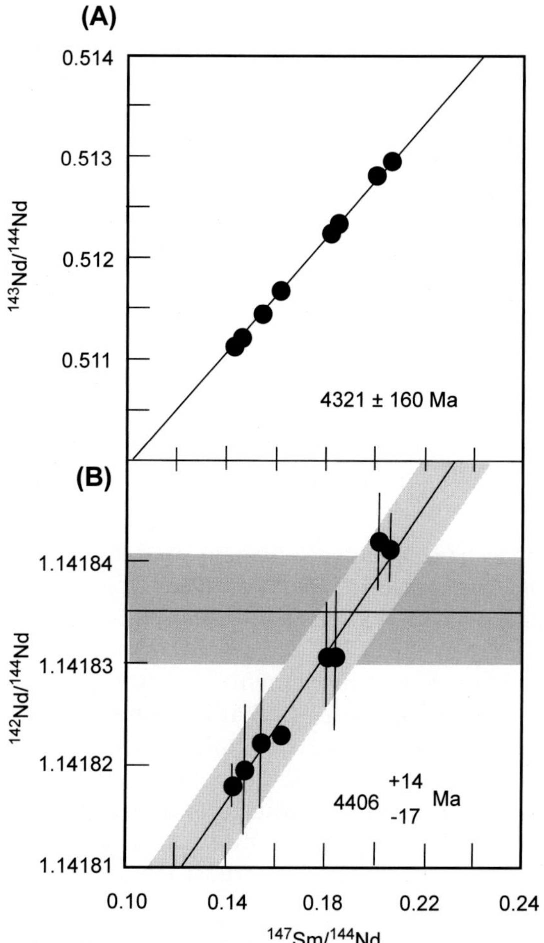

FIGURE 6.3

Isochrons for the Ujaraaluk samples. (A) $^{147}Sm/^{144}Nd$ versus $^{143}Nd/^{144}Nd$ isochron diagram. The *line* shows the best fit line. The number in the figure is the age. (B) $^{147}Sm/^{144}Nd$ versus $^{142}Nd/^{144}Nd$ isochron diagram. The *slanted line* indicates the best fit line, and the shadow shows the error in the regression. The *horizontal line* and *shadow* show the terrestrial standard value and the 4.5 ppm intermediate precision range. The number with the errors is the age.

The figure is modified from O'Neil J, Carlson RW, Paquette J-L, Francis D. Formation age and metamorphic history of the Nuvvuagittuq greenstone belt. Precambrian Res 2012;220–221:23–44.

discovered from Archean metasediments. Four zircons had near-concordant U–Pb ages of approximately 4180 Myr. This result shows that pre-3800 Myr silica-saturated rocks were present on the Earth's crust (Froude et al., 1983).

Compston and Pidgeon (1986) found 4276 ± 6 Ma detrital zircon from the Jack Hills area, also in Western Australia. The frequency of old zircons is 12%, 5 times higher than the Mt. Narryer area. The high U and low Th suggested that they came from the felsic igneous rocks.

Holden et al. (2009) developed the method using SHRIMP II to survey $^{207}Pb/^{206}Pb$ for 5 s, and candidates were analyzed for longer time. They analyzed 10,000 grains from Jack Hills and found 7% were older than 3800 Myr, with the oldest grain being 4372 ± 6 Ma. The oldest population was 4350 Ga.

6.7 FIRST WATER ON THE EARLY EARTH FROM OXYGEN ISOTOPIC DATA OF ZIRCON BY HR-SIMS

The oldest zircons in Mt. Narryer and Jack Hills are of felsic silicate origin, which is similar to the present continental crust. In addition, they are in metamorphosed sedimentary rocks, which indicates liquid water existed at approximately 4350 Ma. The oldest rock (note that "zircon" is only a mineral, but "rock" means the "whole rock," which is an assemblage of various minerals) is gneiss (a metamorphosed sedimentary rock) from the Isua area in Greenland, which also indicates that water already existed at 3800 Ma.

HR-SIMS also gives us oxygen isotopic data of diameters of less than 10 μm. Wilde et al. (2001), in a detailed HR-SIMS study of Jack Hills zircons, discovered a detrital zircon with an age as old as 4404 ± 8 Myr, about 130 million years older than any previously identified zircons on Earth. They found the zircon was zoned for rare earth elements and oxygen isotopic ratios ($\delta^{18}O$ values of 7.4–5.0‰), which implies that it formed from an evolving magmatic source. The high $\delta^{18}O$ value and micro-inclusions of SiO_2 were consistent with growth from a granitic melt with a $\delta^{18}O$ value of 8.5–9.5‰. Magmatic oxygen isotopic ratios indicated the involvement of supracrustal material that underwent low temperature interaction with a liquid hydrosphere. Therefore, this zircon is the earliest evidence for continental crust and oceans on Earth.

A similar study was presented by Mojzsis et al. (2001) in an article in *Nature*. They found detrital zircons from quartzitic rocks in the Murchison District of Western Australia, including 3910–4280-Myr-old zircons that had $\delta^{18}O$ values from 5.4 ± 0.6 to 15 ± 0.4‰. These data indicated that the zircons of approximately 4300 Ma formed from magmas containing a significant component of continental crust, which formed in the presence of water near the Earth's surface. These data are consistent with the presence of a hydrosphere interacting with the crust 4300 Myr ago.

References

Alexander Jr EC, Lewis RS, Reynolds JH, Michel MC. Plutonium-244: confirmation as an extinct radioactivity. Science 1971;172:837–40.

Allegre CJ, Mahnes G, Goppel C. The major differentiation of the Earth at ~4.45 Ga. Earth Planet Sci Lett 2008;267:386–98.

Amelin Y, Krot AM, Hutcheon ID, Ulyanov AA. Lead isotopic ages of chondrules and calcium-aluminum-rich inclusions. Science 2002;297:1678–83.

Amelin Y, Wadhwa M, Lugmair GW. Pb-isotopic dating of meteorites using 202Pb-205Pb double spike: comparison with other high-resolution chronometers. Lunar Planet Sci Conf XXXVII, 2006, #1970.

Burkhardt C, Kleine T, Palme H, Bourdon B, Zipfel J, Friedrich J, Ebel D. Hf-W mineral isochron for Ca, Al-rich inclusions: age pf the solar system and the timing of core formation in planetesimals. Geochim Cosmochim Acta 2008;72:6177–97.

Carlson RW, Borg LE, Gaffney AM, Boyet M. Rb-Sr, Sm-Nd and Lu-Hf isotope systematics of the lunar Mg-suite: the age of the lunar crust and its relation to the time of Moon formation. Phil Trans R Soc A 2014;372:20130246.

Caro G, Bourdon B, Birck J-L, Moorbath S. High-precision $^{142}Nd/^{144}Nd$ measurements in terrestrial rocks: constraints on the early differentiation of the Earth's Mantle. Geochim Cosmochim Acta 2006;70:164–91.

Cates NL, Mojzsis SJ. Pre-3750 Ma supracrustal rocks from the Nuvvuagittuq supracrustal belt, northern Quebec. Earth Planet Sci Lett 2007;255:9–21.

Compston W, Pidgeon RT. Jack Hills, evidence of more very old detrital zircons in Western Australia. Nature 1986;321:766–9.

David J, Godin L, Stevenson R, O'Neil J, Francis D. U–Pb ages (3.8–2.7 Ga) and Nd isotope data from the newly identified Eoarchean Nuvvuagittuq supracrustal belt, Superior Craton, Canada. Geol Soc Am Bull 2009;121:150–63.

Froude DO, Ireland TR, Kinny PD, et al. Ion microprobe identification of 4,100–4,200 Myr-old terrestrial zircons. Nature 1983;616:175–8.

Holden P, Lanc P, Ireland TR, Harrison TM, Foster JJ, Bruce Z. Mass-spectrometric mining of Hadean zircons by automated SHRIMP multi-collector and single-collector U/Pb zircon age dating: the first 100,000 grains. Int J Mass Spectrom 2009;286:53–63.

Holland G, Ballentine CJ. Seawater subduction controls the heavy noble gas composition of the mantle. Nature 2006;441:186–91.

Jeffery PM, Reynolds JH. Origin of excess Xe^{129} in stone meteorites. J Geophys Res 1961;66:3582–3.

Kleine T, Touboul M, Bourdon B, et al. Hf-W chronology of the accretion and early evolution of asteroids and terrestrial planets. Geochim Cosmochim Acta 2009;73:5150–88.

Kuroda PK. Nuclear fission in the early history of the Earth. Nature 1960;137:36.

Marty B. Neon and xenon isotopes in MORB: implications for the earthatmosphere evolution. Earth Planet Sci Lett 1989;94:45–56.

McDonough WF, Sun S-S. Composition of the Earth. Chem Geol 1995;120:223–53.

Mojzsis SJ, Harrison TM, Pidgeon RT. Oxygen-isotope evidence from ancient zircons for liquid water at the Earth's surface 4300 Myr ago. Nature 2001;409:178–80.

Mukhopadhyay S. Early differentiation and volatile accretion in deep mantle neon and xenon. Nature 2012;486:101–10.

O'Neil J, Carlson RW, Paquette J-L, Francis D. Formation age and metamorphic history of the Nuvvuagittuq greenstone belt. Precambrian Res 2012;220–221:23–44.

Patterson C. Age of meteorites and the Earth. Geochim Cosmochim Acta 1956;10:230.

Pepin RO, Porcelli D. Xenon isotope systematics, giant impacts, and mantle degassing on the early Earth. Earth Planet Sci Lett 2006;250:470–85.

Reynolds JH. Determination of the age of the elements. Phys Rev Lett 1960a;4:8–10.

Reynolds JH. Isotopic composition of xenon from enstatite chondrites. Z fur Naturforsch A 1960b;15:1112–4.

Rizo H, Boyet M, Blichert-Toft J, Rosing M. Combined Nd and Hf isotope evidence for deep-seated source of Isua lavas. Earth Planet Sci Lett 2011;312:267–79.

Rudge JF, Kleine T, Bourdon B. Broad bounds on Earth's accretion and core formation constrained by geochemical models. Nat Geosci 2010;3:439–43.

Staudacher T, Allegre CJ. Terrestrial xenology. Earth Planet Sci Lett 1982;60:389–406.

Trieloff M, Kunz J. Isotope systematics of noble gases in the Earth's mantle: possible sources of primordial isotopes and implications for mantle structure. Phys Earth Planet Inter 2005;148:13–38.

Wilde SA, Valley JW, Peck WH, Graham CM. Evidence from detrital zircons for the existence of continental crust and oceans on the Earth 4.4 Gyr ago. Nature 2001;409:175–8.

CHAPTER 7

Life on Mars From the Martian Meteorite?

Chapter Outline

7.1 INTRODUCTION

The space program of the National Aeronautics and Space Association (NASA) focused on searching for life on Mars primarily because of that planet's similarities with Earth. The first exploration of Mars was a flyby mission made by the Mariner 4 probe in 1965. This exploration revealed that the surface of Mars was barren. It also found that the air pressure on Mars is thin, less than 1 kPa compared to 100 kPa on Earth. This suggested that there is no large amount of liquid

Origins of the Earth, Moon, and Life. http://dx.doi.org/10.1016/B978-0-12-812058-3.00007-7

or atmosphere on the Martian surface (Dick, 2006; Martel et al., 2012). The Mariner 4 probe also suggested that Mars does not have a magnetic field, which would protect the planet against deadly solar and cosmic rays (see Section 7.2.6).

After the Mariner 4 mission, two Viking spacecraft landed on Mars in 1976. Although the experiments performed in these missions gave evidence for the presence of oxidizing materials in the Martian soil, absence of organic molecules in the analyzed samples led to the general consensus that there is no life on the surface of Mars (Dick, 2006; Martel et al., 2012).

However, in 1996, a notorious study by D. S. McKay et al. (1996) presented evidence that life existed in the Martian meteorite ALH84001. This meteorite was an igneous orthopyroxenite discovered in Antarctica in 1984. It was believed to have originated from Mars both because of its mineralogy and the isotope composition of the air trapped inside the meteorite (Mittlefehldt, 1994; Gibson et al., 1997). McKay et al. (1996) found traces of polycyclic aromatic hydrocarbons (PAHs), magnetite crystals, carbonate globules, and pseudo-microfossils of miniature bacteria (McKay et al., 1996).

This research caused many researchers to begin searching for extraterrestrial life and spearheaded the creation and expansion of NASA's astrobiology program. Furthermore, this report was famous for inspiring a major scientific controversy about life on other planets and exciting debates about what exactly constitutes life and whether nanobacteria is real life (Arrhenius and Mojzsis, 1996; Grady et al., 1996; Gibson et al., 1997; Hamilton, 2000; Hogan, 2003; Young and Martel, 2010; Martel et al., 2012).

In this chapter, before relating the controversy about finding life in the Martian meteorite, we provide basic knowledge of astrobiology in Sections 7.2.1–7.2.5. These sections can be skipped by those who know biochemistry well. Then the evidence that supported past life in the ALH84001 meteorite is explained in Section 7.3. In the following sections (Sections 7.4 and 7.5), the same evidence is alternatively explained by chemical processes, refuting the claim of life in the meteorite. Morphology, especially, is a poor indicator for life; however, it played an important role again in the discussion of the oldest life of the Apex chert (Section 8.4).

7.2 FUNDAMENTAL KNOWLEDGE OF ASTROBIOLOGY

7.2.1 Amino Acids

Amino acids, especially 2-, alpha- or α-amino acids, are biologically important. This prefix indicates the type of amino acids that have amine ($—NH_2$) and carboxylic acid (—COOH) at the same carbon (the center carbon in Fig. 7.1 attached with amine and carboxylic acid). Therefore, the general chemical formula of the amino acids becomes $H_2NCHRCOOH$, where R is an organic substituent known as a side-chain.

In Fig. 7.1, two types of amino acids are shown. The chemical composition is the same, but the two amino acids cannot be overlapped, because they are mirror images. It is like right and left hands, which are the same, but cannot be

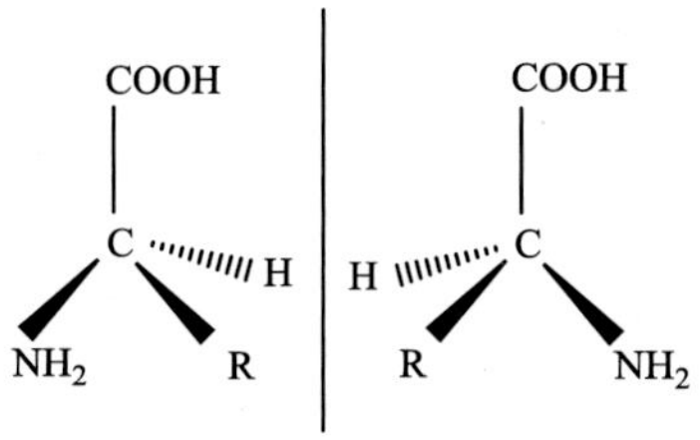

D-type L-type

FIGURE 7.1
Isomers of amino acids. The left and right configurations are D- and L-amino acids, respectively.

overlapped. This is called a chiral image, and either of the two is called an enantiomer. These are named as D-amino or L-amino acids. For amino acids, D or L can be identified by the "CORN" rule.

Practically, COOH, R, and NH_2 are arranged around the chiral center carbon with H away from the viewer. When CO→R→N is counterclockwise, it is the L form. If the rotation is clockwise, it is the D form. All amino acids synthesized or used by living things are L-amino acids, with a few exceptions.

In Fig. 7.2, typical amino acids are shown. Histidine, threonine, isoleucine, tryptophan, leucine, lysine, valine, methionine, and phenylalanine are essential amino acids for humans. The essential amino acids are different for each living thing.

7.2.2 Proteins

Two amino acids with side chains of R_1 and R_2 can combine as follows:

$$
\begin{aligned}
&NH_2 - CH(R_1) - COOH + NH_2 - CH(R_2) - COOH \\
&\rightarrow NH_2 - CH(R_1) - CO - NH - CH(R_2) - COOH + H_2O
\end{aligned}
\tag{7.2.1}
$$

this —CO—NH— bond is called a peptide bond (see Fig. 7.3 for 3D image). The amino acid can combine infinitely.

Proline is not an amino acid, but an imino acid (Box 7.1). However, it is treated as an amino acid. As proline has an unusual ring structure, the peptide bond differs from the norm:

```
C–C–NH  + NH2–CH(R1)–COOH          C–C–NH
|   |                        →     |   |                + H2O
C – C–COOH                         C–C–CO–NH–CH(R1)–COOH          (7.2.2)
```

or

```
NH2–CH(R2)–COOH +  C–C–NH            NH2–CH(R2)–CO
                   |   |        →                |
                   C–C–C–COOH                  C–C–N  + H2O
                                               |   |
                                               C – C–COOH         (7.2.3)
```

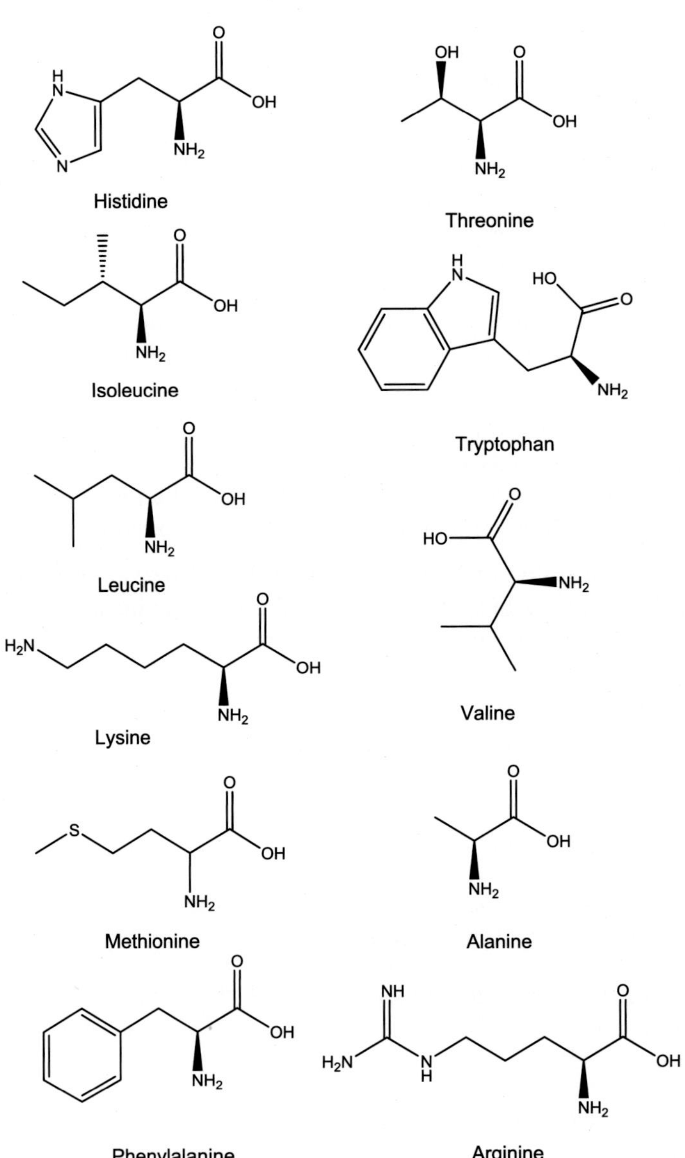

FIGURE 7.2
Amino acids. (Histidine, threonine, isoleucine, tryptophan, leucine, lysine, valine, methionine, and phenylalanine are essential amino acids for human).

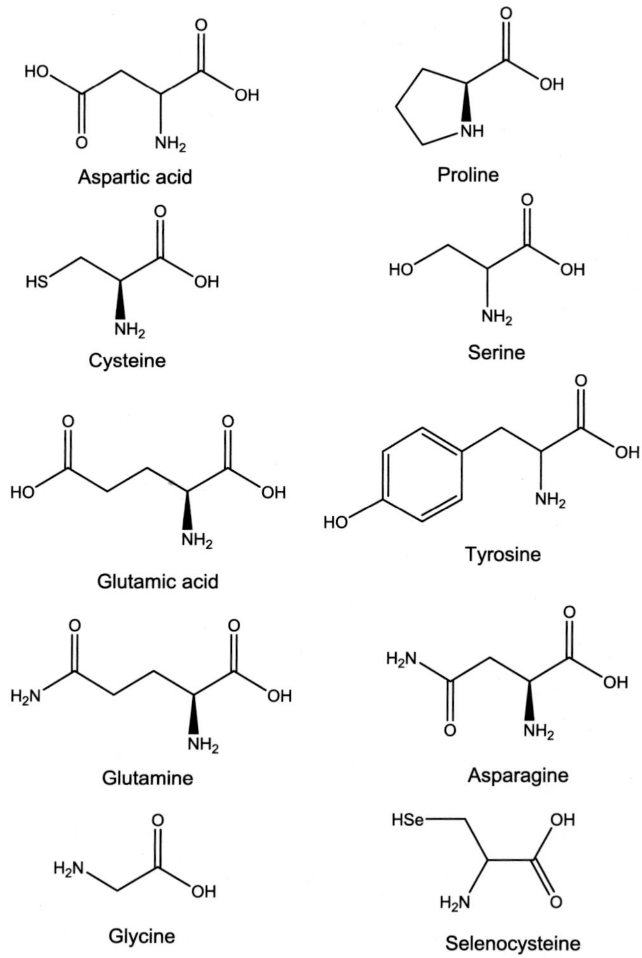

FIGURE 7.2 Cont'd.

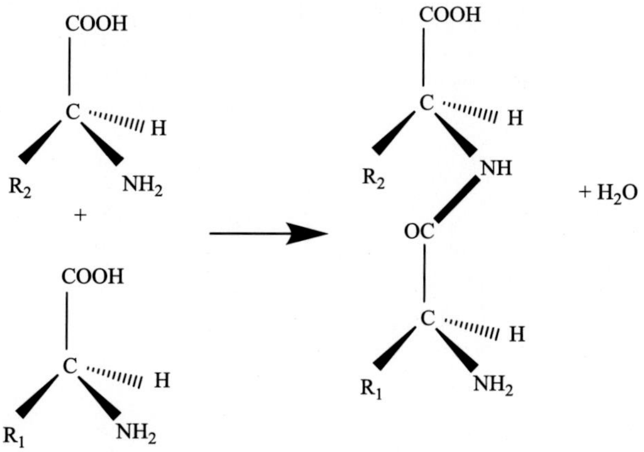

FIGURE 7.3
Combination of two amino acids by a peptide bond.

BOX 7.1 AN AMINO ACID AND AN IMINO ACID

An amino acid has both amino (—NH_2) and carboxyl (—COOH) functional groups. An imino acid has both imino (>C=NH) and carboxyl (—COOH) functional groups.

This polypeptide polymer molecule is called a protein, and some polypeptides have special biological functions. The end of the protein with a free carboxyl group is called C-terminus or carboxy terminus (the amino acid with R_2 in Fig. 7.3), and the end with a free amino group is called N-terminus (or the amino acid with R_1 in Fig. 7.3).

7.2.3 Purines and Pyrimidines

In Fig. 7.4, purines and pyrimidines are shown. Adenine (A) and guanine (G) are purines, and cytosine (C), thymine (T), and urasil (U) are pyrimidines. These are the most important parts in nucleic acid, and genetic information is stored in the sequence of these molecules. The red color nitrogen atom is connected to pentose sugar (see Fig. 7.5A and B).

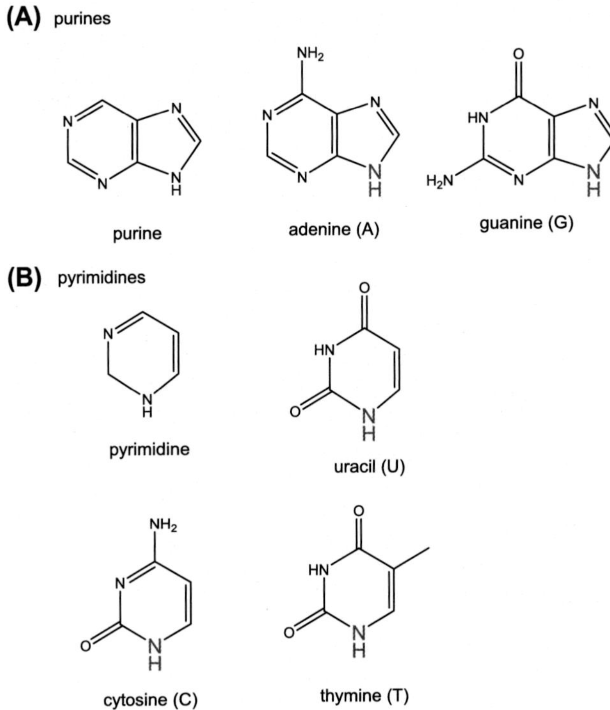

FIGURE 7.4
(A) Purines and (B) pyrimidines.

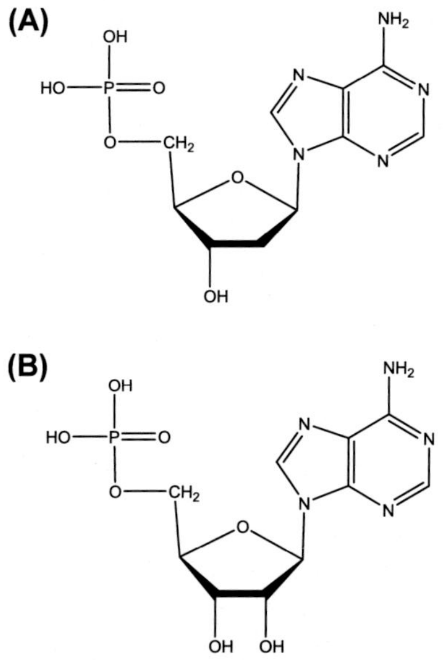

(A)

(B)

FIGURE 7.5
Nucleotides. (A) Adenine-deoxyribonucleic acid-phosphoric acid. (B) Adenine-ribonucleic acid-phosphoric acid. The molecule (B) is also called as adenosine monophosphate (AMP), and was used in the experiment of Fig. 10.2B.

In astrobiology or biochemistry, the abbreviation of A, G, C, T, and U are often used to show the genetic information in deoxyribonucleic acid (DNA, see Fig. 7.5A) and ribonucleic acid (RNA, see Fig. 7.5B). In DNA and RNA, A, G, C, and T, and A, G, C, and U are used, respectively.

The A, T, G, and C are bound with a five-carbon sugar, called deoxyribose, that is connected with one phosphoric acid. This one set of molecules is called a nucleotide. When the five-carbon sugar is deoxyribose (Fig. 7.5A), the polymer of this molecule is called deoxyribonucleic acid (DNA). When the five-carbon sugar is ribose (Fig. 7.5B), the polymer of this molecule is called ribonucleic acid (RNA).

7.2.4 Hydrogen Bond

The five molecules form a constant combination by the hydrogen bond. A combines with T or U in RNA, and G combines with C, respectively. The hydrogen bonds are shown as dotted lines in Fig. 7.6. The hydrogen bond is an electrostatic attraction between hydrogen and polar molecules, such as nitrogen or oxygen. The hydrogen bonds appear in various reactions in living things. The three-dimensional (3D) structure of proteins, which is a hot topic in biochemical research, is defined using hydrogen bonds. (In this area of research, supercomputers are needed to simulate the 3D structure of proteins.)

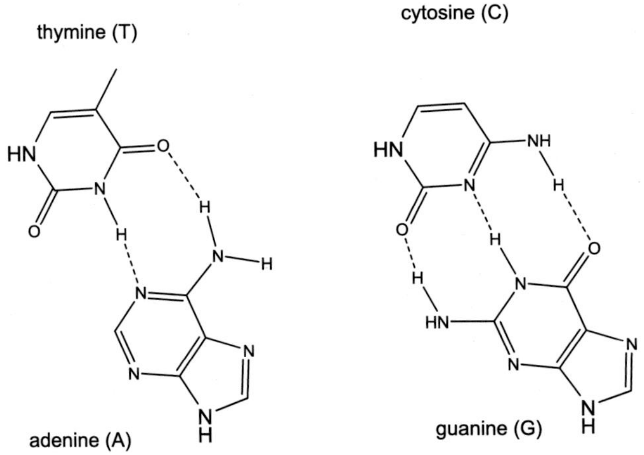

FIGURE 7.6
Hydrogen bonds.

It is astonishing that living things use hydrogen bonds to store genetic information only by A-T and G-C pairs. These bonds are not so strong that they cannot be cut, but strong enough not to be cut easily. Some carcinogens have a similar structure to these molecules, and enter into the A-T or G-C pairs, breaking the genetic information.

7.2.5 Helix and Double Helix

Nucleotides make long polymers, as shown in the left side of Fig. 7.7, which is labeled from 5′ to 3′. This polymer makes a helix with a corresponding right side helix. Finally, the two polymers make the double helix as shown in Fig. 7.7. Two ends of phosphoric acid and deoxyribose are called the 5′ end and 3′ end, respectively. The direction of 5′→3′ is defined to be the normal direction, and when the base direction of the DNA in Fig. 7.7 needs to be shown, it is shown along the normal direction as AGT, not as TGA.

7.2.6 Hydroxyl Alcohols, Hydroxyl Aldehydes, and Hydroxyl Ketones With Linear and Branched Structures

Hydroxyl alcohols, hydroxyl aldehydes and hydroxyl ketones with linear and branched structures are shown in Fig. 7.8A and B. The chemical formulas are shown by the Fisher projection (see Box 7.2). When an aldehyde functional group exists at the first carbon, and the hydroxyl alcohol at the end of the carbon chain, it is called an aldose. When a ketone functional group resides in a carbon atom at positions 2 or 3, it is called a ketose. The five carbon aldoses are called pentoses or monosaccharides (see Box 7.3).

7.2.7 High Energetic Light on a Planet and Life

High-energy light, especially gamma-ray, X-ray and ultraviolet light, are harmful to life on Earth. As there could be life in the universe that uses these kinds of

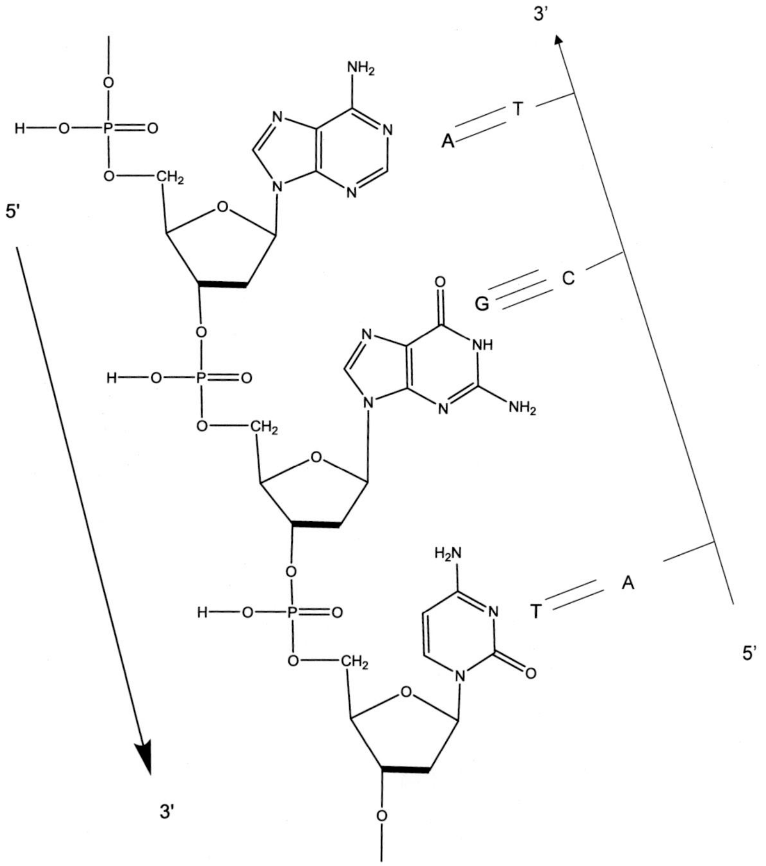

FIGURE 7.7
DNA. Only the one strand is written. The other side is abbreviated as a *right side arrow*.

light, what follows applies specifically to life on Earth. The reason high-energy light is harmful is that it damages DNA (deoxyribonucleic acid) and RNA (ribonucleic acid), which are essential molecules for life. When high-energy light passes through cell fluid, many free radicals are formed that damage DNA and/or RNA. DNA stores the blueprint of proteins to maintain life, and RNA is a copy made from the blueprint. The unicellular organism dies when the damage is more than the level required to maintain life. In multicellular organisms, when these molecules or proteins are damaged and cells does not work properly, they kill themselves (apoptosis). If the number of the apoptotic cells and directly killed cells are exceed the level required to maintain life, the multicellular organisms die (see Box 7.4 for the exceptional animal).

When a planet has an atmosphere, high-energy light is scattered or absorbed. The light energy decreases and the intensity of high-energy light becomes weaker, resulting in a more comfortable environment. Thus the existence of an atmosphere is very important in determining whether the planet is habitable or not.

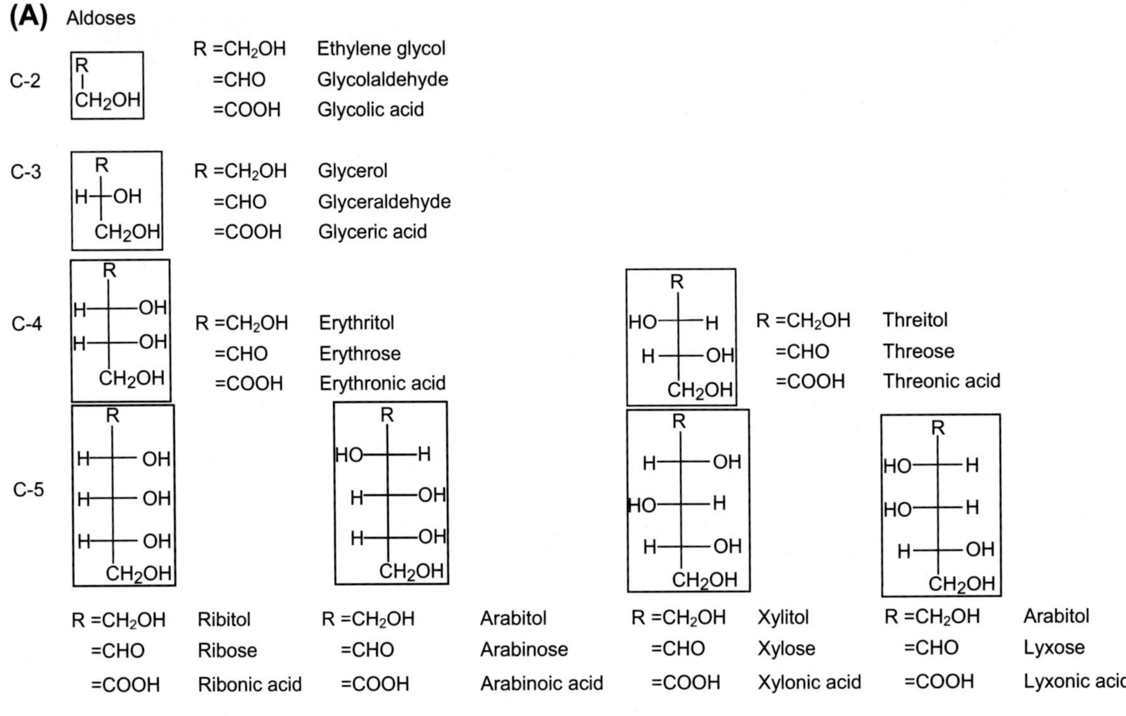

(B) Ketoses

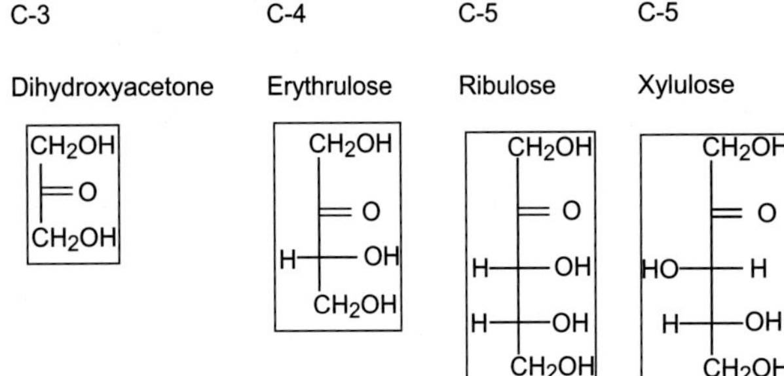

FIGURE 7.8
Structures and names of linear and branched (A) aldoses and (B) ketoses. The Fischer projections (see Box 7.2) indicate D-enantiomer form only. See Section 7.2.1 for "D- and L-enantiomers".

BOX 7.2 FISCHER PROJECTION

The Fischer projection was devised by H. E. Fischer to present a three-dimensional organic molecule in a two-dimensional form. Usually, the Fischer projection is used to represent monosaccharides (simple sugars), whose general formula is $C_nH_{2n}O_n$. All bonds except initial and terminal bonds are indicated as horizontal or vertical lines. The carbon chain is shown as vertical lines, and carbon atoms exist at the center of crossing lines. An example of a Fischer projection of a tetrahedral molecule on a flat plane is shown in Fig. Box 7.1A. A Fischer projection of xylitol, which is a sugar alcohol used as a sweetener, is shown in Fig. Box 7.1B. As the xylitol is positively beneficial for dental health, it is used in chewing gum, lozenges, etc.

BOX 7.2 FISCHER PROJECTION—continued

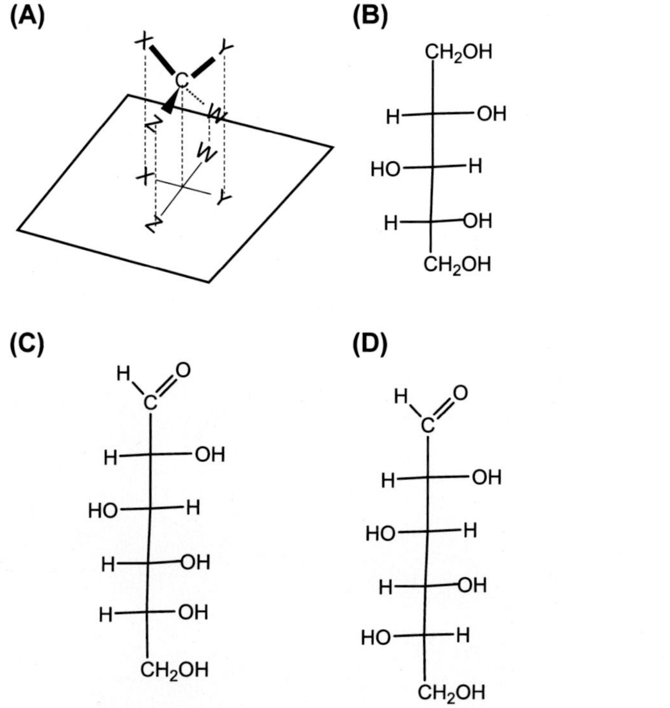

FIGURE BOX 7.1
The Fischer projection. (A) Projection of a tetrahedral molecule on a flat plane. (B) A Fischer projection of xylitol. (C) A Fischer projection of D-glucose. (D) A Fischer projection of L-glucose.

BOX 7.3 PENTOSE (MONOSACCHARIDE)

A pentose is a monosaccharide with five carbon atoms. Pentoses are separated into two groups: aldopentoses and ketopentoses. The aldopentoses have an aldehyde functional group at position 1. The ketopentoses have a ketone functional group in position 2 or 3. As the aldopentoses have three chiral centers, eight (2^3) enantiomers (see Section 7.2.1 and Fig. 7.8A and B) are possible, including D- and L-riboses, arabinoses, xyloses and lyxoses. As the ketopentoses have two chiral centers, four (2^2) enantiomers (see Fig. 7.8A) are possible as D- and L-ribuloses and xyluloses.

BOX 7.4 TARDIGRADES—THE STRONGEST ANIMAL IN EXTREME CONDITIONS

Tardigrades, which are also called water bears or moss piglets (see Fig. Box 7.2), can survive radiation levels 1000 times higher than what other animals can tolerate. They can withstand 5000 Gy (gray) of gamma rays. For humans, 5–10 Gy is lethal. The reason may be that they have a repair system for DNA damage. They can also survive extreme conditions such as high temperature, low pressure, and dehydration, especially when they are in a state of cryptobiosis.

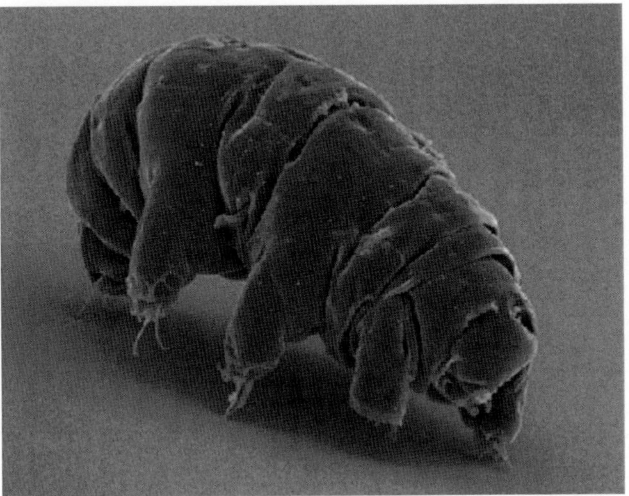

FIGURE BOX 7.2
A scanning electron microscope image of a tardigrade (Milnesium tardigradum).
Copyright: Schokraie E, Warnken U, Hotz-Wagenblatt A, Grohme MA, Hengherr S, Förster F, et al. Comparative proteome analysis of Milnesium tardigradum in early embryonic state versus adults in active and anhydrobiotic state. PLoS ONE 2012;7(9):e45682. http://dx.doi.org/10.1371/journal.pone.0045682. CC BY 2.5, https://commons.wikimedia.org/w/index.php?curid=22716809.

7.2.8 Magnetic Field of a Planet and Life

When a planet has a liquid metallic core, the magnetic field is generated by the so-called "Dynamo effect." When a planet is solidified perfectly, there is no magnetic field around the planet. Therefore, the inside of the planet can be imaged when it has a magnetic field.

A planet with a magnetic field is very advantageous for life to flourish, as it decelerates the fast-moving energetic particles of the solar wind, which are harmful to living things. The energetic particles are forced to move around and along the magnetic field lines, which go from north to south poles. (Please assume a vertically stood magnet which corresponds to a planet. When charged particles come, remember Fleming's left hand rule. The particles finally go to either pole. When the current runs in the ionosphere, it is observed as an aurora.)

7.3 REPORT FOR THE DISCOVERY OF LIFE IN THE MARTIAN METEORITE ALH84001

Before criticizing, we must scrutinize the original report of McKay et al. (1996). They studied an Antarctic meteorite, ALH84001 (Mittlefehldt, 1994), which is an igneous orthopyroxenite composed of coarse-grained orthopyroxene [$(Mg,Fe)SiO_3$] and minor maskelynite ($NaAlSi_3O_8$), olivine [$(Mg,Fe)SiO_4$], chromite ($FeCr_2O_4$), pyrite (FeS_2), and apatite [$Ca_3(PO_4)_2$]. It crystallized at 4.5 Ga. It records at least two shock events separated by a period of annealing. The first shock was at ~4.0 Ga.

The $\delta^{13}C$ values of the carbonate in ALH84001 were −17‰ to +42‰. The value of +42‰ is higher than those of other SNC (shergottite, nakhlite, shassignite) meteorites. This range of $\delta^{13}C$ exceeds the range made by most terrestrial inorganic processes. Alternatively, biogenic processes are known to produce negative values in $\delta^{13}C$ on Earth.

In examining the Martian meteorite ALH84001 by state-of-the-art methods, McKay et al. (1996) concluded that the following pieces of evidence were compatible with the existence of past life on Mars:

1. The igneous Mars rock (of unknown geologic context) was penetrated by a fluid along fractures and pore spaces; then, such places became the sites of secondary mineral formation and possible biogenic activity.
2. A formation age for the carbonate globules is later (i.e., younger) than the age of the igneous rock.
3. Scanning electron microscope (SEM) and transmission electron microscope images of carbonate globules and features resembled terrestrial microorganisms, terrestrial biogenic carbonate structures, or microfossils.
4. Magnetite and iron sulfide particles could have resulted from oxidation and reduction reactions known to be important in terrestrial microbial systems.
5. The presence of polycyclic aromatic hydrocarbons (PAHs), which are explained in Section 7.4.1, is associated with surfaces rich in carbonate globules.
6. Although none of these observations is in itself conclusive for the existence of past life and there are alternative explanations for each of these phenomena taken individually, when they are considered collectively, particularly in view of their spatial association, one cannot help but conclude that they are evidence for primitive life on early Mars.

7.4 DETAILED EXPLANATION IN THE REPORT OF McKAY ET AL. (1996)

7.4.1 Polycyclic Aromatic Hydrocarbons (PAHs)

Polycyclic aromatic hydrocarbons (PAHs; see Fig. 7.9) are molecules of aromatic rings made of carbon and hydrogen. On Earth, PAHs are abundant as fossil molecules in ancient sedimentary rocks, coal, and petroleum. Although PAHs are ubiquitous, they are not produced by living organisms and do not play specific roles in living things.

McKay et al. (1996) analyzed freshly broken fracture surfaces on small chips of ALH84001 for PAHs using a microprobe two-step laser mass spectrometer (μL^2MS). The average PAH concentration on the surfaces was more than 1 ppm. Contamination checks and control experiments indicated that the observed organic material was from ALH84001. No PAHs in ALH84001 with intact fusion crust were present, and the PAH signal increased with increasing depth, becoming constant at approximately 1200 μm within the interior, well away from the fusion crust.

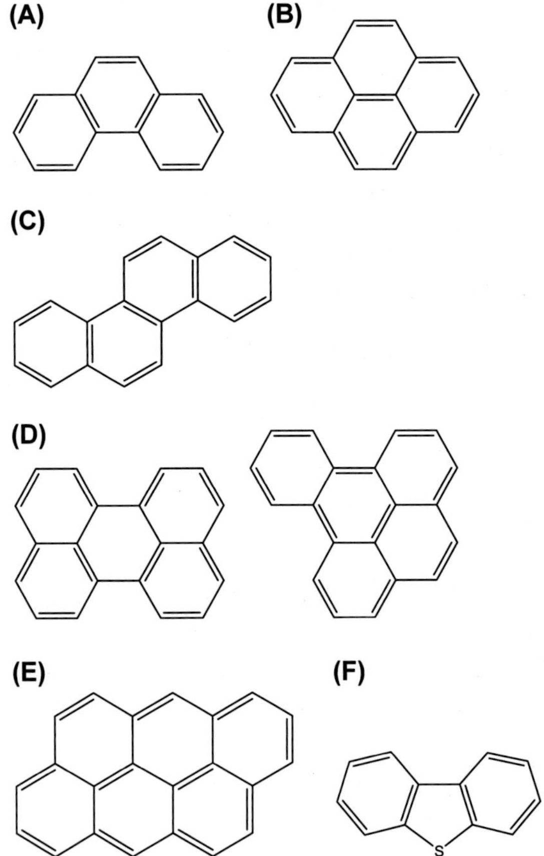

FIGURE 7.9
Chemical structures of polycyclic aromatic hydrocarbons (PAHs) in ALH 84001. (A) Phenanthrene ($C_{14}H_{10}$); (B) pyrene ($C_{16}H_{10}$); (C) chrysene ($C_{18}H_{12}$); (D) perylene or benzopyrene ($C_{20}H_{12}$); (E) anthanthrene ($C_{22}H_{12}$); (F) dibenzothiophene ($C_{12}H_8S$).

The accumulation of PAHs on the Greenland ice sheet over the past 400 years has been studied in ice cores (Kawamura and Suzuki, 1994). The total concentration of PAHs in the cores varies from 10 ppt for preindustrial times to 1 ppb for recent snow deposition. Concentrations of PAHs in Antarctic ice were expected to lie between these two values. Analysis of Antarctic salt deposits on a heavily weathered meteorite (LEW 85320) by μL^2MS did not show the presence of terrestrial PAHs within detection limits, which suggested that the terrestrial contamination of PAHs for ALH84001 is less than 1%.

The freshly broken but pre-existing fracture surfaces rich in PAHs also typically displayed carbonate globules. The globules tend to be disc-shaped and flattened parallel to the fracture surface. Intact carbonate globules appeared orange in visible light and had a rounded appearance; many displaying alternating black and white rims. Under high magnification stereo light microscopy or SEM stereo imaging, some of the globules appear to be quite thin and pancake-like, suggesting that the carbonates formed in the restricted width of a thin fracture. This geometry limited their growth perpendicular to, but not parallel to, the fracture.

The PAHs were highest in the carbonate-rich regions. Two groups of PAHs were identified from averaged mass spectra. A middle-mass group of 178–276 Da dominated, which was composed of simple 3- to 6-ring PAH skeletons. The alkylated homologs were less than 10% of the total signal intensity. Principal peaks at 178, 202, 228, 252, and 278 Da were assigned to phenanthrene ($C_{14}H_{10}$; see Fig. 7.9A), pyrene ($C_{16}H_{10}$; see Fig. 7.9B), chrysene ($C_{18}H_{12}$; see Fig. 7.9C), perylene or benzopyrene ($C_{20}H_{12}$; see Fig. 7.9D), and anthanthrene ($C_{22}H_{12}$; see Fig. 7.9E).

A second weak high-mass group of PAHs of about 300–450 Da was identified. The primary source of PAHs was anthropogenic emissions, which were characterized by the presence of abundant aromatic heterocycles, primarily dibenzothiophene ($C_{12}H_8S$; 184 Da; see Fig. 7.9F). The PAHs in ALH84001 were present at ppm levels but dibenzothiophene was not observed in any of the samples.

7.4.2 Carbonate Globules and Magnetite Crystals

In analysis by an electron microprobe, carbonate globules were observed (Fig. 7.10). The larger globules (>100 μm) had Ca-rich cores surrounded by alternating Fe and Mg-rich bands. Several sharp thin bands were present near the edge of the globule. The first band was rich in iron (Fe) and sulfur (S), the second was rich in magnesium (Mg) with no Fe, and the third was rich in Fe and S again. The Fe-rich rims were composed mainly of fine-grained magnetite ranging in size from approximately 10–100 nm and minor amounts of pyrrhotite (~5 vol%).

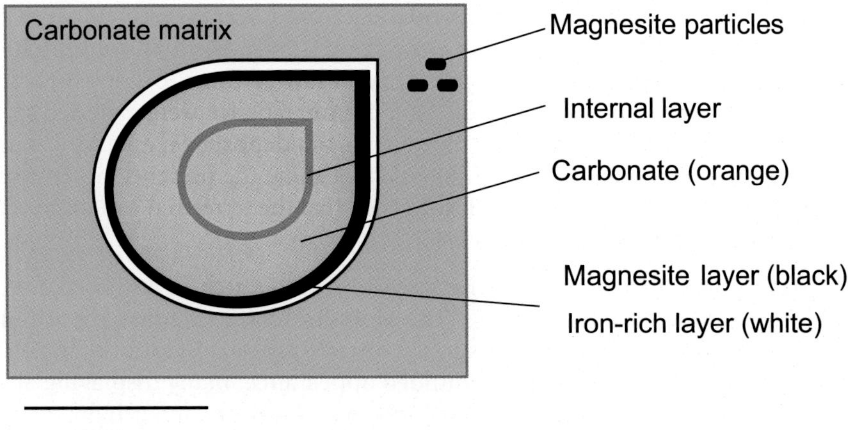

FIGURE 7.10
A schematic diagram of a carbonate globule in a backscattered electron image of ALH84001 after Martel et al. (2012). The colors are artificial colors.

Magnetite crystals were cubic, teardrop, or irregular shaped. Individual crystals had well-preserved structures with no lattice defects. The magnetite and Fe-sulfide were in a fine-grained carbonate matrix. The composition of the fine-grained carbonate matrix matched that of coarse-grained carbonates located adjacent to the rim.

7.4.3 Structures Resembling Terrestrial Life

The occurrence of fine-grained carbonate and magnetite phases could be explained by either inorganic or biogenic processes. Magnetite particles were similar (chemically, structurally, and morphologically) to terrestrial magnetite particles known as magnetofossils, remaining as bacterial magnetosomes. The sketch of extracellular precipitated superparamagnetic magnetite particles produced by terrestrial magnetotactic bacteria is shown in Fig. 7.11A, which resembled some magnetite crystals of ALH84001.

The elongated carbonates formed by connection of tiny ovoid-to-cylindrical particles resembling the magnetite particles in Fig. 7.11A were observed in Fig. 7.11B. The object was about 400 nm in the longest dimension, and about 40 nm across. Such objects were found on the surface of calcite concretions grown from Pleistocene ground water in southern Italy, and were interpreted as nanobacteria.

The origin of these textures on the surface of the ALH84001 carbonates was unclear. One possible explanation is that the textures were erosional remnants of the carbonate that happened to be in the shape of ovoids and elongate forms.

An alternative explanation was that these textures, as well as the nanosize magnetite and Fe-sulfides, were the products of microbiological activity. In general, authigenic secondary minerals in Antarctica are oxidized or hydrated. The lack of PAHs in the other analyzed Antarctic meteorites, the sterility of the sample, and the nearly unweathered nature of ALH84001 argued against an Antarctic biogenic origin.

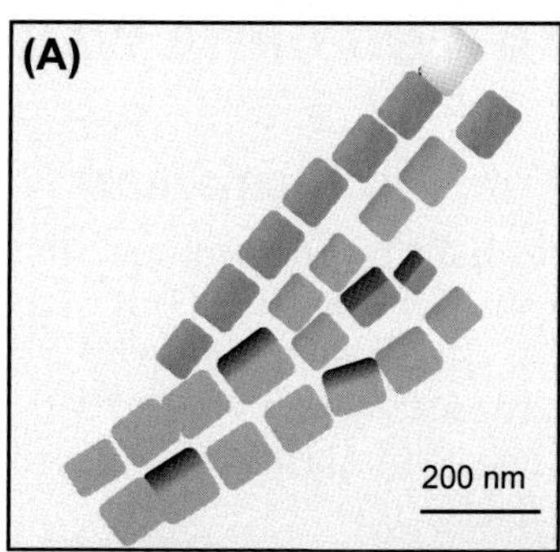

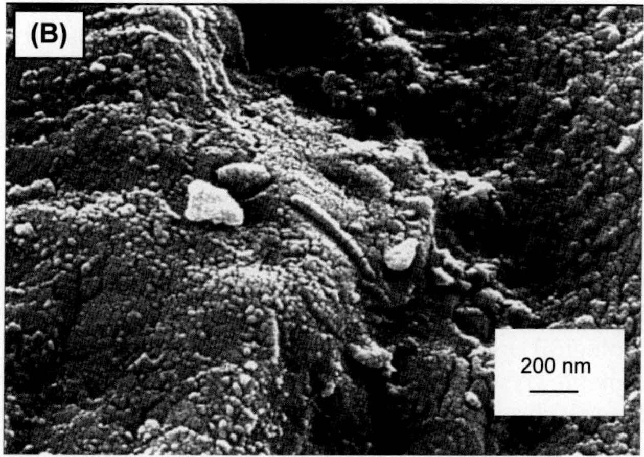

FIGURE 7.11
Schematic diagrams showing morphologies of crystals. (A) A sketch of magnetites observed in terrestrial magnetotactic bacteria. (B) The tube-like structures observed in situ in the ALH84001 meteorite.
(A) This sketch is drawn based on the photographs in Martel J, Young D, Peng H-H, Wu C-Y, Young JD. Biomimetic properties of minerals and the search for life in the martian meteorite ALH84001. Annu Rev Earth Planet Sci 2012;40:167–93. (B) Image credit: NASA/JSC/Stanford University (PIA00288.jpg; the scale is added by the author).

The filamentous features were similar in size and shape to nanobacteria in limestone. The elongate forms resembled some forms of fossilized filamentous bacteria in the terrestrial fossil record. In general, the terrestrial bacteria microfossils were more than an order of magnitude larger than the forms seen in the ALH84001 carbonates.

The carbonate globules in ALH84001 shown in Fig. 7.10 were clearly a key element in the interpretation of this martian meteorite. The origin of these globules was controversial. Harvey and McSween (1996) and Mittlefehldt (1994) argued, on the basis of microprobe chemistry and equilibrium phase relationships, that the globules were formed by high-temperature metamorphic or hydrothermal reactions. Alternatively, Romanek et al. (1992) argued on the basis of isotopic relationships that the carbonates were formed under low-temperature hydrothermal conditions. Finally, they determined that the carbonate globules had a biogenic origin and were likely formed at low temperatures. Pure Mg-carbonate (magnesite) was produced by biomineralization under alkaline conditions.

7.5 LIFE IN THE MARTIAN METEORITE WAS A BIG MISTAKE!

7.5.1 Magnetite Crystals Were Not Biomarkers

On Earth, some species of bacteria contain magnetosomes, which consist of membrane-delineated vesicles with crystals of magnetite (Fe_3O_4) and/or greigite (Fe_3S_4) (Bazylinski and Frankel, 2004; Chang and Kirschvink, 1989). McKay et al. (1996). Studies of one group (Thomas-Keprta et al., 2000, 2001, 2002) reported that the magnetite crystals detected in ALH84001 were similar to those found in terrestrial magnetotactic bacteria.

Several characteristics of the magnetite particles, such as crystal purity, structural perfection, and lack of defects were proposed to be signs of biological processes (Thomas-Keprta et al., 2000). Some authors noted that only one-fourth of the magnetite in ALH84001 had sizes and shapes similar to those of terrestrial magnetotactic bacterial (Buseck et al., 2001; Golden et al., 2004). That is, most magnetite crystals observed in ALH84001 were completely different in size and shapes from those found in magnetotactic bacteria from the Earth. The crystals found in ALH84001 were on average much smaller (Grady et al., 1996).

There were many structural, morphological, and crystallographic variations in the magnetite crystals of the magnetotactic bacteria, which made it difficult to confirm a biological origin for magnetite particles simply by comparing them with the magnetite crystals observed in terrestrial bacteria (Buseck et al., 2001). Based on these results, it became obvious that the features of magnetite crystals such as shapes, sizes, purity, and crystalline structures did not indicate a biogenic origin (Buseck et al., 2001; Golden et al., 2004). Some researchers had also reported that magnetite chains could be produced in an artificial condition (Golden et al., 2004; Nealson and Cox, 2002), insisting against the use of these chains as evidence for biogenic origin.

These results gave doubt to the hypothesis that the magnetite crystals in ALH84001 were of biogenic origin. Several researchers were against the use of magnetite crystals as signs of past life (Barber and Scott, 2002; Buseck et al., 2001; Frankel and Buseck, 2000; Golden et al., 2004). Based on this information, the hypothesis that magnetotactic microorganisms were present in the ancient meteorite ALH84001 became suspect.

7.5.2 Nanobacteria Was Not Life

Many scientists were not convinced of the idea of nanobacteria (Abbott, 1999, 2000; Bradbury, 1998; Day, 1998; Deresinski, 2008; Hamilton, 2000; Hogan, 2003, 2004a, 2004b; Hopkins, 2008; Saey, 2008; Travis, 1998). Whereas other authors (Abbott, 1999; Hogan, 2004b; Hopkins, 2008; Saey, 2008; Urbano and Urbano, 2007) considered the idea of nanobacteria as the difference of thinking grounds, Martel's group (Martel and Young, 2008; Martel et al., 2010, 2012; Wu et al., 2009; Young et al., 2009; Young and Martel, 2010) and others (Barr et al.,

2003; Benzerara et al., 2003; Cisar et al., 2000; Drancourt et al., 2003; Kirkland et al., 1999; Raoult et al., 2008; Vali et al., 2001) tried to prove nanobacteria to be an experimentally-mistaken idea.

In particular, Martel's group showed that the nanobacteria specimens as studied in the many publications were nonliving mineralo-organic nanoparticles. They further demonstrated that the round shape of the bacteria-like forms was due to the presence of an amorphous mineral phase transiently stabilized by mineralization inhibitors that bind the growing mineral and prevent it from forming faceted edges (Martel and Young, 2008; Martel et al., 2012; Wu et al., 2009; Young et al., 2009; Young and Martel, 2010). Amorphous minerals represented random networks which were contrasted with the orderly periodic crystal structure. This randomness caused the mineral to be rounded. The small initial size of the mineral nanoparticles was caused by the presence of mineralization inhibitors that interfered with particle growth.

Proteins can act as mineralization inhibitors (Martel and Young, 2008; Martel et al., 2012; Raoult et al., 2008; Wu et al., 2009; Young et al., 2009), and other biological macromolecules such as phospholipids (Cisar et al., 2000; Young et al., 2009), and inorganic ions such as magnesium (Martel and Young, 2008; Martel et al., 2012; Young et al., 2009). The nanobacteria proponents, such as Kajander and McKay, claimed that the presence of both inorganic and organic inhibitors within a mineral particle was resulted from a normal living process (Ciftcioglu and McKay, 2010; Ciftcioglu et al., 2006; Hjelle et al., 2000; Kajander and Ciftcioglu, 1998; Kajander et al., 1997; Kumar et al., 2006; Lieske, 2008; Shiekh et al., 2009; Sommer and Wickramasinghe, 2005).

In the case of ALH84001, various ions were present during the formation of the carbonate globules because of the presence of magnesite and iron-rich minerals on the surfaces of the globules. Several researchers had made bacteria-like structures similar to the ones found in ALH84001 by the precipitation of calcium carbonate in vitro (Kirkland et al., 1999; Martel et al., 2012; Vecht and Ireland, 2000). In addition, bacteria-like structures similar to the ALH84001 had been observed in other meteorites from the Moon, where life was considered to be unlikely. This also supported the idea that the bacteria-like structures could be produced from non-living processes (Martel et al., 2012; Sears and Kral, 1998).

7.5.3 Similarities and Dissimilarities Between Microcrystals and Microorganisms

It is a very important finding that minerals which precipitate in solution in the presence of mineralization inhibitors can form amorphous phases and structures that are morphologically similar to bacteria (Martel and Young, 2008; Martel et al., 2012; Wu et al., 2009; Young and Martel, 2010; Young et al., 2009). The structures resembling bacteria have particular formations that resemble cellular divisions. These cell-like morphologies have been used in the past as evidence of living processes (Kajander and Ciftcioglu, 1998; Kajander et al., 1997;

Martel et al., 2012). However, these same formations can be attributed to the mixture of mineral nuclei (Martel and Young, 2008; Martel et al., 2012; Young and Martel, 2010; Young et al., 2009).

The bacteria-like structures formed in a particular environment can show a narrow size distribution, which can be attributed to the homogeneous composition of the solution in which precipitation occurred (Martel et al., 2012). Some nanoparticle samples contain both small (20–500 nm) and large (several micrometer) particles, which is similar to what is observed in nanobacteria cultures obtained from human serum (Kajander and Ciftcioglu, 1998; Kajander et al., 1997; Martel et al., 2012). The presence of both small and large particles may be explained by the Ostwald ripening, in which small crystals disappear and form large particles over time owing to the higher stability of the larger particles (Mann, 2001). In addition to forming bacteria-like structures, minerals have other characteristics that resemble those of living microorganisms. For example, minerals can form cellular structures resembling cellular membranes or walls, and minerals can aggregate and produce colonies that are analogous to the ones formed by living terrestrial bacteria (Martel and Young, 2008; Martel et al., 2012). The formation of colonies has been used in the past as an indicator of living organisms (Folk, 1993). Mineral deposition and growth can be similar to the growth or increase in size of living cells.

Minerals can serve as a template for the formation of new crystals of similar compositions. This process is termed secondary nucleation. During secondary nucleation, ions in solution align themselves along the crystal structure and form ionic bonds with the other ions present in the crystal in a process that may form a new crystal nucleus on top of the first crystal. Following formation of the crystal nucleus, these nascent crystals may detach from the template crystal and then grow to the size of a mineral nanoparticle through further ion deposition and crystal growth (Mann, 2001). This nucleation process can mimic the proliferation of a microorganism that divides by binary fission or budding. The nucleation process is made more complicated by the presence of mineralization inhibitors that keep the minerals in their amorphous or semicrystalline state, allowing them to grow while also allowing them to retain their spherical or bacteria-like morphology (Martel et al., 2012).

The processes of nucleation and crystal growth are limited by the availability of ions that can get incorporated into the crystal structure (Mann, 2001). They are also modulated by the presence of nucleation inhibitors in the medium. Thus, the secondary nucleation process is doomed to slow down and finally stop once the ions in solution are consumed (Martel et al., 2012). Crystal growth also stops when the crystal-growing sites are blocked by mineralization inhibitors and the accumulation of crystal defects (Mann, 2001). This process of slow particle proliferation and its seeming saturation also mimic the exponential growth and plateau observed during the culture of living microorganisms.

The growth curves of bacteria in a batch culture are usually characterized by these four phases: (1) the lag phase, (2) the exponential phase, (3) the stationary phase, and (4) the death phase (Martel et al., 2012). The proliferation of nonliving biomimetic nanoparticles in medium often follows growth characteristics similar to those of a bacterial batch culture. The exponential nature of nanoparticle proliferation is one of its many biomimetic properties. However, there are certain distinctions between the growth curves of bacterial culture and those of biomimetic nanoparticles. Biomimetic nanoparticle growth does not experience a death phase. Rather, a sustained growth phase extends a long time beyond reasonable bacterial growth and death cycles. It is observed that continuous accumulation of precipitating nanoparticles aggregate after the initial growth phase (Martel et al., 2012; Wu et al., 2009; Young et al., 2009).

The properties of mineral precipitation can also resemble those of the subculture of microorganisms that are transferred to a freshly cultured medium (Cisar et al., 2000; Young et al., 2009). For instance, when an aliquot containing several crystals or amorphous minerals is diluted into a fresh culture medium and the solution is incubated for a period of days to weeks, the number of nanoparticles in the solution increases. This process can be attributed to secondary nucleation that occurs on the seeded crystals or amorphous minerals. In this case, the mineral seeds serve as templates for the formation and growth of new crystals in a manner similar to crystal proliferation as described above (Martel et al., 2012).

7.5.4 Ineffectiveness of Morphological Similarities

The nanobacteria hypothesis is formed on morphological grounds, as no chemical evidence is found by any laboratory except the initial studies advanced by Kajander and the nanobacteria camp (Martel et al., 2012). The lack of specific biomarkers other than morphology necessitates great care when searching for life in unknown specimens (Martel et al., 2012).

In studies of morphological signatures of terrestrial fossils and extraterrestrial life, Garcia-Ruiz et al. have shown that morphology alone is a poor and ambiguous indicator of biogenicity (Garcia-Ruiz et al., 2002, 2003; Hyde et al., 2003). These authors demonstrated that the morphologies of common inorganic minerals such as barium carbonate and silica that precipitate in alkaline environments resemble primitive organisms and complex biological structures. Notably, they propose a mechanism to explain the formation of the smooth and curved surfaces in waves (Garcia-Ruiz et al., 2009). The waves curl and form structures like cauliflowers. These cauliflower-like forms are reminiscent of the colonies of nanobacteria observed in human body fluids and the ALH84001 meteorite.

In the case of the ALH84001 studies, some researchers argued that the bacteria-like structures could represent gold coating in sample preparation, which is used for electron microscopy to make the samples electronically conducive (Bradley et al., 1997). These authors showed that the formation of parallel lamellae on surface fractures as well as appropriate sample orientation can give the impression of bacteria-like formations in ALH84001 samples. However, using atomic force microscopy (AFM), another group showed that even without using gold coating, bacteria-like structures were apparent in ALH84001 samples, thus concluding that the bacteria-like morphologies are not just the result of sample preparation (Steele et al., 1998; Martel et al., 2012). Using various forms of optical and electron microscopy, abundant bacteria-like structures in mineral precipitates that were not coated were observed (Martel and Young, 2008; Martel et al., 2010; Peng et al., 2011; Wu et al., 2009; Young et al., 2009). Coating artifacts are unlikely to explain all the bacteria-like structures found in the ALH84001 meteorite. Instead, they are more likely the result of simple chemical processes (Martel et al., 2012).

These observations indicate that the mere fact that bacteria-like structures are observed in geological specimens cannot be considered convincing evidence of past life. The bacteria-like structures are more likely to arise from chemical processes that are ubiquitous and that take place whenever minerals precipitate from supersaturated solutions in the presence of mineralization inhibitors (Martel et al., 2012).

7.5.5 Conclusion: Morphology Is Not a Decisive Factor

In the discussions of Sections 7.5.1–7.5.4 we have learned the morphology is not a decisive factor for determination of living things. Researchers, especially geologists, sometimes forget this simple rule. Thus confusion sometimes occur, as shown in Chapter 8, Sections 8.4, 8.8 and 8.9. We must always pay attention to morphological discussions.

References

Abbott A. Battle lines drawn between 'nanobacteria' researchers. Nature 1999;401:105.

Abbott A. Researchers fail to find signs of life in 'living' particles. Nature 2000;408:394.

Arrhenius G, Mojzsis S. Extraterrestrial life: life on Mars – then and now. Curr Biol 1996;6:1213–6.

Barber DJ, Scott ERD. Origin of supposedly biogenic magnetite in the Martian meteorite Allan Hills 84001. Proc Natl Acad Sci USA 2002;99:655–6.

Barr SC, Linke RA, Janssen D, et al. Detection of biofilm formation and nanobacteria under lon-term cell culture conditions in serum samples of cattle, goats, cats, and dogs. Am J Vet Res 2003;64:176–82.

Bazylinski DA, Frankel RB. Magnetosome formation in prokaryotes. Nat Rev Microbiol 2004;2:217–30.

Benzerara K, Menguy N, Guyot F, Dominici C, Gillet P. Nanobacteria-like calcite single crystals at the surface of the Tatouine meteorite. Proc Natl Acad Sci USA 2003;100:7438–42.

Bradbury J. Nanobacteria may lie at the heart of kidney stones. Lancet 1998;352:121.

Bradley JP, Harvey RP, McSween Jr HY. No 'nanofossils' in Martian meteorite. Nature 1997;390:454–6.

Buseck PR, Dunin-Borkowski RE, Devouard B, Frankel RB, McCartney MR. Magnetite morphology and life on Mars. Proc Natl Acad Sci USA 2001;98:13490–5.

Chang SR, Kirschvink JL. Magnetofossils, the magnetization of sediments, and the evolution of magnetite biomineralization. Annu Rev Earth Planet Sci 1989;17:169–95.

Ciftcioglu N, McKay DS, Mathew G, Kajander EO. Nanobacteria: fact of fiction? Characteristics detection, and medical importance of novel self-replicating, calcifying nanoparticles. J Investig Med 2006;54:385–94.

Ciftcioglu N, McKay DS. Pathological calcification and replicating calcifying-nanoparticles: general approach and correlation. Pediatr Res 2010;67:490–9.

Cisar JO, Xu DQ, Thompson J, Swaim W, Hu L, Kopecko DJ. An alternative interpretation of nanobacteria-induced biomineralization. Proc Natl Acad Sci USA 2000;97:11511–5.

Day M. Mean microbes-hard little bugs could cause everything from tumours to dementia. New Sci 1998;159:11.

Deresinski S. Nan(non)bacteria. Clin Infect Dis 2008;47:v–i.

Dick SJ. NASA and the search for life in the universe. Endeavour 2006;30:71–5.

Drancourt M, Jacomo V, Lepidi H, et al. Attempted isolation of *Nanobacterium* sp. microorganisms from upper urinary tract stones. J Clin Microbiol 2003;41:368–72.

Frankel RB, Buseck PR. Magnetite biomineralization and ancient life on Mars. Curr Opin Chem Biol 2000;4:171–6.

Folk RL. SEM imaging of bacteria and nannobacteria in carbonate sediments and rocks. J Sediment Res 1993;63:990–9.

Garcia-Ruiz JM, Carnerup A, Christy AG, Welham NJ, Hyde ST. Morphology: an ambiguous indicator of biogenicity. Astrobiology 2002;2:353–69.

Garcia-Ruiz JM, Hyde ST, Carnerup AM, et al. Self-assembled silica-carbonate structures and detection of ancient microfossils. Science 2003;302:1194–7.

Garcia-Ruiz JM, Melero-Carcia E, Hyde ST. Morphogenesis of self-assembled nanocrystalline materials of barium carbonate and silica. Science 2009;323:362–5.

Gibson Jr EK, McKay DS, Thomas-Keprta K, Romanek CS. The case for relic life on Mars. Sci Am 1997;27:58–65.

Golden DC, Ming DW, Morris RV, Brearley AJ, Lauer Jr HV. Evidence for exclusively inorganic formation of magnetite in Martian meteorite ALH84001. Am Mineral 2004;89:681–95.

Grady M, Wright I, Pillinger C. Opening a martian can of worms? Nature 1996;382:575–6.

Hamilton A. Nanobacteria: gold mine or minefield of intellectual enquiry? Microbiol Today 2000;27:182–4.

Harvey R, McSween Jr HP. A possible high-temperature origin for the carbonates in the martian meteorite ALH84001. Nature 1996;382:49–51.

Hjelle JT, Miller-Hjelle MA, Poxton IR, et al. Endotoxin and nanobacteria in polycystic kidney disease. Kidney Int 2000;57:2360–74.

Hogan J. 'Microfossils' made in the laboratory. New Sci 2003;(2422):14–5.

Hogan J. Are nanobacteria alive or just strange crystals? New Sci 2004a;(2448):6–7.

Hogan J. Nanobacteria revelations provoke new controversy. New Sci 2004b. Available from: http://www.newscientist.com/article/d5009-nanobacteria-revelations-provoke-new-controversy.html.

Hopkins M. Nanobacteria theory takes a hit. Nat News 2008. http://dx.doi.org/10.1038/news.2008.762.

Hyde ST, Carnerup AM, Larsson AK, Christry AG, Garcia-Ruiz JM. Self-assembry of carbonate-silica colloids: between living and non-living form. Phys A 2003;339:24–33.

Kajander EO, Kuronen I, Akerman KK, Pelttari A, Ciftcioglu N. Nanobacteria from blood: the smallest culturable autonomously replicating agent on earth. Proc SPIE 1997;3111:420–8.

Kajander EO, Ciftcioglu N. Nanobacteria: an alternative mechanism for pathogenic intra-and extracellular calcification and stone formation. Proc Natl Acad Sci USA 1998;95:8274–9.

Kawamura K, Suzuki I. Ice core record of polycyclic aromatic hydrocarbons over the past 400 years. Naturwissenschaften 1994;81:502–5.

Kirkland BL, Lynch FL, Rahnis MA, Folk RL, Molineux IJ, McLean RJC. Alternative origins for nannobacteria-like objects in calcite. Geology 1999;27:347–50.

Kumar V, Farell G, Yu S, Harrington S, Fitzpatrick L, et al. Cell biology of pathologic renal calcification contribution of crystal transcytosis, cell-mediated calcification, and nanoparticles. J Investig Med 2006;54:412–24.

Lieske JC. Can biologic nanoparticles initiate nephrolithiasis? Nat Clin Pract Nephrol 2008;4:308–9.

Mann S. Chemical control of biomineralization. In: Biomineralization: principles and concepts in bioinorganic materials chemistry. Oxford: Oxford Univ. Press; 2001. p. 38–67.

Martel J, Young JD. Purported nanobacteria in human blood as calcium carbonate nanoparticles. Proc Natl Acad Sci USA 2008;105:5549–54.

Martel J, Wu CY, Young JD. Critical evaluation of gamma-irradiated serum used as feeder in the culture and demonstration of putative nanobacteria and calcifying nanoparticles. PLoS One 2010;5:e10343.

Martel J, Young D, Peng H-H, Wu C-Y, Young JD. Biomimetic properties of minerals and the search for life in the martian meteorite ALH84001. Annu Rev Earth Planet Sci 2012;40:167–93.

McKay DS, Gibson Jr EK, Thomas-Keprta KL, et al. Search for past life on Mars: possible relic biogenic activity in martian meteorite ALH84001. Science 1996;273:924–30.

Mittlefehldt DW. ALH84001, a cumulate orthopyroxenite member of the martian meteorite clan. Meteoritics 1994;29:214–21.

Nealson KH, Cox BL. Microbial metal-ion reduction and Mars: extraterrestrial expectations? Curr Opin Microbiol 2002;5:296–300.

Peng H-H, Martel J, Lee YH, Ojcius DM, Young JD. Serum-derived nanoparticles: de novo generation and growth in vitro, and internalization by mammalian cells in culture. Nanomedicine 2011;6:643–58.

Raoult D, Drancourt M, Azza S, et al. Nanobacteria are mineralo fetuin complexes. PLoS Pathog 2008;4:e41.

Romanek CS, Grossman EL, Morse JW. Carbon isotopic fractionation in synthetic aragonite and calcite: effects of temperature and precipitation rate. Geochim Cosmochim Acta 1992;56:419–30.

Saey TH. Rest in peace nanobacteria, you were not alive after all. Sci News 2008;173:6–7.

Sears DW, Kral TA. Martian "microfossils" in lunar meteorites? Meteorit Planet Sci 1998;33:791–4.

Shiekh FA, Miller VM, Lieske JC. Do calcifying nanoparticles promote nephrolithiasis? A review of the evidence. Clin Nephrol 2009;71:1–8.

Sommer AP, Wickramasinghe NC. Functions and possible provenance of primordial proteins – Part II: microorganism aggregation in clouds triggered by climate change. J Proteome Res 2005;4:180–4.

Steele A, Goddard D, Beech IB, et al. Atomic force microscopy imaging of fragments from the Martian meteorite ALH84001. J Microsc 1998;189:2–7.

Thomas-Keprta KL, Bazylinski DA, Kirschvink JL, Clemett SJ, McKay DS. Elongated prismatic magnetite crystals in ALH84001 carbonate globules: potential Martian magnetofossils. Geochim Cosmochim Acta 2000;64:4049–81.

Thomas-Keprta KL, Clemett SJ, Bazylinski DA, Kirschvink JL, McKay DS. Truncated hexaoctahedral magnetite crystals in ALH84001: presumptive biosignatures. Proc Natl Acad Sci USA 2001;4:2164–9.

Thomas-Keprta KL, Clemett SJ, Bazylinski DA, Kirschvink JL, McKay DS. Magnetofossils from ancient Mars: a robust biosignature in the Martian meteorite ALH84001. Appl Environ Microbiol 2002;68:3663–72.

Travis J. The bacteria in the stone: extra-tiny microorganisms may lead to kidney stones and other diseases. Sci News 1998;154:75–7.

Urbano P, Urbano F. Nanobacteria: facts or fancies? PLoS Pathog 2007;3:e55.

Vali H, McKee MD, Ciftcioglu N, et al. Nanoforms: a new type of protein-associated mineralization. Geochim Cosmochim Acta 2001;65:63–74.

Vecht A, Ireland TG. The role of vaterite and aragonite in the formation of pseudo-biogenic carbonate structures: implication for Martian exobiology. Geochim Cosmochim Acta 2000;64:2719–25.

Wu CY, Martel J, Young D, Young JD. Fetuin-A/albumin-mineral complexes resembling serum calcium granules and putative nanobacteria: demonstration of a dual inhibition-seeding concept. PLoS One 2009;4:e8058.

Young JD, Martel J, Young L, Wu CY, Young A, Young D. Putative nanobacteria represent physiological remnants and culture by-products of normal calcium homeostasis. PLoS One 2009;4:e4417.

Young JD, Martel J. The rise and fall of nanobacteria. Sci Am 2010;302:52–9.

CHAPTER 8

The Hadean and Archaean Atmosphere and the Oldest Records of Life as Micro- or Chemofossils

Chapter Outline

8.1 INTRODUCTION

In the first part of Chapter 8, the origin and composition of the initial atmosphere of the Earth in the Hadean and Archean eons is investigated and discussed. ("Hadean" is the time interval between the Earth's formation and the beginning of the Archean eon.) Currently, most water is considered to have come by the late heavy bombardment (LHB) in the Archean eon. However, this is only theoretically assumed, and there is no actual evidence for the Hadean atmosphere. Then, the oldest geological record of life is related and discussed.

Searching for the oldest geological record on Earth is important, because the clues to the earliest life could be found there. The choice of geological sample and locality is very important. The most famous Archean areas are the Pilbara area in

Origins of the Earth, Moon, and Life. http://dx.doi.org/10.1016/B978-0-12-812058-3.00008-9

Western Australia and the Isua Supracrustal Belt (ISB) in Western Greenland. In this chapter, discoveries of microfossils and morphological evidence from these areas are shown. However, the determination that such microfossils are in fact the remnants of living things reminds us of the case of the big mistake about Martian life as shown in Chapter 7. Please remember that the morphology or the shape of microfossils are not solely conclusive evidence for the existence of life, but rather increase the risk of making the wrong decision.

Thus, in the last part of Chapter 8, in addition to efforts trying to find microfossils in the oldest geological records from Archean rocks, searches for the chemofossils (graphite) as evidence of the earliest life are conducted. The long stretch of time from the Archean eon to today should have destroyed the microfossils by metamorphism, weathering, etc. Nowadays, carbon isotopic ratios ($\delta^{13}C$) of the chemofossils in graphitic form using secondary ion mass spectrometric techniques (SIMS; see Box 2.6) give us definitive information.

8.2 THE PERSPECTIVE OF ATMOSPHERIC EVOLUTION FROM THE HADEAN TO THE ARCHEAN EARTH

As the sun formed from its molecular cloud, it was accompanied by disk material that consisted of gas and small dust particles. Over several tens of millions of years, these dust particles formed the planets. This process occurred in several stages in the terrestrial planet zone, including moon-forming impacts on the proto-Earth (Canup and Asphaug, 2001).

A Hadean atmosphere containing N_2 and CO_2 and a Hadean ocean containing H_2O seems to have formed as a natural consequence of planetary accretion in the terrestrial planet region. The atmosphere which had weak reducing potential with relatively high partial pressure of CO_2 should have formed (Holland, 1984; Walker, 1985; Kasting, 1993; Ferus et al., 2015; Furukawa et al., 2015). In this condition the important biological precursor compounds for life were synthesized. It should be noted that high partial pressure of CO_2 in the early Hadean atmosphere can be presumed. Such gas is like intestinal gas, so it is reasonable to suppose that life may have fermented.

Atmospheric O_2 levels rose naturally and gradually, but not immediately, occurring by photosynthesis and organic carbon burial. At the same time, the concentrations of CO_2 and other greenhouse gases did not compensate for the brightening sun. The Earth's relatively stable climate was a result of the negative feedback between atmospheric CO_2, surface temperature, and the weathering rate of silicate rocks (Kasting, 1993).

This atmospheric evolution implies that Earth is not a unique planet. If planets exist around other stars, some of them could reside in orbits where the illumination is similar to that received by the Earth. Planetary climates are buffered by the carbonate–silicate cycle. Therefore, the habitable zone around late F to mid K stars (see the top scale of Fig. Box 1.4; the Hertzsprung–Russel diagram) may be wider, and other habitable planets may exist. If the origin of life was not a fortuitous event, many of these planets could be inhabited and on some, intelligent life may even have evolved. Both of these speculations can be tested: the first by spectroscopic investigations from large, space-based telescopes; and the second by monitoring microwave and radio emissions from space (Kasting, 1993).

Morbidelli et al. (2000) suggested that the most plausible sources of the water accreted by the Earth were in the outer asteroid belt, in the giant planet regions, and in the Kuiper Belt. It is plausible that the Earth accreted water from the early phases when the solar nebula was still present to the late stages of gas-free scattered planetesimals. Asteroids and comets from the Jupiter-Saturn region were the first water deliverers, when the Earth was less than half its present mass. The bulk water presently on the Earth was carried by a few planetary embryos, originally formed in the outer asteroid belt and accreted by the Earth at the final stage of its formation. (see Fig. 2.10: The blue planetesimals are water rich, which could be a source of water for Earth).

Finally, a late veneer (this could be the same as or different from the late veneer for the highly siderophile elements discussed in Chapter 4), accounting for at most 10% of the present water mass, occurred due to comets from the Uranus-Neptune region and from the Kuiper Belt. The net result of accretion from these several reservoirs is that the D:H ratio of water on Earth is essentially the typical water condensed in the outer asteroid belt. This is in agreement with the observation that the D:H ratio in the oceans is very close to the mean value of the D:H ratio of water inclusions in carbonaceous chondrites.

8.3 TRANSPORTATION OF MATERIALS ON THE ARCHAEAN EARTH BY LATE HEAVY BOMBARDMENT

Following the solidification of the Moon approximately 4.5 Ga, the initially heavy impacter flux declined (Koeberl et al., 2000) and increased again during the late heavy bombardment (LHB). As discussed in Chapter 4, a lot of materials fell on the Moon and of course on the Earth as well. The LHB occurred from approximately 4 to 3.85 Ga, after the Hadean and in the Archean eon, and is recorded in lunar craters (Culler et al., 2000; Hartmann et al., 2000; Gomes et al., 2005; De Niem et al., 2012). This event should have delivered large amounts of materials including biologically important molecules, such as water or organic materials, to the early Earth. The cause of the LHB is linked to a dynamic instability of the outer solar system in the context of the so-called "Nice model" (see Section 2.9 and Fig. 2.10), when Jupiter's orbit changed as a result of resonances with Saturn and small cometary bodies (Tsiganis et al., 2005; Nesvorny and Morbidelli, 2012). These changes released the impacters from previously stable asteroidal and cometary reservoirs on to the planets.

The impactor flux on the Earth is believed to be approximately 10 times higher during the LHB than in the period immediately preceding the LHB, and slowly decreased afterward (Koeberl, 2006; Morbidelli et al., 2012; Geiss and Rossi, 2013).

8.4 IS THE APEX CHERT, FROM THE PILBARA AREA, WESTERN AUSTRALIA, THE OLDEST MICROFOSSILS?

Based on the environment with water and initial atmosphere, life SOMEHOW started. It is unknown when and how the first life appeared on Earth. This chapter describes when life potentially started based on geological records. The earliest record of life may be microfossils, but they should have experienced intense metamorphism, which would have obliterated any fragile microfossils contained therein. Schopf (1993) thought the most promising area for searching for microfossils was the Pilbara Block of northwestern Australia. The region is underlain by a 30-km thick sequence of relatively well preserved sedimentary and volcanic rocks of from approximately 3000 to 3500 Ma.

He detected a diverse assemblage of filamentous microbial fossils in the early Archean eon (~3465 Ma by the zircon dating). He believed to have discovered 11 taxa (including eight undescribed species) of cellularly preserved filamentous microbes in a bedded chert unit of the Early Archean Apex Basalt of northwestern Western Australia, which survived intense metamorphism. This prokaryotic assemblage establishes that cyanobacterium-like microorganisms were extant and morphologically diverse at least as early as approximately 3465 Ma and suggests that O_2-producing photoautotrophs may have already evolved by this early stage in biotic history (see Fig. 8.1), which contains some sketches of microfossils reported by Schopf (1993, 1994).

However, the validity of these microfossils is challenged later.

8.5 CARBON ISOTOPIC FRACTIONATION

Carbon has two stable isotopes of ^{12}C and ^{13}C, and one radioactive isotope of ^{14}C (half-life is 5730 years). As the half-life of ^{14}C is appropriately short, ^{14}C is used for dating up to approximately 30,000 years. The carbon isotopes of ^{12}C and ^{13}C are used for the fractionation of carbon as:

$$\delta^{13}C\ [‰] = \left[\left({}^{13}C/{}^{12}C\right)_{sample}/\left({}^{13}C/{}^{12}C\right)_{standard} - 1\right] \times 1000 \quad (8.5.1)$$

Usually, the Pee Dee Belemnite (PDB) is used as the carbon isotope standard, which is based on a Cretaceous marine fossil from the Pee Dee Formation in South Carolina. All $\delta^{13}C$ data without notices are $\delta^{13}C_{PDB}$ in this book.

Carbon isotope fractionation is useful there is a significant difference between that of organic carbon and inorganic carbon. As shown in Fig. 8.2, the organic carbon is light ($\delta^{13}C_{PDB}$ is −35‰ to −15‰; e.g., Schopf and Kudryavtsev, 2014), whereas the inorganic carbon is heavy ($\delta^{13}C_{PDB}$ is −10–0‰; e.g., Hoefs, 1973). Therefore, the carbon isotopic ratio is a good indicator to discriminate carbon in a fossil-like object derived from organic or inorganic origins. Furthermore, carbon isotope ratios of micrometer-scale microfossils can be determined when high-resolution secondary ion mass spectrometry (HR-SIMS) is applied (see Box 2.6).

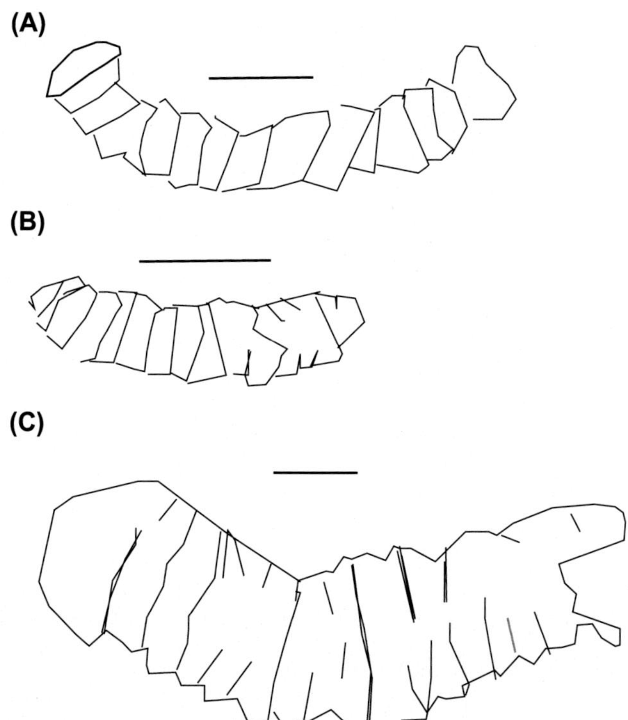

FIGURE 8.1
A carbonaceous microfossil in thin section of the Early Archean Apex chert of Western Australia. (A) *Primaevifilum delicatulum* Schopf (1993). The most popular microbial taxum in the Apex chert. (B) *Primaevifilum conicoteminatum* Schopf (1993). (C) *Archaeoscillatoriopsis maxima*, n. gen., n. sp. (holotype). Each bar is a scale bar of 10 μm.
The sketches are simplified from interpretive drawings of Schopf JW. Microfossils of the early Archean Apex chert: new evidence of the antiquity of life. Science 1993;260:640–6.

8.6 THE EVIDENCE OF LIFE OLDER THAN 3800 MA AT THE ISUA SUPRACRUSTAL BELT AND AKILIA ISLAND, WEST GREENLAND

Mojzsis et al. (1996) reported the carbon isotopic fractionation of carbonaceous inclusions within apatite (calcium phosphate) grains by high resolution (HR)-SIMS (see Box 2.6) from the oldest known sediment sequences of approximately 3800 Ma banded iron formation from the ISB in West Greenland and a similar formation from nearby Akilia island that may be older than 3850 Ma.

The carbon in carbonaceous inclusions was isotopically light (see Fig. 8.2), indicative of biological activity; no known abiotic process can explain the data. Unless some unknown abiotic process exists that is able both to create such isotopically light carbon and then selectively incorporate it into apatite grains, their results provide the evidence for emergence of life on Earth by at least 3800 Ma.

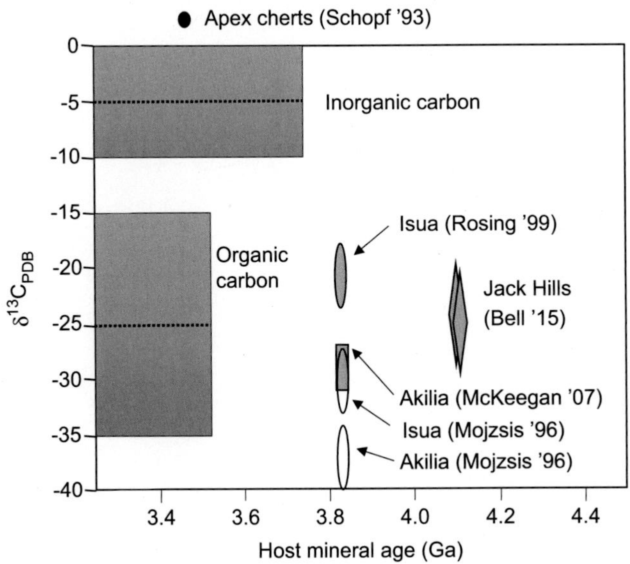

FIGURE 8.2
$\delta^{13}C$ for carbon samples measured by SIMS versus host mineral age compared with inorganic and organic carbon.
Organic and inorganic carbon values from Schopf JW, Kudryavtsev AB. Biogenicity of Earth's earliest fossils. In: Dilek Y, Fumes H, editors. Evolution of Archean crust and early life. Modern Approaches in Solid Earth Sciences, vol. 7. Dordrecht: Springer; 2014. p. 333–49 and Hoefs J. Stable isotope geochemistry. Berlin: Springer; 1973, respectively.

Ueno et al. (2002) also measured carbon isotopic compositions of graphite in seven metasediments [Metasediment is altered sediment (mudstone, sandstone, chert, etc.) by metamorphism.] and two carbonate rocks from approximately 3.8 Ga from ISB, West Greenland by HR-SIMS. The $\delta^{13}C$ values of graphite globules in the metasediments and the carbonate rocks were −18‰ to −2‰ and −7‰ to −3‰, respectively. The highest $\delta^{13}C$ value of graphite globules in the metasediment rose from −14‰ to −5‰ as the metamorphic grade increased. Thus, they concluded that the existing light carbon is modified by metamorphism and decomposition of carbonate or exchanged with ^{13}C-enriched reservoir. Therefore, the light carbon found by Mojzsis et al. (1996) was probably premetamorphic origin. This supported the existence of life at Isua at approximately 3.8 Ga.

McKeegan et al. (2007) constructed three-dimensional (3D) molecular-structural images of apatite grains and associated minerals embedded in a banded quartz-pyroxene-magnetite from supracrustal rocks from Akilia, south western Greenland by using confocal Raman spectroscopy (see Box 8.1). The sample was the same rock used by Mojzsis et al. (1996), which contained isotopically light graphite inclusions in apatite. The graphite inclusions were perfectly contained within apatite without cracks. The carbon isotopic composition of such inclusion was isotopically light ($\delta^{13}C = -29 \pm 4$‰), which was determined by

HR-SIMS in agreement with earlier analyses. This result is consistent with the hypothesis that graphite contained in apatite grains of the >3830 Ma Akilia metasediments are chemofossils of the early life. (Although the original form of life is perfectly lost, the constituent carbon remained in an apatite grain, being transformed into graphite. Thus it can be called a chemofossil.)

BOX 8.1 CONFOCAL RAMAN SPECTROSCOPY

When the monochromatic light is lit on a sample molecule, the light is reflected, absorbed or scattered. In some cases, Raman scattering occurs, in which lights of shorter and longer wavelengths (Stokes and anti-Stokes Raman scattering lights) are emitted from the sample molecule. Raman scattering is dependent on the very weak vibrational states of the molecule ($\times 10^{-7}$). To enhance the Raman scattering lights, a larger number of photons in the incident light is required. Thus, laser light generally is used.

Confocal microscopy is an optical imaging technique to enhance resolution and contrast. There is a spatial pinhole at the confocal plane of the lens to remove out-of-focus light in the Z-direction. The Raman spectroscopy instrument is used with this confocal microscope so that not only the small X–Y area (>20 nm area) but also the Z-directional image can be analyzed, which means an X-Y-Z Raman scattering light image, or 3D image, can be obtained.

8.7 ^{13}C-DEPLETED CARBON MICROPARTICLES IN >3700 MA SEA-FLOOR SEDIMENTARY ROCKS FROM WEST GREENLAND

Rosing (1999) reported that turbiditic and pelagic sedimentary rocks (see Box 8.2) from ISB in western Greenland (3779 ± 81 Ma by the ^{147}Sm–^{143}Nd isochron of the sediment and amphibolite samples) might contain reduced carbon of biogenic origin. The carbon was in 2–5 μm graphite globules and had an isotopic composition of $\delta^{13}C$ about −19‰. These data and the occurrence indicated that the graphite represented biogenic detritus, which was perhaps derived from planktonic organisms.

The largest difference from the study of Mojzsis et al. (1996) is that Rosing (1999) chose samples with well-preserved sedimentary structures, that showed clear depositional contact with the basaltic Garbenschiefer Formation (they became amphibolites by recent metamorphism), with occasional pillow structures. The sedimentary rocks can be traced approximately 100 m. In the least deformed samples, mica and chlorite defined a schistosity parallel to the sedimentary bedding. The black slaty units with thicknesses of less than 10 cm containing graphite are finely laminated by dark layers (greywacke) also containing graphite.

In more deformed rocks, all sedimentary structures have been obliterated, and up to 1 cm biotite and garnet porphyroblasts (round crystals of garnet) are present. In such recrystallized samples, graphite is absent. These relations indicate

BOX 8.2 SOME COMMON KNOWLEDGE ON SEDIMENTARY ROCKS, METAMORPHISM AND DEEP SEA VOLCANISM

"Turbiditic" sediments are sediments formed near continents. Therefore, original grain sizes are relatively large compared to the pelagic sediments. "Pelagic" sediments are sediments formed far from a continent; therefore, the original grain sizes are fine. The pelagic sediments sometimes form alternate layers with cherts, which have a composition of almost 100% silica. Cherts is considered to be of pure chemical or biogenic origin. Most cherts today are of biogenic origin.

Amphibolites are metamorphic rocks that are considered to have been volcanic rock, such as basalt, before metamorphism. The pillow structures are typical of lavas erupted in the deep sea. The pillow structure is typically observed in mid-oceanic ridge basalts (MORBs). Even after metamorphism, the pillow structure can sometimes remain in amphibolites. When the pillow structure is observed, it is clear evidence that the basalt was erupted under the sea as MORBs or the sea existed at the eruption.

When garnet is observed in metamorphic rocks, it means the metamorphic temperature and pressure were relatively high. When the metamorphism is going in the direction of higher temperature and pressure, it is called progressive metamorphism. In contrast, when the metamorphism goes to lower temperature and pressure, it is called retrogressive metamorphism.

Foliation is observed in metamorphic rocks, especially when sheet-like minerals like mica or chlorite align to one planar or wavy direction. Schistosity is a mode of foliation in metamorphism that is planar (sheet-like) and rather strong. When metamorphic rocks show strong foliation or strong schistosity, a strong parallel force in the direction of the original bedding was applied to the rocks.

that the graphite globules formed before the earliest growth of garnet and biotite disappeared during progressive metamorphism.

Sedimentological and geochemical evidence indicates that the carbon forming the graphite globules is of biogenic origin. In analogy to modern oceanic pelagic shales, the precursor organic detritus of the graphite globules could have been derived from planktonic organisms that sedimented from surface waters. Thus the organism should be photoautotrophic.

8.8 QUESTIONING THE EVIDENCE FOR THE EARTH'S OLDEST FOSSILS IN APEX CHERTS

Structures resembling remarkably preserved bacterial and cyanobacterial microfossils from 3465 Ma Apex cherts of the Warrawoona Group in Western Australia (see Fig. 8.1) have provided the oldest morphological evidence for life on Earth and had been taken to support an early beginning for oxygen-producing photosynthesis.

However, Brasier et al. (2002) newly researched recollected materials using optical and electron microscopy, digital image analysis, micro-Raman spectroscopy and other geochemical techniques. They reinterpreted the observed microfossil-like structure as secondary artifacts formed from amorphous graphite within multiple generations of metalliferous hydrothermal vein chert and volcanic glass. Although there is no support for primary biological morphology, a Fischer-Tropsch-type synthesis of carbon compounds (see Box 8.3) and carbon isotopic fractionation were inferred for one of the oldest known hydrothermal systems on Earth.

BOX 8.3 FISCHER–TROPSCH REACTION AND CARBON ISOTOPIC FRACTIONATION

A Fischer–Tropsch reaction or process is a chemical reaction that converts a mixture of carbon monoxide and hydrogen into hydrocarbons:

$$n\,CO + (2n+1)\,H_2 \rightarrow C_nH_{2n+2} + n\,H_2O$$

The Fischer–Tropsch reaction can also make oxidized carbon:

$$2n\,CO + (n+1)\,H_2 \rightarrow C_nH_{2n+2} + n\,CO_2$$

Lancet and Anders (1970) first measured carbon isotopic fractionation in the Fischer–Tropsch reaction. They observed fractionation of −50‰ to −100‰ at 400K, which was similar to that of carbonaceous chondrites. When hydrated silicates are formed, the reaction of:

$$\underset{\text{Olivines}}{4(Mg,Fe)_2SiO_4} + 4H_2O + 2CO_2 \rightarrow \underset{\text{Carbonates}}{2\,(Mg,Fe)\,CO_3} + \underset{\text{Serpentines}}{2(Mg,Fe)_3Si_2O_5(OH)_4}$$

occurs. This is similar to a serpentinization reaction.

8.9 OBJECTION TO THE EARLIEST LIFE ON AKILIA ISLAND

Hayes (1996) critically commented on the difficulty of the SIMS analysis and why graphite inclusions in apatites were used (Section 8.6) in "News and Views" of the same issue of Nature. If the data are obtained as the whole rock analysis, the data might be acceptable.

The $\delta^{13}C$ values of the Isua graphite have been interpreted by previous investigators (e.g., Schidlowski and Aharon, 1992; Mojzsis et al., 1996) as metamorphic modification of organic matter with initial $\delta^{13}C$ values of approximately −30‰. However, Naraoka et al. (1996) proposed a two-component mixing model in which a major component had $\delta^{13}C$ values of approximately −12‰ and a minor component of approximately −25‰. The minor component might be biogenic, but it was not certain whether it was an Archean biological product or a more recent one.

Their thermodynamic consideration of the temperature and composition of metamorphic fluids and carbon isotope calculations suggested that the major component of graphite with $\delta^{13}C$ values around −12‰ was formed by an inorganic process. The graphite appears at temperatures between 700 and 400°C by one or both of the following two processes:

$$CO_2 + CH_4 \rightarrow 2C + 2H_2O \tag{8.9.i}$$

during cooling of fluids with a CO_2:CH_4 molar ratio of approximately 1.

$$4FeO + CO_2 \rightarrow 2Fe_2O_3 + C \tag{8.9.ii}$$

where FeO comes from olivine in ultramafic rocks.

The $\delta^{13}C_{Total\text{-}carbon}$ values of the fluids for both Eq. (8.9.i) and (8.9.ii) were between −12‰ and −5‰, suggesting that the carbon-bearing fluids could be derived from the mantle.

Similarly, earlier studies (e.g., Schidlowski and Aharon, 1992; Mojzsis et al., 1996) on isotopic characteristics of graphite occurring in rocks of the approximately 3.8 Ga ISB in southern West Greenland (see Section 8.6) were criticized by van Zuilen et al. (2002). They showed that graphite occurred abundantly in secondary carbonate veins in the ISB that were formed at depth in the crust by injection of hot fluids reacting with older crustal rocks. During these reactions, graphite formed from Fe(II)-bearing carbonates at high temperature. These metasomatic rocks, which had no biological relevance, were earlier thought to be of sedimentary origin and this graphite provided the basis for early life.

Mojzsis et al. (1996) reported the carbon isotope ratios in grains of apatite. Sano et al. (1999) directly measured the age of the apatite using U–Pb and Pb–Pb dating by HR-SIMS (see Box 2.6). They obtained 1504 ± 336 and 1459 ± 160 Ma, respectively. This threw doubt on the conclusion of Mojzsis et al. (1996, 1999), who replied that by the metamorphic events the carbonaceous inclusions were not affected but the host apatites were.

Mojzsis et al. (1996) interpreted a quartz-pyroxene rock as a banded iron formation (BIF) from the island of Akilia, southwest Greenland, containing ^{13}C-depleted graphite that had been evidence for the earliest (3850 Myr ago) life on Earth (see Section 8.6). However, Fedo and Whitehouse (2002) observed that field relationships on Akilia recorded multiple intense deformation events that had resulted in parallel transposition of early Archean rocks, the tails of which commonly formed the banding in the quartz-pyroxene rock. Geochemical data possessed distinct characteristics consistent with an ultramafic igneous, not BIF, protolith for this lithology and the adjacent schists. Later metasomatic silica and iron introduction have merely resulted in the rock that superficially resembled a BIF. Thus Fedo and Whitehouse (2002) concluded that an ultramafic igneous origin did not support the claims of Mojzsis et al. (1996) that the carbon isotopic composition of graphite inclusions represented evidence for life at the time of crystallization.

8.10 BACK TO THE ISUA SUPRACRUSTAL BELT (ISB), WESTERN GREENLAND

Ohtomo et al. (2014) analyzed the geochemistry and structure of the ^{13}C-depleted graphite in ISB, Western Greenland. Raman spectroscopy and geochemical analyses indicated that the schists were formed from clastic marine sediments that contained ^{13}C-depleted carbon at the time of their deposition. Transmission electron microscope (TEM) observations showed that the grapmahite in the schist formed nanoscale polygonal and tube-like grains. In contrast, abiotic graphite in carbonate veins exhibits a flaky morphology. The graphite grains in the schist contained distorted crystal structures and disordered stacking of sheets of graphene. The observed morphologies were consistent with pyrolyzed and pressurized organic compounds during metamorphism. Therefore, they concluded that the graphite contained in the Isua metasediments represents traces of early life that flourished in the oceans at least 3.7 Ga.

Fedo and Whitehouse (2002) wondered what evidence could be used. The Apex chert was refuted by Brasier et al. (2002). The Akilia island chert was refuted by themselves. The graphite particles in ISB in Western Greenland in 3700–3800 Ma of Rosing (1999) might be solely the oldest record of life, as it has not yet been refuted.

8.11 GEOLOGICAL EVIDENCE OF RECYCLING OF ALTERED CRUST IN THE HADEAN EON

Using SIMS, Cavosie et al. (2005) measured $\delta^{18}O$ in 4400–3900 Ma igneous zircons from the Jack Hills, Western Australia, which gave a record of the oxygen isotope composition of magmas in the earliest part of the Archean eon. The main finding of Cavosie et al. (2005) is that evidence for the recycling of altered crust was preserved in igneous zircons from Jack Hills with magmatic $\delta^{18}O$ values as high as 6.5‰ by 4325 Ma, and up to 7.3‰ by 4200 Ma. The $\delta^{18}O$ values in these zircons were indistinguishable from the range of igneous zircons found from the Archean eon, and demonstrate from an oxygen isotope perspective that no fundamental, planetary-scale change in magmatic process was recorded in zircon from at least 4325 to 2500 Ma.

The detrital magmatic zircons possibly represent many igneous rocks and are clearly not from the same magmatic event. The magmatic $\delta^{18}O$ values obtained from microvolumes of zircon are interpreted to preserve igneous chemistry, including U–Pb concordance, Th/U chemistry, and internal zoning. Although some grains have internal zoning patterns that are ambiguous, both the lower (5.3‰) and upper (7.3‰) limits of magmatic $\delta^{18}O$ values for the zircons in the study are defined by multiple grains, which are an average of many growth zoning patterns. The range of magmatic $\delta^{18}O$ of the zircons observed is within previously reported ranges for detrital zircons of similar age from Jack Hills.

Valley et al. (2002) proposed the hypothesis that the early Earth was cool. No known rocks have survived from the first 500 My of Earth's history, but studies of single zircons suggest that some continental crust formed as early as 4.4 Ga, 160 My after accretion of the Earth, and that surface temperatures were low enough for liquid water. Surface temperatures are inferred from high $\delta^{18}O$ values of zircons. The range of $\delta^{18}O$ values is constant throughout the Archean eon (4.4–2.6 Ga), suggesting uniformity of processes and conditions. The hypothesis of a cool early Earth suggests long intervals of relatively temperate surface conditions from 4.4 to 4.0 Ga that were conducive to liquid-water oceans and possibly life.

8.12 ORIGIN OF LIFE BACK TO 4.1 BILLION YEARS AGO

Some of the detrital zircons from Jack Hills have ages up to nearly 4.4 Ga. From a population of over 10,000 Jack Hills zircons, Bell et al. (2015) identified one more than 3.8-Ga zircon that contains primary graphite inclusions. These inclusions were judged as primary, because (1) their enclosure in a crack-free host were shown by transmission X-ray microscopy using the Stanford Synchrotron Radiation Lightsource (SSRL); and (2) the graphite crystal habit. Then they measured carbon isotopic ratios in inclusions in a concordant, 4.10 ± 0.01-Ga zircon using HR-SIMS. They obtained $\delta^{13}C_{PDB}$ of −24 ± 5‰, which is consistent with a biogenic carbon and may be evidence that life appeared by 4.1 Ga, or approximately 300 My earlier than has been previously proposed.

References

Bell EA, Boehnke P, Harrison TM, Mao WL. Potentially biogenic carbon preserved in a 4.1 billion year-old zircon. Proc Natl Acad Sci USA 2015;112:14518–21.

Brasier MD, Green OR, Jephcoat AP, et al. Questioning the evidence for Earth's oldest fossils. Nature 2002;416:76–81.

Canup RM, Asphaug E. Origin of the Moon in a giant impact near the end of the Earth's formation. Nature 2001;412:708–11.

Cavosie AJ, Valley JW, Wilde SA. Magmatic delta O-18 in 4400-3900 Ma detrital zircons: a record of the alteration and recycling of crust in the Early Archean. Earth Planet Sci Lett 2005;235:663–81.

Culler TS, Becker TA, Muller RA, Renne PR. Lunar impact history from $^{40}Ar/^{39}Ar$ dating of glass spherules. Science 2000;287:1785–8.

De Niem D, Kuehrt E, Morbidelli A, Motschmann U. Atmospheric erosion and replenishment induced by impacts upon the Earth and Mars during a heavy bombardment. Icarus 2012;221:495–507.

Fedo CM, Whitehouse MJ. Metasomatic origin of quartz-pyroxene rock, Akilia, Greenland, and implications for Earth's earliest life. Science 2002;296:1448–52.

Ferus M, Nesvorný D, Sponer J, et al. High-energy chemistry of formamide: a unified mechanism of nucleobase formation. PNAS 2015;112:657–62.

Furukawa Y, Nakazawa H, Sekine T, Kobayashi T, Kakegawa T. Nucleobase and amino acid formation through impacts of meteorites on the early ocean. Earth Planet Sci Lett 2015;429:216–22.

Geiss J, Rossi AP. On the chronology of lunar origin and evolution: implications for Earth, Mars and the Solar System as a whole. Astron Astrophys Rev 2013;21:1–54.

Gomes R, Levison HF, Tsiganis K, Morbidelli A. origin of the cataclysmic late heavy bombardment period of the terrestrial planets. Nature 2005;435:466–9.

Hartmann WK, Ryder G, Dones L, Grinspoon D. The time-dependent intense bombardment of the primordial Earth/Moon system. In: Canup RM, Righter K, editors. Origin of the Earth and Moon. Tucson: The University of Arizona Press; 2000. p. 493–512.

Hayes JM. The earliest memories of life on Earth. Nature 1996;384:21–2.

Hoefs J. Stable isotope geochemistry. Berlin: Springer; 1973.

Holland HD. The chemical evolution of the atmosphere and oceans. Princeton: Princeton University Press; 1984.

Kasting JF. Earth's early atmosphere. Science 1993;259:920–6.

Koeberl C, Reimold WU, McDonald I, Rosing M. Impacts and the early Earth. In: Gilmour I, Koeberl C, editors. Lecture notes in Earth sciences. Berlin: Springer; 2000. p. 73–97.

Koeberl C. Impact processes on the early Earth. Elements 2006;2:211–6.

Lancet MS, Anders E. Carbon isotope fractionation in the Fischer-Tropsch synthesis and in meteorites. Science 1970;170:980–2.

McKeegan KD, Kudryavtsev AB, Schopf JW. Raman and ion microscopic imagery of graphitic inclusions in apatite from older than 3830 Ma Akilia supracrustal rocks, West Greenland. Geology 2007;35:591–4.

Mojzsis SJ, Arrhenius G, McKeegan KD, Harrison TM, Nutman AP, Friend CRL. Evidence for life on Earth before 3,800 million years ago. Nature 1996;384:55–9.

Mojzsis SJ, Harrison TM, Arrhenius G, McKeegan KD, Grove M. Reply-Origin of life from apatite dating? Nature 1999;400:127–8.

Morbidelli A, Chambers J, Lunine JI, et al. Source regions and timescales for the delivery of water to the Earth. Meteorit Planet Sci 2000;35:1309–20.

Morbidelli A, Marchi S, Bottke WF, Kring DA. A sawtooth-like timeline for the first billion years of lunar bombardment. Earth Planet Sci Lett 2012;355:144–51.

Naraoka H, Ohtake M, Maruyama S, Ohmoto H. Non-biogenic graphite in 3.8-Ga metamorphic rocks from the Isua district, Greenland. Chem Geol 1996;133:251–60.

Nesvorny D, Morbidelli A. Statistical study of the early solar system's instability with four, five, and six giant planets. Astron J 2012;144:20–68.

Ohtomo Y, Kakegawa T, Ishida A, Nagase T, Rosing MT. Evidence for biogenic graphite in early Archaean Isua metasedimentary rocks. Nat Geosci 2014;7:25–8.

Rosing MT. ^{13}C-depleted carbon microparticles in >3700-Ma sea-floor sedimentary rocks from west Greenland. Science 1999;283:674–6.

Sano Y, Terada K, Takahashi Y, Nutman AP. Origin of life from apatite dating? Nature 1999;400:127.

Schidlowski M, Aharon P. Carbon cycle and carbon isotope record: geochemical impact of life over 3.8 Ga of earth history. In: Schidolowski M, Kimberley S, Golubic MM, McKirdy DM, Trudinger PA, editors. Early organic evolution: implications for mineral and energy resources. Berlin: Springer; 1992. p. 147–75.

Schopf JW. Microfossils of the early Archean Apex chert: new evidence of the antiquity of life. Science 1993;260:640–6.

Schopf JW. Disparate rates, differing fates: tempo and mode of evolution changed from the Precambrian to the Phanerozoic. Proc Natl Acad Sci USA 1994;91:6735–42.

Schopf JW, Kudryavtsev AB. Biogenicity of Earth's earliest fossils. In: Dilek Y, Fumes H, editors. Evolution of Archean crust and early life. Modern Approaches in Solid Earth Sciences, vol. 7. Dordrecht: Springer; 2014. p. 333–49.

Tsiganis K, Gomes R, Morbidelli A, Levison HF. Origin of the orbital architecture of the giant planets of the Solar System. Nature 2005;435:459–61.

Ueno Y, Yurimoto H, Yoshioka H, Komiya T, Maruyama S. Ion microprobe analysis of graphite from ca. 3.8 Ga metasediments, Isua supracrustal belt, West Greenland: relationship between metamorphism and carbon isotopic composition. Geochim Cosmochim Acta 2002;66:1257–68.

Valley JW, Peck WH, King EM, Wilde SA. A cool early Earth. Geology 2002;30:351–4.

van Zuilen M, Lepland A, Arrhenius G. Reassessing the evidence for the earliest traces of life. Nature 2002;418:627–30.

Walker JCG. Carbon-dioxide on the early earth. Orig Life Evol Biosph 1985;16:117–27.

CHAPTER 9

How Did Initial Life-Related Molecules Appear on Earth?

Chapter Outline

Origins of the Earth, Moon, and Life. http://dx.doi.org/10.1016/B978-0-12-812058-3.00009-0

9.1 INTRODUCTION

In this chapter, the origin of life-related molecules is discussed. Life-related molecules are divided into two groups. One group consists of molecules made of carbon (C), hydrogen (H), and oxygen (O), such as aldoses, ketoses, and sugars, including ribose (see Fig. 7.8A and B). In this chapter, they are presented as CHOs. Another group consists of C, H, O, and nitrogen molecules, such as formamide, nucleobases, and amino acids. In this chapter, they are expressed as CHONs. The origins of CHOs and CHONs are separately discussed in this book, whereas inorganic materials are not included.

There are several proposals on the origin of CHOs and CHONs on Earth:

1. High energy such as an electrical spark or ultraviolet light activated organic synthesis of CHONs on Earth, as was proposed by Miller (1953) (Section 9.2).
2. CHOs and CHONs in comets and asteroids fell on the Earth during the late heavy bombardment (LHB) without decomposition, as carbonaceous chondrites containing them fell to Earth (Section 9.3).
3. Strong UV light produced CHOs in space, and fell to Earth (Section 9.4).
4. Impact-shocks (icy comet–icy comet or icy comet–Earth) in space or on Earth with extremely high energy synthesized CHONs (Sections 9.5–9.8).

To prove these hypotheses, various experiments have been performed for synthesizing CHOs and CHONs. This chapter presents a range of experiments, from classic to up-to-date research.

9.2 CLASSIC EXPERIMENTS

9.2.1 The Primordial Soup Theory

A.I. Opalin (1894–1980) was a noted biochemist in Soviet Union. He was very much interested in how life began. He thought that the early Earth was in reducing conditions which contain CH_4, NH_3, H_2, and water vapor. He proposed the hypothesis that life was developed by gradual chemical evolution of carbon molecules in Earth's primordial soup (Oparin, 1938).

J.B.S. Haldane (1892–1964) was a British-Indian biologist and a mathematician who published an article explaining the primordial soup theory called "The Origin of Life" in 1929. He described the primitive ocean as a "vast chemical laboratory" and a "hot dilute soup" containing inorganic compounds. The sunlight generated a variety of organic compounds from the oxygenless atmosphere of CO_2, NH_3, and water vapor. The first molecules reacted with each other and produced more complex compounds of CHONs and finally cellular components.

In 1924, Oparin presented a similar idea in Russia, and in 1936 he introduced it to the English-speaking world.

9.2.2 The Miller Experiment (or the Miller-Urey Experiment)

To prove the Oparin–Haldane hypothesis, S. Miller conducted the famous Miller experiment (or Miller-Urey experiment) under the supervision of H. Urey in 1952 (Miller, 1953). In Fig. 9.1, the apparatus which Miller used is indicated.

Two flasks were connected with two pipes. One flask with pure water was heated to boiling to make water vapor. The water vapor was led through one pipe to another reaction flask containing methane, ammonia, and hydrogen gases, in

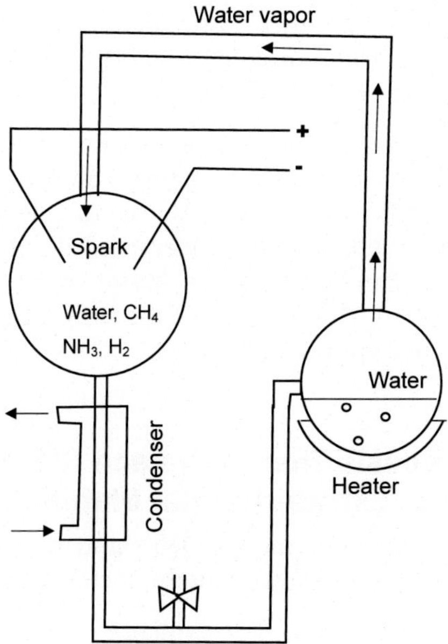

FIGURE 9.1
The Miller experiment.

which electrical sparks were supplied. The water vapor was cooled by a condenser, and the liquid water was returned to the boiling flask through another pipe.

The water in the flask became pink, and deep red after a week. Most turbidity was due to colloidal silica from the glass (they do not seem to have used a "silica glass flask" but a "soda glass flask"). At the end of the run, the boiling flask was removed and $HgCl_2$ was added to prevent the growth of living organisms. Then the sample solution was combined with $Ba(OH)_2$ and evaporated in vacuum by adding H_2SO_4, and then neutralized with $Ba(OH)_2$, filtered, and concentrated in vacuum. The paper chromatogram showed glycine, α-alanine, β-alanine, aspartic acid, and unidentified organic compounds.

Although the synthesized amino acids were racemic, it was proved that some amino acids (CHONs) can be synthesized by such a simple condition.

9.3 COSMIC ORIGIN OF CHOS AND CHONS

9.3.1 CHOs and CHONs in Carbonaceous Chondrites and Comets

In a collision between pure ice comets, no organic materials are formed because the impacters are composed only of ice (water). However, various amino acids and nucleobases (CHONs) are found in carbonaceous chondrites (Kvenvolden et al., 1970; Cronin and Moore, 1971; Stoks and Schwartz, 1979, 1981; Martins et al., 2008; Callahan et al., 2011). In addition, amino acid and precursors (ammonia, methanol, and carbonyl compounds, which correspond to CHOs) were observed in actual carbonaceous chondrites (Crovisier and Bockelée-Morvan, 1999; Bockeléee-Morvan et al., 2000; Ehrenfreund and Charnley, 2000; Ehrenfreund et al., 2002; Mumma et al., 2003; Festou et al., 2005; DiSanti et al., 2013).

The amino acid precursors (such as ammonia, methanol, and carbonyl compounds) have been observed in actual comets; for example, Halley, Hyakutake, Tempel-1, Giacobini-Zinner, Hartley-2, and Hale-Bopp (Crovisier and Bockelée-Morvan, 1999; Bockelée-Morvan et al., 2000; Ehrenfreund and Charnley, 2000; Ehrenfreund et al., 2002; Mumma et al., 2003; Festou et al., 2005; DiSanti et al., 2013). Glycine, one of the simplest proteins, was detected on comet 81P/Wild from samples returned by NASA's Stardust mission (Elsila et al., 2009).

9.3.2 The Hypothesis That CHOs and CHONs on Earth Came From Extraterrestrial Materials

In this simple hypothesis, CHOs and CHONs came from extraterrestrial materials such as carbonaceous meteorites or interplanetary dust particles (IDPs).

This simple idea has been supported by evidence involving the identification of CHOs and CHONs in carbonaceous chondrites. Amino acid precursors, such as ammonia, methanol and carbonyl compounds have also been observed in comets.

9.3.3 Transportation of CHOs and CHONs to Earth by the Late Heavy Bombardment

Major counterarguments are that organic matter cannot survive the extremely high temperature (>10^4K) reached at impact, which atomize the meteoritic bodies. Only small particles of 10^{-12}–10^{-6}g, which are gently decelerated by the atmosphere, can deliver the organic matter without decomposition on the way to Earth. However, we can actually obtain hand specimens of carbonaceous chondrites.

Anders (1989) estimated the amount of such soft-landed organic carbon to be about 20 g cm^{-2} in the few hundred Myrs during the late heavy bombardment (LHB). It may have included some biologically important compounds of CHOs and CHONs that were not formed by abiotic synthesis on Earth.

It was speculated that the Earth accreted important prebiotic organic molecules for the origins of life from impacts of carbonaceous asteroids and comets during the LHB period. However, in comet or asteroid interactions with the atmosphere, surface impact, and resulting organic pyrolysis, it was demonstrated that organic materials will not survive impacts at velocities greater than approximately 10 km s^{-1} and that even comets and asteroids as small as 100 m in radius cannot survive. However, for plausible dense early atmospheres (CO_2 pressure of 1×10^6 Pa), Chyba et al. (1990) found that Earth 4.5 Gyr ago could accrete intact cometary organics at a rate of at least approximately 10^6–10^7 kg per year, a flux that then declined with a half-life of 10^8 years. These results may have increased terrestrial oceanic and total biomasses to 3×10^{12} kg and 6×10^{14} kg, respectively.

Chyba and Sagan (1992; see Box 9.1 for Sagan) proposed that the sources of organic molecules on the early Earth are divided into three categories:

1. Delivery by extraterrestrial objects, which is discussed in this section.
2. Organic synthesis by another form of energy (such as an electrical spark or ultraviolet light) on the Earth, as was conducted by the Miller experiment (Section 9.2);
3. Organic synthesis driven by impact shocks on the Earth afterward, which will be discussed in Sections 9.4–9.7.

Estimates of the early terrestrial atmosphere suggest that quantities of organics either delivered by (1) or produced by (2) in the LHB were comparable to those produced by other energy sources as (3).

BOX 9.1 CARL E. SAGAN (1934–96)

Carl Sagan was a famous American astrobiologist, astronomer, astrophysicist, cosmologist, and author. He wrote more than 20 books as author, co-author or editor, and published more than 600 scientific papers. The TV program, "Cosmos" made him very famous and enlightened many Americans.

9.3.4 Counterarguments to the Cosmic Origin of CHOs and CHONs

There are some counterarguments to the cosmic origin of CHOs and CHONs. First of all, the amounts of CHONs in IDPs are unclear (Glavin et al., 2004; Maurette, 2006). Second, organic molecules such as CHONs in carbonaceous chondrites and/or IDPs may be very fragile.

Experiments to synthesize CHONs were continued by Oró's group (Oró, 1961; Oró and Kimball, 1961). The organic synthesis of CHONs by glow-discharge was continued by Harada and Suzuki (1977). Ferris et al. (1978) proposed the importance of hydrogen cyanide (HCN) for the synthesis of purines, which are an important block of DNA (see Section 7.2). Shapiro (1999) synthesized cytosine (see Fig. 7.4). Miyakawa et al. (2002a,b) proposed the cold origin of life.

Alongside these terrestrial origins of life, impact shock origin of CHOs and CHONs became strong counterarguments. This is the third proposal of Chyba and Sagan (1992) and Anders (1989): that the synthesizing energy is obtained by the impact shocks of meteorites. The largest difference from the cosmic origin is that the carbons of organic molecules were supplied from the terrestrial atmosphere, because partial pressure of CO_2 was high enough in the Hadean and Archean eons. Syntheses of CHONs by shock of meteoritic impact goes back to the experiments of the Sagan and Bar-Nun groups (Bar-Nun et al., 1970; Bar-Nun and Shaviv, 1975).

9.3.5 The Possible Model for CHONs and Phosphorus

Pasek and Lauretta (2008) calculated and concluded that meteorites and comets may have been an important source of prebiotic C, N, and P (called CHONPs) on the early Earth. They also concluded that life should have originated shortly after the LHB, because concentrations of organic compounds and reactive phosphorus were enough for life to start. This work quantifies the sources of potentially prebiotic extraterrestrial CHONPs. In addition, they showed that the fluxes of CHONPs correlate with total iridium fluxes, and estimated the effect of the atmosphere on the survival of material. They concluded that:

1. IDPs were a better source of organic compounds to the surface of the Earth than noncarbonaceous chondrites;
2. extraterrestrial metallic material was much more abundant on the early Earth, and delivered reactive P in the form of phosphide minerals to Earth's surface; and
3. large impacts provided substantial local enrichment of potentially prebiotic reagents.

9.4 INTRODUCTION TO IMPACT-SHOCK EXPERIMENTS FOR SYNTHESES OF CHOs AND CHONs

The most probable way to synthesize CHOs and CHONs considered today is the impact-shock between an icy comet and another icy comet in space or between an icy comet and the Earth when the comet enters the Earth's Archean atmosphere and then falls into the Archean Ocean. Three cases were easily imaginable, and were experimentally simulated. The first case is an icy comet hitting another icy comet in space (Martins et al., 2013; Meinert et al., 2016; Sections 9.5 and 9.6 for CHOs and CHONs, respectively). The second case covers both reactions of ice–ice and ice-Earth collisions because only the energy drives the chemical reactions (Ferus et al., 2015; Section 9.7). Therefore, the plasma made by the high-energy laser does not matter as the source of the collisional energy. The final experiment simulated an icy comet falling on the Earth's early ocean (Furukawa et al., 2015; Section 9.8), which noted the starting materials. Hence, Section 9.5 covers the synthesis of CHOs, whereas Sections 9.6–9.8 target the syntheses of CHONs.

The simulated reaction, starting materials, irradiation medium, experimental condition, target molecules and their detection methods in each modern experiment, are summarized in Table 9.1.

Table 9.1 Summary of Experimental Conditions for Synthesizing Life-Related Organic Molecules

Researcher	Simulated Reaction	Starting Materials	Irradiation	Experimental Condition	Target Molecules	Detection
Meinert et al. (2016)	Icy grains at molecular cloud stage	Ice, Methanol, NH_3	Ultraviolet light	10^{-5} Pa, 78K	CHOs	GC × GC–TOFMS
Martins et al. (2013)	Icy comet/ icy comet collision	NH_3 solution, Dry ice (CO_2), Methanol	Gas gun	>50 GPa	CHONs	GC–MS
Ferus et al. (2015)	Icy comet/ icy comet or Earth collision (LHB)	Formamide ($HCONH_2$), Olivine chondrite	High energy laser	>4500K	CHONs	GC–MS
Furukawa et al. (2015)	Icy comet fell into seawater	Fe, Ni (Fe_3O_4), NH_4HCO_3 (Mg_2SiO_4), H_2O, N_2 gas	Gas gun		CHONs	UHPLC–MS/MS

9.5 SYNTHESES OF CHOs BY UV IRRADIATION ON INTERSTELLAR ICES

Meinert et al. (2016) published an experiment in which CHOs (aldoses and ketoses) were made by ultraviolet irradiation of interstellar ice composed of water (H_2O), methanol (CH_3OH), and ammonia (NH_3).

The largest difference of Meinert's experiment from the others in Table 9.1 is that this experiment aimed to synthesize aldoses and ketoses, and to finally produce ribose, which is a backbone of ribonucleic acid (RNA). RNA is a basis of the RNA world (see Section 10.6.3), that will evolve into the DNA world.

In the experiment of Meinert et al. (2016), samples were prepared after Nuevo et al. (2007). Briefly, the starting material gases (H_2O, $^{13}CH_3OH$, and NH_3 in proportions of 10:3.5:1) were deposited on an MgF_2 substrate cooled at 78K in 10^{-5} Pa as a thin film of ices. The MgF_2 substrate is transparent from visible to UV light. Then the gas-deposited substrate was irradiated with UV photons for 142h using an H_2 discharge lamp providing essentially Lyman-α photons (see Box 1.6) at 122nm with a tail including an H_2 recombination line at approximately 160nm and a continuum down to the visible range. The ratio of ultraviolet photons to deposited molecules was approximately 1:1.

The substrate was further irradiated at room temperature with right-hand circularly polarized synchrotron radiation (CPSR) at 10.2eV for 2h at beamline DESIRS of the SOLEIL (Source Optimisée de Lumière d'Énergie Intermédiaire du Lure in French) synchrotron, the French national synchrotron facility. They initially intended to induce asymmetric photochemical reactions. No stereochemical effects induced by CPSR could be identified. Photon–molecule interactions occurred predominantly.

The sample was washed with water, derivatized, and analyzed by two-dimensional gas chromatography with time-of-flight mass spectrometer after Meinert and Meierhenrich (2012). Briefly, two gas-chromatographic columns are connected for a two-dimensional measurement with a reflectron TOFMS for recording retention times and mass spectra (GC^2-TOFMS).

The ice samples contained aldoses and ketoses shown in Fig. 9.2A and B (see also Fig. 7.8A and B), respectively. Erythrose(*) and threose(**) were below the quantification limit and detection limit, respectively. When the background noise is shown as σ, signal intensities of the quantification and detection limits were expressed as less than 10 σ and less than 3σ, respectively. In other words, almost all aldoses and ketoses with a carbon number from 2 to 5 were synthesized more than the quantification levels. They detected the monosaccharides ribose, arabinose, xylose, and lyxose, which belong to aldopentoses, and ribuloses and xylulose, which belong to ketopentoses.

The diversity of sugar molecules can be explained by the framework of a photochemically initiated formose-type reaction. Reactant and intermediate species of the formose reaction are formaldehyde and glycolaldehyde, which undergo aldol condensations to make hydroxyl aldehydes and hydroxyl ketones with linear and branched structures (Fig. 9.3). These photo-synthesized products in Fig. 9.3 add up to more than 3.5% by mass, therefore, it can be said that these sugars, sugar alcohols, and sugar acids are synthesized in major constituent levels. Ribose is also synthesized to major constituent levels; thus, this result is a direct link between cosmochemistry and astrobiology.

It should also be noticed that synthesized organic materials are CHOs, although the nitrogen source is added as ammonia. The reactions shown in Section 9.6.3 should have occurred. To make CHON materials, another starting material such as formamide would be required (Section 9.7), or higher energy would be needed.

(A) Aldoses

		ppm (w/w)		ppm (w/w)
C-2	Ethylene glycol	550	C-5	
	Glycolaldehyde	2390	Ribitol	560
	Glycolic acid	6330	Ribose	260
C-3	Glycerol	2860	Ribonic acid	82
	Glyceraldehyde	302	Arabitol	1150
	Glyceric acid	2440	Arabinose	200
C-4	Erythritol	5070	Arabinoic acid	165
	Erythrose	*	Xylitol	630
	Erythronic acid	960	Xylose	240
	Threitol	7200	Xylonic acid	67
	Threose	**	Arabitol	1150
	Threonic acid	840	Lyxose	145
			Lyxonic acid	140

(B) Ketoses

		ppm (w/w)
C-3	Dihydroxyacetone	540
C-4	Erythrulose	37
C-5	Ribulose	2010
C-5	Xylulose	470

FIGURE 9.2
Structures, names and abundances of linear and branched (A) aldoses and (B) ketoses detected in the experiment products. The concentration is ppm (weight/weight). Erythrose(*) and threose(**) were below quantification limit and below detection limit, respectively. Identified C-6 analytes are not shown.

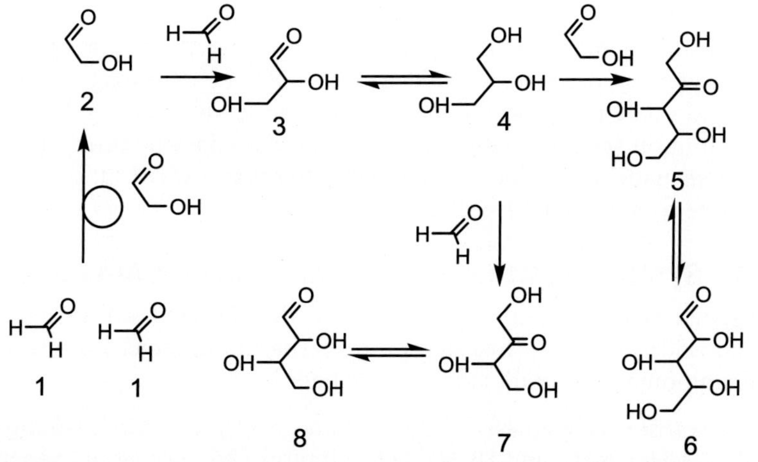

FIGURE 9.3
The formose reaction. Formaldehyde **1** reacts under autocatalytic reaction kinetics to form glycolaldehyde **2**, which undergoes an aldol reaction by forming glyceraldehyde **3**. Dihydroxyacetone **4** is formed by aldoseketose isomerization of **3** and reacts with **2**, making pentulose **5**, which isomerizes to an aldopentose **6** such as ribose. In another pathway, dihydroacetone **7** and formaldehyde **1** reacts forming ketotetrose **7** and aldotetrose **8** (Breslow, 1959).

9.6 THE EXPERIMENT WHERE THE ICY COMET HITS ANOTHER ICY COMET IN SPACE

In this experiment of Martins et al. (2013), the case inwhich the icy comet hits another icy comet in space is assumed. Using a gas gun, life-related molecules of CHONs are synthesized in this condition.

9.6.1 Details of the Gas-Gun Experiment

Both carbon dioxide and ammonia are reasonably available as starting materials in the impact environments for prebiotic organic synthesis. Methanol (CH_3OH) is one of the simplest compounds of carbon, and is also available in space. Thus, ice was made from a mixture of ammonia solution, CO_2, and methanol composed of 9.1:8:1. Two experiments were performed. This ice was impacted twice at 7.15 and 7.00 km s^{-1} by a steel projectile. The two pieces (one piece is ~100 g) of the ice were used in one collision experiment (one is a control), but only less than 1 mg was at the peak pressure of more than 50 GPa. The ice piece was recovered, dried, and analyzed by gas chromatography-mass spectrometry (GC–MS). The detection limits for amino acids were approximately 10 pg.

9.6.2 Run Products by Impact Shock Experiments

Impact shock by ice–ice collision produced several amino acids, including linear and methyl α-amino acids. The linear α-amino acids detected ranged from C_2 to C_5 and included glycine, D- and L-alanine, α-aminobutyric acid (α-ABA), and D- and L-norvaline (see Fig. 9.4). Methyl α-amino acids include the nonprotein amino acids isovaline and α-aminoisobutyric acid (α-AIB) (see Fig. 9.4).

A racemic mixture of alanine was detected, with a D:L ratio of 0.99 ± 0.05 (target ice sample no. 1) and 0.99 ± 0.02 (target ice sample no. 2). A racemic mixture of norvaline was also detected, with a D:L ratio of 0.97 ± 0.04 (target ice sample no. 1) and 0.97 ± 0.02 (target ice sample no. 2). This clearly indicates that there was no contribution (contamination) of terrestrial materials, because all terrestrial samples are made of L-amino acids, resulting in a very low D:L ratio if the experiments were contaminated (see Section 7.2).

9.6.3 Synthetic Pathway for Production of α-Amino Acids

A suggested synthetic pathway to produce the detected linear and methyl α-amino acids includes a two-step process. The initial target ice mixture contained ammonia, carbon dioxide, and methanol.

The first reaction is synthesis of the α-amino acid precursors (carbonyl compounds R–CHO, NH_3, and HCN). Oxidation of methanol would generate the α-amino acid precursor carbonyl compounds (such as formaldehyde), which could then synthesize formic acid.

$$CH_3OH \text{ (methanol)} \rightarrow HCHO \text{ (formaldehyde)} \rightarrow HCOOH \text{ (formic acid)}$$

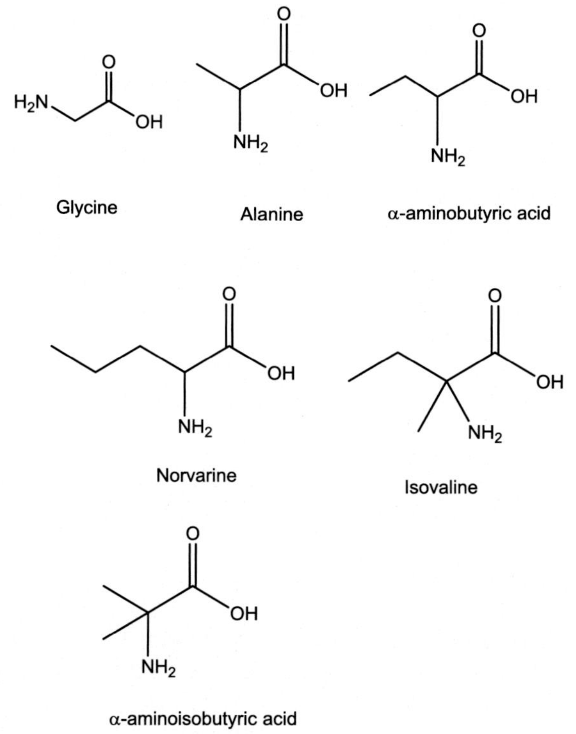

FIGURE 9.4
Compounds formed by the ice–ice collision experiment.

The impact shock waves generate heat, producing additional organic compounds. When heated to a high temperature, and in the presence of ammonia, formic acid forms formamide which then decomposes to HCN plus water.

$$HCOOH\ (\text{formic acid}) + NH_3 \rightarrow H{-}CN\ (\text{hydrogen cyanide}) + 2H_2O$$

The second reaction is Strecker-cyanohydrin synthesis, in which:

$$R{-}CHO + H{-}CN + NH_4OH \rightarrow R{-}C(NH_2){-}CN \rightarrow R{-}C(NH_2){-}COOH$$
(amino acid)

Further support for these reactions is indicated by an ice mixture containing only ammonia and carbon dioxide without methanol. Even with an impact velocity of $7.12\,km\,s^{-1}$, no detectable quantities of amino acids were produced, indicating the importance of the initial methanol.

9.6.4 Experimental Timescale

It is important to note that the impact shock experimentally occurs on timescales of nanoseconds to milliseconds (depending on the size of the impacter), which is much a shorter timescale than that required for the two-step Strecker-cyanohydrin process. A high shock pressure results in the formation of ions and radicals, which will then be involved in the post-shock reaction to form amino acids or CHONs.

9.6.5 Relevance of Experimental Results to the Synthesis of CHONs for the Origin of Life

These results present a significant step forward in our understanding of the origin of the building blocks of life. It is certain that ice–ice collisions occur in space; it is also known that comets contain significant quantities of the compounds used in this study, and that these compounds are found on the impacted surfaces of the icy bodies in the outer solar system. Therefore, it is highly probable that there are conditions on the surfaces of Saturnian bodies in which ammonium compounds, simple alcohols, CO_2, and water ice coexist in an intimately mixed solid form. An icy body with high velocity (5–20 $km\,s^{-1}$) would impart enough energy to promote shock synthesis of more complex CHONs from these ices.

These impact-shock experiments support a revival of the hypothesis of the role of comets in exogenous delivery to the early Earth. These results are also applicable to the chemical evolution of CHONs in primitive bodies, such as the parent body of carbonaceous meteorites. As high concentrations of amino acids (Kvenvolden et al., 1970) and NH_3 (Pizzarello et al., 2011) have been detected in CR2 chondrites, these carbonaceous meteorites and comets may share a common condition for the formation of CHONs. Scaling between laboratory scales (millimeters) and planetary scales (kilometers) should be considered, as planetary-scale impacts would lead to a significant, long-term, melting and evaporative loss of volatiles that may affect the final abundance of CHONs, and the results should be treated with care. However, as there is no atmosphere to stop impacts from millimeter-sized projectiles, the experimental study may be directly applicable and provide the formation mechanism of CHONs.

BOX 9.2 THE HERSCHEL SPACE OBSERVATORY

The Herschel Space Observatory (see Fig. Box 9.1) was built by the European Space Agency and was operated from 2009 to 2013 to observe far infrared and submillimeter wavelengths (55–672 μm). This space observatory was placed on the second Lagrangian point (L2; see Box 2.12) of the Sun and Earth.

Existence of molecular oxygen was first confirmed by the Herschel Space Observatory. It was also suggested that the water on Earth could have initially come from cometary impacts. The water vapor on the dwarf planet Ceres was detected and published in 2014 (Küppers et al., 2014). The observatory was in operation until 2013.

BOX 9.2 THE HERSCHEL SPACE OBSERVATORY—continued

FIGURE BOX 9.1
An artist's rendition of the Herschel Space Observatory.
Image credit: ESA. http://www.esa.int/spaceinimages/Images/2007/09/Herschel.

9.6.6 Definitive Evidence Against Comets as the Source of Earth's Water

The Herschel Space Observatory (see Box 9.2) has shown that 67PCG has a similar deuterium to hydrogen ratio (a D:H ratio) as Earth's water using spectroscopic observation (Hartogh et al., 2011).

For the direct determination of D:H ratio of the comet, a spacecraft Rosetta (see Box 9.3) was launched targeting a coma (nucleus) of comet 67P/Churyumov_ Gerasimenko in 2004. Rosetta carried a mass spectrometer for the determination of the D:H ratio and an enantiomer-sensitive GC–MS designed to measure concentrations of chiral organic molecules (Thiemann and Meierhenrich, 2001). In addition, if such molecules are found (including CHONs) then impact shock could be an important factor in the origin of CHONs.

BOX 9.3 ROSETTA AND PHILAE

Rosetta (Fig. Box 9.2A) is an ESA spacecraft. It targeted a coma (nucleus) of comet 67P/Churyumov_Gerasimenko (abbreviated as 67PCG). Rosetta was launched in 2004, and the spacecraft reached 67PCG in 2014. It successfully approached the coma of 67PCG, and sent photographs (see Fig. Box 9.3). Philae (see Fig. Box 9.2B) is spacecraft attached to Rosetta whose purpose was to land on the coma of 67PCG. It was detached from Rosetta in 2014 and landed on the surface of 67PCG. After the first touchdown, Philae rebounded on the surface of 67PCG several times, and seemed to settle in the dark cleavage of the coma where there is not enough sunshine to generate sufficient energy to communicate with Rosetta.

(A)

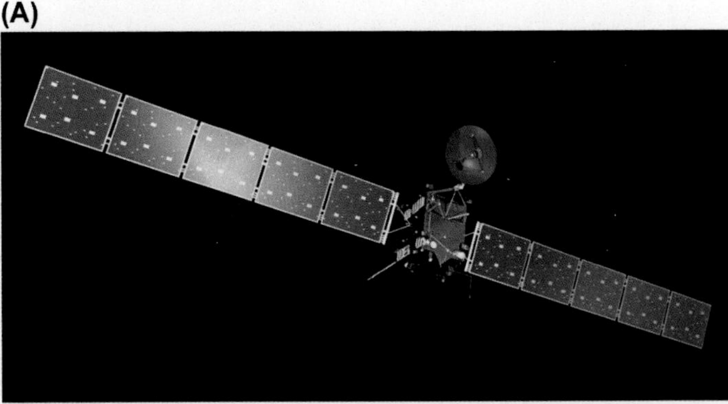

(B)

FIGURE BOX 9.2
Artist's impression of (A) Rosetta and (B) Philae. Rosetta is the spacecraft that targeted a coma of the comet 67P/Churyumov_Gerasimenko (67PCG). Philae is the lander on the coma of 67PCG.
Image credit: (A) ESA/ATG medialab. https://www.flickr.com/photos/europeanspaceagency/11206647984/. (B) ESA. http://sci.esa.int/jump.cfm?oid=53559.

BOX 9.3 ROSETTA AND PHILAE—continued

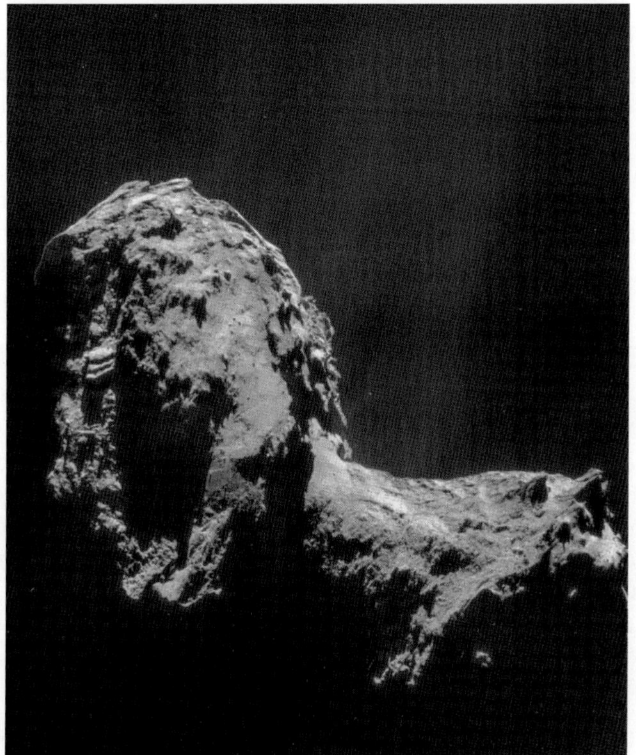

FIGURE BOX 9.3
A photograph of comet 67P/Churyumov_Gerasimenko. It was taken 31 km from the center of the comet. The resolution is 3 m/pixel. The jet activities of water and carbon dioxide gases are seen. As the jet activities are stronger in the center than in other locations, the comet became a dumbbell-like figure. The surface seems to be made of hard organic materials and ice. The inside should be porous and amorphous ice. "Deep fried ice cream" is a good analogy.
Image credit: ESA/Rosetta/NAVCAM. http://www.nasa.gov/sites/default/files/thumbnails/image/pia18899.jpg.

It was very surprising that the mass spectrometer on Rosetta reported 67PCG's very high D:H ratio of approximately 5×10^{-4}, which is very different from those of Earth ($\sim 1.6 \times 10^{-4}$), those of Jupiter-Saturn-Uranus-Neptune ($2–6 \times 10^{-5}$), or other comets, possibly from Oort clouds ($2–4 \times 10^{-4}$). As the D:H ratios of asteroids were $1.3–3 \times 10^{-4}$, it was suggested that the Earth's ocean water came from asteroids, not from comets.

9.7 THE ICY PLANET AND ICY PLANET/ARCHEAN EARTH COLLISIONS

9.7.1 Introduction

Icy planet and icy planet/Archean Earth collisions are assumed to be the origin of the source of the biogenic molecules, or CHONs. To test this assumption and simulate the collision, Ferus et al. (2015) heated a sample with a high-power laser and analyzed run products by GC–MS. In this Section, the experiments of Ferus et al. (2015) are introduced, and analytical results are shown and discussed in detail. It is very interesting that results and implications of the laser heating contradict those of the shock-recovery experiments (Furukawa et al., 2015). It seems that we will approach the origin of life by solving this contradiction.

9.7.2 Birth of the Sun and the Earth, LHB, and CHONs

Ferus et al. (2015) summarized the Earth's history from the birth of the Sun to the LHB. The Sun was accompanied by a disk of material that consisted of gas and small dust particles. Over several tens of Myrs, these dust particles accumulated and formed the planets. This process occurred in several stages including moon-forming impacts on the proto-Earth (see Chapter 2). Then, following the solidification of the Moon approximately 4.5 Ga, the initially heavy impacter flux declined and increased again during the LHB at around 4–3.8 Ga (see Sections 2.9 and 8.3). The origin of the LHB was linked to a dynamic instability of the outer solar system in the Nice model (Section 2.8; Tsiganis et al., 2005; Nesvorny and Morbidelli, 2012).

The theoretical constraints indicated that the impactor flux on Earth became approximately 10 times higher at the LHB than in the period before the LHB, and that this flux slowly decayed afterward (Koeberl, 2006; Morbidelli et al., 2012; Geiss and Rossi, 2013). At its peak, the LHB most likely attained an impact frequency of 10^9 tons of material per year. Typical impact speeds were estimated to have increased from approximately 9 to approximately 21 $km\,s^{-1}$ once the LHB began.

The ratio of the gravitational cross-sections of the Earth and the Moon is approximately 17:1. Thus, for every lunar basin, such as Orientale or Imbrium, basins approximately 17 times larger should have formed on Earth (Bottke et al., 2012).

The atmosphere was partly eroded and transformed (Ferus et al., 2009; De Niem et al., 2012), and the hydrosphere was enriched by water (Morbidelli et al., 2000; Cavosie et al., 2005). These impact-related processes also transformed biomolecules and their precursors on Earth's surface, which would have relevant consequences for the origin of life (Chyba et al., 1990; Chyba and Sagan, 1992).

Although the impact energies were most likely not large enough to produce ocean evaporation or globally sterilizing events (Valley, 2005), they could have served as local energy sources for biomolecule syntheses (Ferus et al., 2012, 2014; Jeilani et al., 2013). The formation of the nucleobases, as well as purine and glycine, were reported during the dielectric breakdown induced by the high-power laser Asterix in the presence of catalytic materials (meteorites, TiO_2, clay)

(Ferus et al., 2012, 2014). Therefore, the high-impact activity may not have been harmful for the formation of biomolecules and the first living structures. In fact, it may have been the source of energy required to initiate chemical reactions, such as the synthesis of biomolecules (Zahnle and Sleep, 2006).

One recent advance in prebiotic chemistry is that formamide could be the parent compound of the components of the first informational polymers (Saladino et al., 2001). Saladino et al. (2012a,b, 2013) extensively studied the formamide-based chemistry that can lead to the synthesis of nucleobases and nucleotides and their metabolic products.

By choosing the appropriate catalyst, purine, adenine, guanine, and cytosine (catalyzed by limestone, Kaolin, silica, alumina, or zeolite), thymine (irradiated by sunlight and catalyzed by TiO_2), and hypoxanthine and uracil (in the presence of montmorillonites) were obtained (Saladino et al., 2012a,b,c, 2013; Senanayake and Idriss, 2006).

Formamide-based synthesis in the high-density energy event (impact plasma) can solve the enigma of the simultaneous formation of four nucleobases. The main objective of Section 9.7 is to demonstrate a mechanism for the formation of the nucleobases through the reaction of formamide and its dissociation products in a high-energy impact event relevant to the LHB.

The plasma formed by the impact of an extraterrestrial body was simulated using the high-power chemical iodine Prague Asterix Laser System. During the dielectric breakdown in gas (laser-induced dielectric breakdown, or LIDB) generated by a laser pulse of energy of 150 J (time interval of ~350 ps, wavelength of 1.315 μm, and output density of 10^{14}–10^{16} W cm^{-2}), the outcomes for a high-energy density event occurred. The shock caused a temperature of 4500K (Babankova et al., 2006), and generated secondary hard radiation (UV and X-ray).

The unstable radicals produced in the formamide dissociation have been identified and quantified using time-resolved discharge emission spectroscopy (Civiš et al., 2012, 2006).

9.7.3 Experiments of Laser-Induced Dielectric Breakdown

Saladino et al. (2006a,b) proposed the prebiotic scenario that formamide could be concentrated in lagoons and was exposed to hard UV radiation at high temperatures and in the presence of various catalysts. After the Nice model, during the LHB, such a lagoon must have been exposed to plasma during an extraterrestrial body impact event, which could initiate a cascade of chemical reactions, eventually leading to the formation of nucleobases.

In the LIDB plasma experiments by Ferus et al. (2015), formamide is first decomposed into radicals. ("•" indicates a radical, which is a free bond. For example, NH• shows a state that one electron of a lone electron pair of N is removed. Therefore, this nitrogen is very reactive.) CN• and NH• are the most abundant species. The absorption gas-phase spectra showed the presence of HCN, CO, NH_3, CO_2, CH_3OH, and N_2O.

Ferus et al. (2014) reported that total yields of CN·, NH·, and stable CO were 55%, 4%, and 41% in the impact plasma. Due to the very rigid structure of CN· from the strong triple bond, this molecule is able to adopt a series of excited electron configurations. These properties make this transient species an ideal reactant in the high-energy plasma environment (Horka et al., 2004; Civis et al., 2008; Ferus et al., 2011a,b). Therefore, CN· were also discovered in interstellar space in the envelopes of giant stars (Shimizu, 1973). Ferus et al. (2012) showed that the stepwise addition of CN· to formamide with atomic H can give rise to the formation of 2,3-diaminomaleonitrile (DAMN) in a highly exergonic reaction. DAMN (see Fig. 9.5) is generally considered to be the common precursor of all nucleobases.

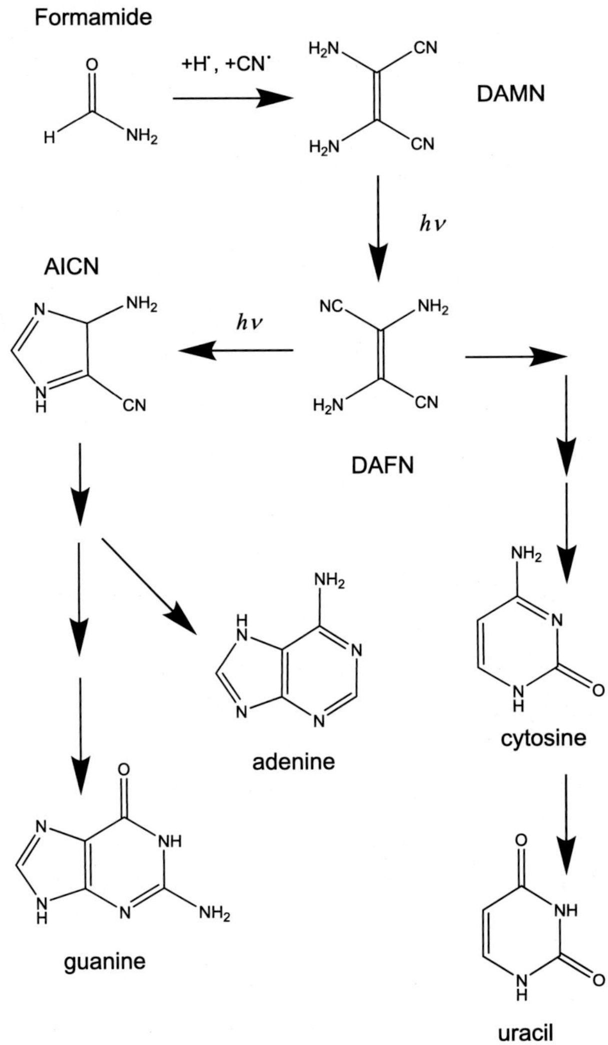

FIGURE 9.5
Flow of high energy synthesis from formamide to four canonical nucleobases after Ferus et al. (2015).

9.7.4 Laser-Induced Dielectric Breakdown Synthesis From Formamide to Four Canonical Nucleobases

All canonical nucleobases except cytosine were detected in the samples of formamide exposed to LIDB. Irradiation of formamide and DAMN using a high-power laser produced adenine, guanine, and uracil, whereas DAMN suspensions produced all of the bases.

When formamide was irradiated in the presence of clay, all four bases were detected. The role of clay is to protect the adsorbed cytosine against deamination to uracil in a further reaction step.

To demonstrate this, formamide was irradiated in the presence of the olivine chondrite meteorite Northwest Africa 6472 (NWA 6472). Olivine is a silicate mineral with very low or negligible absorption capacity. Indeed, the results of the irradiation with and without a chondrite meteorite are roughly similar and clearly show that cytosine was not formed in the presence of olivine, whereas clay supports formation of all studied bases.

Thus the following general model is made for the high-energy synthesis of nucleobases from formamide (Fig. 9.5). The synthesis is initiated by a reaction of formamide with a CN· forming several intermediates. This part of the reaction pathway leads to DAMN.

The photoisomerization of DAMN produces 2,3-diaminofumaronitrile (DAFN), which binds to another CN· to readily cyclize into a trisubstituted pyrimidinyl radical. This moiety serves as the precursor for cytosine and uracil.

Another pathway includes an additional reaction step in which DAFN cyclizes to 4-amino-5-cyanoimidazole (AICN). This synthetic pathway may lead either directly to adenine or the precursor of guanine.

9.7.5 Conclusion of Laser-Induced Dielectric Breakdown Experiments

Ferus et al. (2015) demonstrated that during the era of the LHB, nucleobases may have been synthesized in an impact plasma via reactions of the dissociation products of formamide, such as CN· and NH·, with the formamide parent molecule without a catalyst. Their proposal extends the original idea of Saladino et al. (2006a,b), who suggested formamide to be the precursor of nucleobases in a prebiotic environment. Ferus et al. (2015) suggest that during the LHB, the environment influenced by extremely frequent impact events was potentially favorable for nucleobase synthesis. The first biosignatures of life are dated to roughly coincide with the LHB or near the end of it. In conclusion, it is suggested that the emergence of life is not the result of an accident but a direct consequence of the condition of the primordial Earth.

9.8 THE ICE PLANET THAT FELL ON EARTH'S OCEAN

9.8.1 Introduction

In the previous Sections 9.6 and 9.7, we learned how to build the blocks of life; for example, amino acids and nucleobases or CHONs can be synthesized in ice–ice collisions with high speed. If the CHONs were made at a high rate and fell onto Earth as IDPs, the origin of life can mostly be attributed to the extraterrestrial. Because, as discussed in Section 9.3.3, larger particles can be too hot when entering the Earth, resulting in decomposing, burning, and sterilizing of amino acids, nucleobases, and CHONs (Peterson et al., 1997; Ross, 2006).

However, these impacts cause chemical reactions among meteoritic materials, the ocean, and the atmosphere. Formation of reduced volatiles from inorganic materials has been reported in simulations of post-impact reactions on the early Earth (Fegley et al., 1986; Mukhin et al., 1989; Gerasimov et al., 2002; Schaefer and Fegley, 2010; Kurosawa et al., 2013; Furukawa et al., 2014).

Furthermore, Furukawa's group has investigated such post-impact reactions with experimental simulations and demonstrated the formation of glycine and aliphatic carboxylic acids from inorganic carbon in meteorites (Nakazawa, 2008; Furukawa et al., 2009). Amino acid formation in impacts involving simulated cometary ice composed of ammonia, methanol, and carbon dioxide has also been proposed (Goldman et al., 2010). These studies support the importance of impact-induced reactions as a mechanism for providing the building blocks of life on the early Earth.

Furukawa's group synthesized glycine using solid amorphous carbon as the carbon source (Furukawa et al., 2009). They presumed high partial pressure of CO_2 in the early Hadean atmosphere (Holland, 1984; Walker, 1985; Kasting, 1993; see Section 8.2), and therefore, dissolution of large quantities of CO_3^{2-} in the early oceans should have occurred (Morse and Mackenzie, 1998). Hence, large amounts of carbon would have been available in the post-impact plumes. The huge carbon reservoir on the early Earth might have been used in impact-induced reactions to form various kinds of organic compounds important for life.

Ammonia can be synthesized through the reduction of terrestrial nitrogen species in the ocean, crust, and impact plumes (Summers and Chang, 1993; Brandes et al., 1998; Nakazawa et al., 2005; Smirnov et al., 2008; Schaefer and Fegley, 2010; Furukawa et al., 2014). Thus, both CO_3^{2-} and ammonia were easily available in the impact environments, and could have been carbon and nitrogen sources for prebiotic synthesis on the early Earth.

The purpose of Section 9.8 is to investigate what kind of nucleobases and amino acids are synthesized from CO_3^{2-} and ammonia in simulating the fall of meteorites, following Furukawa et al. (2015).

9.8.2 Shock-Recovery Experiments

Details of the sample container, flyer, and the alignment of samples in the container are described elsewhere (Furukawa et al., 2009). The sample container was made of low carbon stainless steel (SUS 304L). Gaseous nitrogen was introduced into the head space of the sample container. The shock experiments were conducted using a single-stage propellant gun (Sekine, 1997).

For the shock-recovery experiment, forsterite (Mg_2SiO_4), metallic iron, magnetite (Fe_3O_4) and metallic nickel, ^{13}C-labeled ammonium bicarbonate solution (representative of ocean), and gaseous nitrogen (representative of atmosphere) were mixed to be representative of simplified meteorite components (see Table 9.2). The mixtures of IMx, OCx and CCx (x = 1 or 2) represent an iron meteorite, an ordinary chondrite, and a carbonaceous chondrite, respectively.

Table 9.2 Composition of Starting Materials and Impact Velocity (Furukawa et al., 2015)

Type		IM2	IM1	CC1	CC2	OC1	OC2
Starting materials (mg)	Fe	200	300	50	50	100	100
	Fe_3O_4	0	0	100	100	0	0
	Ni	20	30	15	15	30	30
	Mg_2SiO_4	0	0	200	300	200	200
	NH_4HCO_3	170	170	170	40	170	30
	H_2O	130	130	130	150	130	150
	N_2 (gas)	Filled	Filled	Filled	Filled	Filled	Filled
Impact velocity	$km\,s^{-1}$	0.82	0.86	0.89	0.86	0.86	0.87

9.8.3 Run Products of the Impact Experiments

Ultra-high performance liquid chromatography (UHPLC) coupled with tandem mass spectrometry (MS/MS) was used to detect ^{13}C-labeled nucleobases. The ^{13}C-labeling ensured accurate identification of products by removing contamination. The combination of UHPLC and MS/MS facilitated the identification of the specific products. Run products are summarized in Table 9.3.

Various amino acids were detected including glycine (Gly), alanine (Ala), serine (Ser), aspartic acid (Asp), glutamic acid (Glu), valine (Val), leucine (Leu), isoleucine (Ile), proline (Pro), β-alanine (β-Ala), sarcosine (Sar), α-amino-n-butyric acid (α-ABA) and β-aminoisobutyric acid (β-AIBA), all of which were ^{13}C-labeled. The experimental results were the first observation that simultaneous formation of various amino acids and nucleobases were synthesized in the single shock experiment.

Table 9.3 **Run Products of the Impact Experiments (Furukawa et al., 2015)**

Products			IM2	IM1	CC1	CC2	OC1	OC2
Nucleobase	(n mol)	Cytosine	8.8	5.3	tr.	tr.	0.11	0.16
		Urascil	0.096	0.023	BD	BD	BD	BD
Amino acids	(n mol)	Gly	520	2900	350	34	370	55
		Ala	48	210	12	0.21	13	0.59
		Ser	0.3	536	0.2	BD	0.51	BD
		Asp	0.9	1.9	BD	BD	BD	BD
		Glu	BD	0.9	BD	BD	BD	0.3
		Val	BD	0.9	BD	BD	BD	BD
		Ile	BD	tr.	BD	tr.	BD	BD
		Leu	BD	tr.	tr.	BD	BD	BD
		Pro	BD	tr.	BD	BD	BD	BD
		Sar	1.9	140	2	0.13	2.2	0.1
		β-Ala	19	86	13	0.11	11	tr.
		α-ABA	19	86	5.1	BD	4.5	BD
		β-AIBA	2.5	22	0.6	BD	0.6	BD
	Total		610	3500	380	35	400	56
	C conversion rate (%)		0.062	0.35	0.037	0.018	0.039	0.03
Amines	(n mol)	Methlamine	3900	20,000	1700	2500	1000	4200
		Ethylamine	460	1400	100	40	52	81
		Propylamine	34	240	15	2.5	6.2	6.1
		Butylamine	0	37	5.1	BD	1.1	2
	Total		4400	22,000	1800	2500	1100	4300

BD and tr. represent "below detection limit" and "detected in trace amounts," respectively.The C conversion rate is from NH_4HCO_3 into amino acids and nucleobases (atom %).

β-Ala, Sar, α-ABA and β-AIBA are non-proteinogenic amino acids (see Fig. 9.6. "Non-proteinogenic" means that the amino acid is NOT used in the formation of naturally existing proteins); therefore, this further excluded the contamination. Gly was produced in the largest amount. Asp, Glu, and Val were produced only in the IM1 experiment. The ^{13}C-labeled primary amines (methylamine, ethylamine, propylamine, and butylamine) decreased as the length of the alkyl chain increased.

Uracil was formed only in the IM compositions, whereas cytosine was formed in all experiments. When the yields of the amino acids were normalized to the initial amount of $NH_4H^{13}CO_3$, the yields of amino acids decreased in the order of IM1, IM2, OC1, and CC1, depending on the amounts of metallic iron and nickel in the starting materials. The yields also reduced on the initial amounts of NH_4C. The conversion rates were 3.5×10^{-1}–1.8×10^{-2} mol%, which were far higher than those using solid amorphous carbon of 1.0×10^{-6} mol% (Furukawa et al., 2009). This meant that the molecular carbon ($H^{13}CO_3^-$) was more reactive than solid carbon.

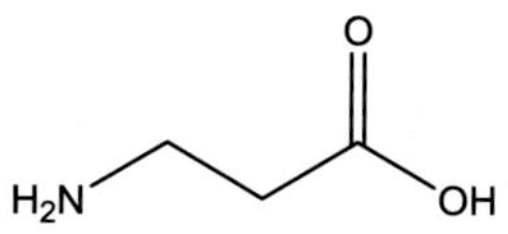

β-alanine (β-Ala)

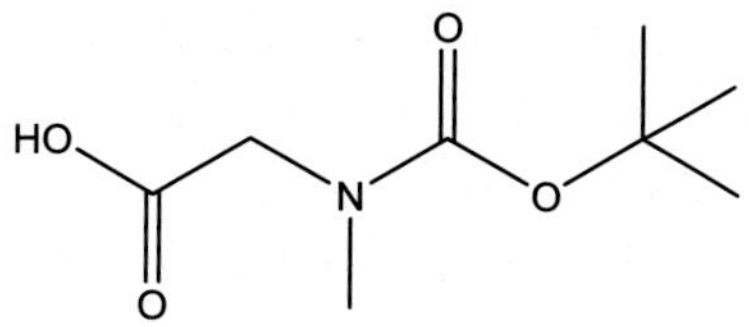

sarcosine (Sar)

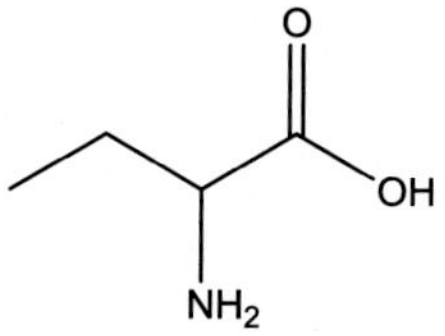

α-amino-n-butyric acid
(α-ABA)

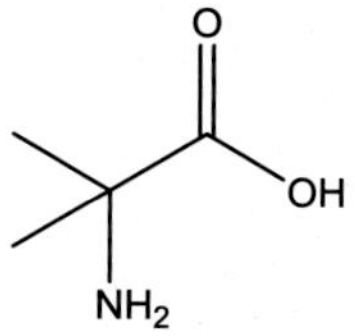

β-aminoisobutyric acid
(β-AIBA)

FIGURE 9.6
Non-proteinogenic amino acids synthesized by the impact experiments.

9.8.4 Implication for the Prebiotic Earth

In this experiment, the impact velocities were lower (~0.9 km s^{-1}) than typical impact velocities of large extraterrestrial materials (~20 km s^{-1}), which are limited simply by resistance of the container against the speed. The container burst at higher velocity impacts. It is estimated that the higher velocity would cause higher temperature and therefore would result in higher production of organic compounds by carbonate reduction reactions. It should be noted that the labile molecules, such as nucleobases and amino acids, were not formed at peak temperatures; instead, they were formed in post-impact conditions with lower temperature-pressure as in this experiment.

As nucleobases and amino acids are generated from a carbon reservoir, such organics can be formed in a CO_2-rich atmosphere. Ammonia could be formed from nitrogen oxides in an N_2 atmosphere by lightning and meteorite impacts, and subsequent reduction by ferrous iron in the ocean or iron sulfides from hydrothermal vents (Summers and Chang, 1993; Brandes et al., 1998; Summers, 1999; Summers et al., 2012). Summers (1999) calculated the amount of ammonia in the prebiotic ocean formed by reduction of nitrogen oxides by ferrous iron to be 7×10^{-5}–8×10^{-7} mol L^{-1}. Brandes et al. (1998) also estimated the reduction rate of nitrogen oxides to ammonia was dependent on the iron sulfide catalysts. Nakazawa et al. (2005) proposed that meteorite impacts reduced atmospheric nitrogen to ammonia, providing NH_3-rich areas in the ocean.

Hartmann et al. (2000) and Valley et al. (2002) estimated impact rates during LHB of 4.0–3.8 Ga to be 10^3–10^5 times higher than those of today. Thus, even if ammonia and bicarbonate concentrations in the Archaean ocean were low, the accumulation of bombardment could make large amounts of nucleobases or amino acids. Anders (1989) estimated that hypervelocity projectiles should have delivered intact organic materials 276 times greater than those meteorites did.

It is interesting that, in contrast to this study, carbonaceous chondrites contain more purine than pyrimidine (Shapiro, 1999; Callahan et al., 2011). In other words, pyrimidine bases were preferentially formed in this study.

9.9 SUMMARY FOR THE FORMATION OF CHOS AND CHONs

Ferus et al. (2015) proposed that formation of the prebiotic nucleobases through pure formamide ($HCONH_2$) during impacts is the key reaction. However, Miyakawa et al. (2002a,b) showed that formamide accumulation in oceans would not be high enough to facilitate the nucleobase synthesis. The presence of formamide-rich lakes offer one solution; however, the geological evidence from that time indicates that the land areas were small (Armstrong and Harmon, 1981; McCulloch and Bennett, 1994).

The continuous formation of CHONs from atmospheric CO_2 and N_2 in the early ocean environment, as well as impact-induced reactions using terrestrial carbon sources seems possible. However, an experiment by Meinert et al. (2016) has again shown the importance of CHO syntheses on ices in space. Thus the most feasible origin of CHOs and CHONs in the early Earth should be the hybrid of space (ice–ice collision in space) and terrestrial (impact of ice on Earth, especially on the ocean) synthetic sources of CHOs and CHONs.

9.10 SELECTION OF CHO AND CHON ENANTIOMERS

The next step is how to explain the selective use of enantiomers of CHOs and CHONs by living things, referred to as "homochirality". Glucose is one type of CHO and the main sugar on Earth with a chemical formula of $C_6H_{12}O_6$. All sugars such as glucose that are utilized by living things are a D-isomer. (The Fisher projection of D-glucose and L-glucose are shown in Fig. Box 7.1C and D, respectively.)

A typical example of a CHON is an amino acid, which is a major constituent of proteins. The most biologically important molecules are made of proteins. The amino acid also has two types of enantiomers, but all living things use L-amino acid, as related in Section 7.2. (Another amino acid is D-amino acid.)

The selection of enantiomers should have started in the early history of living things. In this chapter, this issue has not been discussed (except Section 9.7.3), because it has not been resolved. The LIDB should have affected the L:D ratio; however, as noted, no effects on LIDB were observed.

Three reasons can be easily proposed:

1. Life should have started from selected enantiomers because the abundance of the selected enantiomers was high whether synthesized in space or on Earth.

2. Early life should have used selected enantiomers because of presently unknown advantages;
3. The enantiomer-favored living thing won the early stages of the battle of survival for life by chance, and the present selectivity for enantiomers is not related to either abundance or advantage.

Hence, the next step is to successfully explain the selectivity for enantiomers of CHOs and CHONs, which might give us clues for the constraints on the initial life-related molecules. Although many researchers have noted this, the problem has yet to be solved.

9.11 DISCOVERY OF CHIRAL MOLECULE IN SPACE

McGuire et al. (2016) discovered the interstellar chiral molecule propylene oxide (CH_3CHCH_2O), which is shown in Fig. 9.7. The chiral molecule was detected by radio astronomy aimed at the Sagittarius B2 North [SgrB2(N)] molecular cloud, the preeminent source for new complex-molecular detections in the interstellar medium. The propylene oxide was observed using data from the publicity available Prebiotic Interstellar Molecular Survey project at the Green Bank Telescope, which gives high-resolution, high-sensitivity spectral survey data toward SgrB2(N) from 1 to 50 GHz. The chirality was separated by the radio telescope data of a rotational excitation temperature of approximately 5K at a molecular density of $N_T = 1 \times 10^{13}\,cm^{-2}$ and a velocity of $64\,km\,s^{-1}$. However, by this method, which molecule in Fig. 9.7 is enriched cannot be determined.

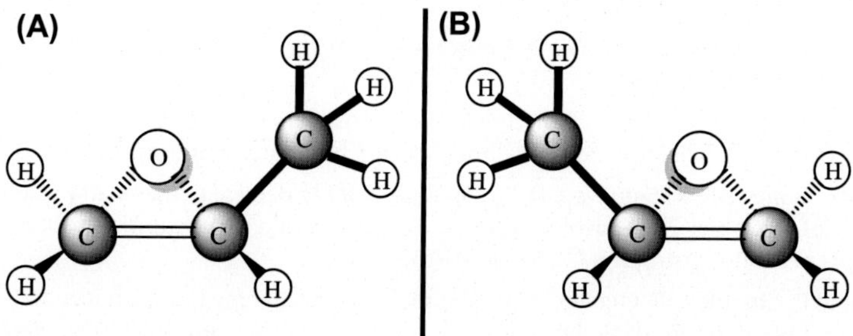

FIGURE 9.7
The molecular structure of (A) S-propylene oxide and (B) R-propylene oxide. *Spheres* mean hydrogen (H), carbon (C), and oxygen (O) atoms.

References

Anders E. Pre-biotic organic-matter from comets and asteroids. Nature 1989;342:255–7.

Armstrong RL, Harmon RS. Radiogenic isotopes: the case for crustal recycling on a near-steady-state no-continental-growth earth and discussion. Philos Trans R Soc A 1981;301:443–72.

Babankova D, Civis S, Juha L, et al. Optical and X-ray emission spectroscopy of high-power laser-induced dielectric breakdown in molecular gases and their mixtures. J Phys Chem A 2006;110:12113–20.

Bar-Nun A, Bar-Nun N, Bauer SH, Sagan C. Shock synthesis of amino acids in simulated primitive environments. Science 1970;168:470–3.

Bar-Nun A, Shaviv A. Dynamics of the chemical evolution of earth's primitive atmosphere. Icarus 1975;24:197–210.

Bockelée-Morvan D, Lis DC, Wink JE, et al. New molecules found in comet C/1995 O1(Hale-Bopp). Investigating the link between cometary and interstellar material. Astron Astrophys 2000;353:1101–14.

Bottke WF, Vokrouhlicky D, Minton D, et al. An Archaean heavy bombardment from a destabilized extension of the asteroid belt. Nature 2012;485:78–81.

Brandes JA, Boctor NZ, Cody GD, et al. Abiotic nitrogen reduction on the early Earth. Nature 1998;395:365–7.

Breslow R. On the mechanism of the formose reaction. Tetrahedron Lett. 1959;1:22–6.

Callahan MP, Smith KE, Cleaves HJ, et al. Carbonaceous meteorites contain a wide range of extraterrestrial nucleobases. Proc Natl Acad Sci USA 2011;108:13995–8.

Cavosie AJ, Valley JW, Wilde SA. Magmatic delta O-18 in 4400-3900 Ma detrital zircons: a record of the alteration and recycling of crust in the Early Archean. Earth Planet Sci Lett 2005 ;235:663–81.

Chyba CF, Thomas PJ, Brookshaw L, Sagan C. Cometary delivery of organic molecules to the early. Earth Planet Sci Lett 1990;249:366–73.

Chyba C, Sagan C. Endogenous production, exogenous delivery and impact-shock synthesis of organic molecules: an inventory for the origins of life. Nature 1992;355:125–32.

Civiš S, Kubat P, Nishida S, Kawaguchi K. Time-resolved Fourier transform infrared emission spectroscopy of H_3^+ molecular ion. Chem Phys Lett 2006;418:448–53.

Civis S, Sedivcova-Uhlikova T, Kubelik P, Kawaguchi K. Time-resolved Fourier transform emission spectroscopy of $A^2\Pi$–$X^2\Sigma^+$ infrared transition of the CN radical. J Mol Spectrosc 2008;250:20–6.

Civiš S, Kubelík P, Ferus M. Time-resolved Fourier transform emission spectroscopy of He/CH4 in a positive column discharge. J Phys Chem A 2012;116:3137–47.

Cronin JR, Moore CB. Amino acid analyses of Murchison, Murray, and Al-lende carbonaceous chondrites. Science 1971;172:1327–9.

Crovisier J, Bockelée-Morvan D. Remote observations of the composition of cometary volatiles. Space Sci Rev 1999;90:19–32.

De Niem D, Kuehrt E, Morbidelli A, Motschmann U. Atmospheric erosion and replenishment induced by impacts upon the Earth and Mars during a heavy bombardment. Icarus 2012;221:495–507.

DiSanti MA, Bonev BP, Villanueva GL, Mumma MJ. Highly depleted ethane and mildly depleted methanol in Comet 21P/Giacobini-Zinner: application of a new empirical 2-band model for CH_3OH near 50 K. Astrophys J 2013;763:1–15.

Ehrenfreund P, Charnley SB. Organic molecules in the interstellar medium, comets, and meteorites: a voyage from dark clouds to the early Earth. Annu Rev Astron Astrophys 2000;38:427–83.

Ehrenfreund P, Irvine W, Becker L, et al. Astrophysical and astrochemical insights into the origin of life. Rep Prog Phys 2002;65:1427–87.

Elsila JE, Glavin DP, Dworkin JP. Cometary glycine detected in samples returned by Stardust. Meteorit Planet Sci 2009;44:1323–30.

Fegley B, Prinn RG, Hartman H, Watkins GH. Chemical effects of large impacts on the Earth's primitive atmosphere. Nature 1986;319:305–7.

Ferris JP, Joshi PC, Edelson EH, Lawless JG. HCN: a plausible source of purines, pyrimidines and amino acids on the primitive earth. J Mol Evol 1978;11:293–311.

Ferus M, Matulkova I, Juha L, Civis S. Investigation of laser-plasma chemistry in CO-N_2-H_2O mixtures using O-18 labeled water. Chem Phys Lett 2009;472:14–8.

Ferus M, Kubelík P, Civiš S. Laser spark formamide decomposition studied by FTIR spectroscopy. J Phys Chem A 2011a;115:12132–41.

Ferus M, Kubelik P, Kawaguchi K, Dryahina K, Spanel P, Civis S. HNC/HCN ratio in acetonitrile, formamide, and BrCN discharge. J Phys Chem A 2011b;115:1885–99.

Ferus M, Civis S, Mladek A, Sponer J, Juha L, Sponer JE. On the road from formamide ices to nucleobases: IR-spectroscopic observation of a direct reaction between cyano radicals and formamide in a high-energy impact event. J Am Chem Soc 2012;134:20788–96.

Ferus M, Michalcikova R, Shestivska V, Sponer J, Sponer JE, Civis S. High-energy chemistry of formamide: a simpler way for nucleobase formation. J Phys Chem A 2014;118:719–36.

Ferus M, Nesvorný D, Sponer J, et al. High-energy chemistry of formamide: a unified mechanism of nucleobase formation. Proc Natl Acad Sci USA 2015;112:657–62.

Festou M, Uwe-Keller H, Weaver HA. Comets-II. Tucson: University Arizona Press; 2005.

Furukawa Y, Sekine T, Oba M, Kakegawa T, Nakazawa H. Biomolecule formation by oceanic impacts on early Earth. Nat Geosci 2009;2:62–6.

Furukawa Y, Samejima T, Nakazawa H, Kakegawa T. Experimental investigation of reduced volatile formation by high-temperature interactions among meteorite constituent materials, water, and nitrogen. Icarus 2014;231:77–82.

Furukawa Y, Nakazawa H, Sekine T, Kobayashi T, Kakegawa T. Nucleobase and amino acid formation through impacts of meteorites on the early ocean. Earth Planet Sci Lett 2015;429:216–22.

Geiss J, Rossi AP. On the chronology of lunar origin and evolution implications for Earth, Mars and the Solar System as a whole. Astron Astrophys Rev 2013;21:1–54.

Gerasimov MV, Dikov YP, Yakovlev OI, Wlotzka F. Experimental investigation of the role of water in impact vaporization chemistry. Deep-Sea Res Part 2 2002;49:995–1009.

Glavin DP, Matrajt G, Bada JL. Re-examination of amino acids in Antarctic micrometeorites. Adv Space Res 2004;33:106–13.

Goldman N, Reed EJ, Fried LE, Kuo IFW, Maiti A. Synthesis of glycine containing complexes in impacts of comets on early Earth. Nat Chem 2010;2:949–54.

Harada K, Suzuki S. Formation of amino acids from ammonium bicarbonate or ammonium formate by contact glow-discharge electrolysis. Naturwissenschaften 1977;64:484.

Hartmann WK, Ryder G, Dones L. Grinspoon: the time-dependent intense bombardment of the primordial Earth/Moon system. In: Canup RM, Righter K, editors. Origin of the Earth and Moon. Tucson: The University of Arizona Press; 2000. p. 493–512.

Hartogh PD, Lis C, Bockelee-Morvan D, et al. Ocean-like water in the Jupiter-family comet 103P/Hartley 2. Nature 2011;478:218–20.

Holland HD. The chemical evolution of the atmosphere and oceans. Princeton: Princeton University Press; 1984.

Horka V, Civis S, Spirko V, Kawaguchi K. The infrared spectrum of CN in its ground electronic state. Collect Czech Chem Commun 2004;69:73–89.

Jeilani YA, Nguyen HT, Newallo D, Dimandja J-MD, Nguyen MT. Free radical routes for prebiotic formation of DNA nucleobases from formamide. Phys Chem Chem Phys 2013;15:21084–93.

Kasting JF. Earth's early atmosphere. Science 1993;259:920–6.

Koeberl C. Impact processes on the early Earth. Elements 2006;2:211–6.

Küppers M, O'Rourke L, Bockelee-Morvan D, et al. Localized sources of water vapour on the dwarf planet (1) Ceres. Nature 2014;505:525–7.

Kurosawa K, Sugita S, Ishibashi K, et al. Hydrogen cyanide production due to mid-size impacts in a redox-neutral N_2-rich atmosphere. Orig Life Evol Biosph 2013;43:221–45.

Kvenvolden K, Lawless J, Pering K, et al. Evidence for extraterrestrial amino-acids and hydrocarbons in the Murchison meteorite. Nature 1970;228:923–6.

Martins Z, Botta O, Fogel ML, et al. Extraterrestrial nucleobases in the Murchison meteorite. Earth Planet Sci Lett 2008;270:130–6.

Martins Z, Price MC, Goldman N, Sephton MA, Burchell MJ. Shock synthesis of amino acids from impacting cometary and icy planet surface analogues. Nat Geosci 2013;6:1045–9.

Maurette M. Micrometeorites and the mysteries of our origins. New York: Springer; 2006.

Meinert C, Meierhenrich UJ. A new dimension in separation science: comprehensive two-dimensional gas chromatography. Angew Chem Int Ed Engl 2012;51:10460–70.

Meinert C, Myrgorodska I, de Marcellus P, et al. Ribose and related sugars from ultraviolet irradiation of interstellar ice analogs. Science 2016;352:208–12.

McCulloch MT, Bennett VC. Progressive growth of the Earth's continental crust and depleted mantle: geochemical constraints. Geochim Cosmochim Acta 1994;58:4717–38.

McGuire BA, Carroll PB, Loomis RA, et al. Discovery of the interstellar chiral molecule propylene oxide (CH_3CHCH_2O). Science 2016;352:1449–52.

Miller SL. A production of amino acids under possible primitive earth conditions. Science 1953;117:528–9.

Miyakawa S, Cleaves HJ, Miller SL. The cold origin of life: A. Implications based on the hydrolytic stabilities of hydrogen cyanide and formamide. Orig Life Evol Biosph 2002a;32:195–208.

Miyakawa S, Cleaves HJ, Miller S. The cold origin of life: B. Implications based on pyrimidines and purines produced from frozen ammonium cyanide solutions. Orig Life Evol Biosph 2002b;32:209–18.

Morbidelli A, Chambers J, Lunine JI, et al. Source regions and timescales for the delivery of water to the Earth. Meteorit Planet Sci 2000;35:1309–20.

Morbidelli A, Marchi S, Bottke WF, Kring DA. A sawtooth-like timeline for the first billion years of lunar bombardment. Earth Planet Sci Lett 2012;355:144–51.

Morse JW, Mackenzie FT. Hadean ocean carbonate geochemistry. Aquat Geochem 1998;4:301–19.

Mukhin LM, Gerasimov MV, Safonova EN. Origin of precursors of organic molecules during evaporation of meteorites and mafic terrestrial rocks. Nature 1989;340:46–8.

Mumma MJ, Disanti MA, Russo ND, et al. Remote infrared observations of parent volatiles in comets: a window on the early solar system. Adv Space Res 2003;31:2563–75.

Nakazawa H, Sekine T, Kakegawa T, Nakazawa S. High yield shock synthesis of ammonia from iron, water and nitrogen available on the early Earth. Earth Planet Sci Lett 2005;235:356–60.

Nakazawa H. Origin and evolution of life: endless ordering of the Earth's light elements. In: Okada H, Mawatari SF, Suzuki N, Gautum P, editors. International symposium on origin and evolution of natural diversity. Sapporo: Hokkaido University; 2008. p. 13–9.

Nesvorny D, Morbidelli A. Statistical study of the early solar system's instability with four, five, and six giant planets. Astron J 2012;144:20–68.

Nuevo M, Meierhenrich UJ, d'Hendecourt L, et al. Enantiomeric separation of complex organic molecules produced from irradiation of interstellar/circumstellar ice analog. Adv Space Res 2007;39:400–4.

Oparin AI. The origin of life. Macmillan; 1938.

Oró J. Mechanism of synthesis of adenine from hydrogen cyanide under possible primitive earth conditions. Nature 1961;191:1193–4.

Oró J, Kimball AP. Synthesis of purines under possible primitive earth conditions. I. Adenine from hydrogen cyanide. Arch Biochem Biophys 1961;94:217–27.

Pasek M, Lauretta D. Extraterrestrial flux of potentially prebiotic C, N, and P to the early Earth. Orig Life Evol Biosph 2008;38:5–21.

Peterson E, Horz F, Chang S. Modification of amino acids at shock pressures of 3.5 to 32 GPa. Geochim Cosmochim Acta 1997;61:3937–50.

Pizzarello S, Williams LB, Lehman J, Holland GP, Yarger JL. Abundant ammonia in primitive asteroids and the case for a possible exobiology. Proc Natl Acad Sci USA 2011;108:4303–6.

Ross DS. Cometary impact and amino acid survival: chemical kinetics and thermochemistry. J Phys Chem A 2006;110:6633–7.

Saladino R, Crestini C, Costanzo G, Negri R, Di Mauro E. A possible prebiotic synthesis of purine, adenine, cytosine, and 4(3H)-pyrimidinone from formamide: implications for the origin of life. Bioorg Med Chem 2001;9:1249–53.

Saladino R, Crestini C, Ciciriello F, Di Mauro E, Costanzo G. Origin of informational polymers: differential stability of phosphoester bonds in ribomonomers and ribooligomers. J Biol Chem 2006a;281:5790–6.

Saladino R, Crestini C, Ciciriello F, Costanzo G, Di Mauro E. About a formamide based origin of informational polymers: syntheses of nucleobases and favourable thermodynamic niches for early polymers. Orig Life Evol Biosph 2006b;36:523–31.

Saladino R, Botta G, Pino S, Costanzo G, Di Mauro E. From the one-carbon amide formamide to RNA all the steps are prebiotically possible. Biochimie 2012a;94:1451–6.

Saladino R, Botta G, Pino S, Costanzo G, Di Mauro E. Genetics first or metabolism first? The formamide clue. Chem Soc Rev 2012b;41:5526–65.

Saladino R, Crestini C, Pino S, Costanzo G, Di Mauro E. Formamide and the origin of life. Phys Life Rev 2012c;9:84–104.

Saladino R, Botta G, Delfino M, Di Mauro E. Meteorites as catalysts for prebiotic chemistry. Chem Eur J 2013;19:16916–22.

Schaefer L, Fegley B. Chemistry of atmospheres formed during accretion of the Earth and other terrestrial planets. Icarus 2010;208:438–48.

Sekine T. Shock wave chemical synthesis. Eur J Solid State Inorg Chem 1997;34:823–33.

Senanayake SD, Idriss H. Photocatalysis and the origin of life: synthesis of nucleoside bases from formamide on TiO_2(001) single surfaces. Proc Natl Acad Sci USA 2006;103:1194–8.

Shapiro R. Prebiotic cytosine synthesis: acritical analysis and implications for the origin of life. Proc Natl Acad Sci USA 1999;96:4396–401.

Shimizu M. Interstellar dust and related topics. In: Greenberg JM, van de Hulst HC, editors. International astronomical union symposium. Dordrecht: International Astronomical Union; 1973. p. 405–12.

Smirnov A, Hausner D, Laffers R, Strongin DR, Schoonen MAA. Abiotic ammonium formation in the presence of Ni-Fe metals and alloys and its implications for the Hadean nitrogen cycle. Geochem Trans 2008;9:5.

Stoks PG, Schwartz AW. Uracil in carbonaceous meteorites. Nature 1979;282:709–10.

Stoks PG, Schwartz AW. Nitrogen-heterocyclic compounds in meteorites: significance and mechanisms of formation. Geochim Cosmochim Acta 1981;45:563–9.

Summers DP, Chang S. Prebiotic ammonia from reduction of nitrite by iron (II) on the early Earth. Nature 1993;365:630–2.

Summers D. Sources and sinks for ammonia and nitrite on the early Earth and the reaction of nitrite with ammonia. Orig Life Evol Biosph 1999;29:33–46.

Summers DP, Basa RCB, Khare B, Rodoni D. Abiotic nitrogen fixation on terrestrial planets: reduction of no to ammonia by FeS. Astrobiology 2012;12:107–14.

Thiemann WHP, Meierhenrich U. ESA mission ROSETTA will probe for chirality of cometary amino acids. Orig Life Evol Biol 2001;31:199–200.

Tsiganis K, Gomes R, Morbidelli A, Levison HF. Origin of the orbital architecture of the giant planets of the Solar System. Nature 2005;435:459–61.

Valley JW, Peck WH, King EM, Wilde SA. A cool early Earth. Geology 2002;30:351–4.

Valley JW. A cool early Earth? Sci Am 2005;293:58–65.

Walker JCG. Carbon-dioxide on the early earth. Orig Life Evol Biosph 1985;16:117–27.

Zahnle KJ, Sleep NH. Impacts and the early evolution of life. In: Thomas PJ, Hicks RD, Chyba CF, McKay CP, editors. Comets and the origin and evolution of life. 2nd ed. Berlin: Springer; 2006. p. 207–52.

CHAPTER 10

From Life-Related Molecules to Life

Chapter Outline

10.1 INTRODUCTION

After the late veneer, the Earth cooled and life-related molecules were formed by the means described in Chapter 9. Then life was born. Of course, this is too hasty! A few sentences are not nearly enough to explain how evolution occurred over a few billion years.

This chapter should be most interesting and even exciting! Following this chapter, we could create life! This may require discussing ethical issues. Unfortunately, (or fortunately), the question of how life-related molecules evolved into life has not been resolved; therefore, there is no need to worry about ethics for the moment.

First we learn about the modern cell and current life, and compare them with initial life (protocell). Then some plausible models for the origin of the protocell are explained. Some models concentrate on just a few functions (duplication of cell membranes), whereas others concentrate on all of these requirements (reproduction models of the protocell).

Origins of the Earth, Moon, and Life. http://dx.doi.org/10.1016/B978-0-12-812058-3.00010-7

BOX 10.1 SIX KINGDOMS OF EUKARYOTA

Eukaryota is composed of six kingdoms: Excavata, Amoebozoa, Opisthokonta, Rhizaria, Chromalveolata, and Archaeplastida.

Excavata consist of various flagellate protozoa.

Amoebozoa includes most lobose amoeboids and slime molds.

Opisthokonta includes animals, fungi, choanoflagellates, etc.

Rhizaria consists of foraminifera, radiolarian, and various other amoeboid protozoa.

Chromalveolata include stramenopiles (brown algae, diatoms, etc.), haptophyte, cryptophyta, and alveolata.

Archaeplastida includes land plants, green algae, red algae, and glaucophytes.

10.2 CHARACTERIZATION OF PRESENT LIFE

In biological classification, life is divided into eight hierarchies (domain, kingdom, phylum, class, order, family, genus, and species).

All life on Earth is divided into three domains: Bacteria, Archaea and Eukaryota.

Archaea have neither cell nucleus nor any other membrane-bound organelles in their cells.

Bacteria are a few micrometers in length, and have various shapes, ranging from spheres to rods and spirals.

Eukaryota cells contain a nucleus and other organelles enclosed within membranes. For example, all plants and animals are eukaryotes. Eukaryota is composed of six kingdoms (see Box 10.1).

Archaea and Bacteria have a similar cell structure, but the cell composition of Archaea is essentially different from that of Bacteria and Eucaryota. Cell membranes of Bacteria and Eucaryota are phospholipids, which consist of a phosphate head and glycerol-ester lipids (see Fig. 10.1). In contrast, Archaea cell membranes are made of phosphate head and glycerol-ether lipids (see Fig. 10.2). This apparently relates to their survival in a hostile environment, because the ether bond is stronger at high temperature and low pH than the ester bond.

10.3 THE SCHEMATIC STRUCTURE OF THE EUKARYOTE CELL

The schematic structure of the eukaryote cell is shown in Fig. 10.3. The cell is contained in the cell membrane, which is made of a double layer (see Fig. 10.4B) of phospholipids (schematically depicted in Fig. 10.4A), also called a phospholipid bilayer. This membrane is embedded with many kinds of proteins to protect the cell and cytoplasm in hostile conditions such as high osmotic pressure (see Fig. 10.4B).

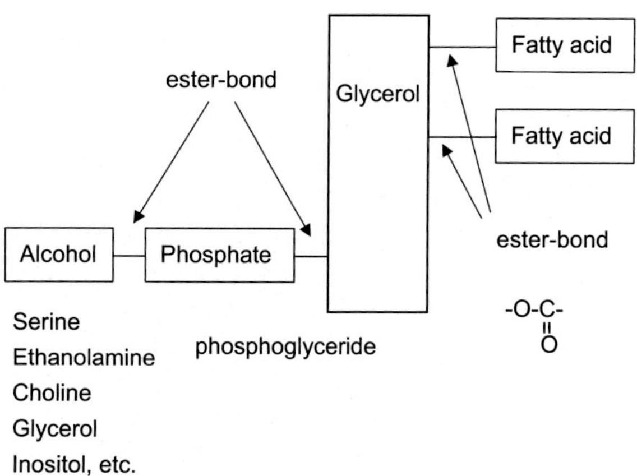

FIGURE 10.1
Schematic structure of phospho-ester-lipids in a membrane of Bacteria and Eukaryota.

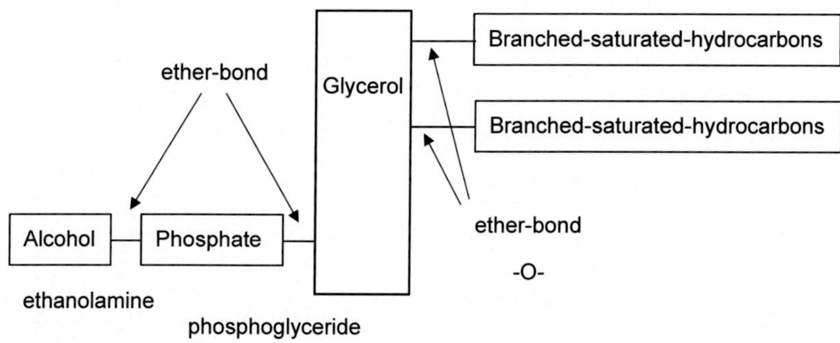

FIGURE 10.2
Schematic structure of phospho-ether-lipids in a membrane of Archaea.

The DNA is contained in chromosomes in the nucleus, which is clearly separated by the phospholipid bilayer inside the cell (see Fig. 10.3). In contrast, Archaea and Bacteria cells, called the prokaryote cells, have no nucleus. Mitochondria, which generate energy efficiently, are also separated by the phospholipid bilayer. Mitochondria also do not exist in prokaryote cells. Endoplasmic reticulum (ER), which is a transport network for molecules, and ribosomes, which consist of RNA and protein molecules, also exist. The origin and evolution of these microorganisms in the eukaryote cell are discussed in Section 10.10.

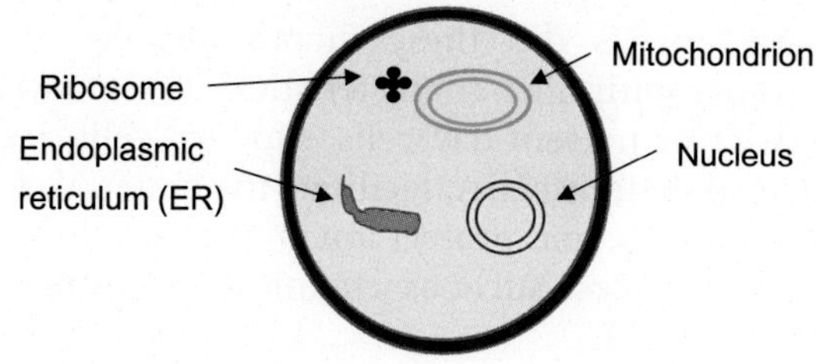

FIGURE 10.3
The schematic structure of the eukaryote cell.

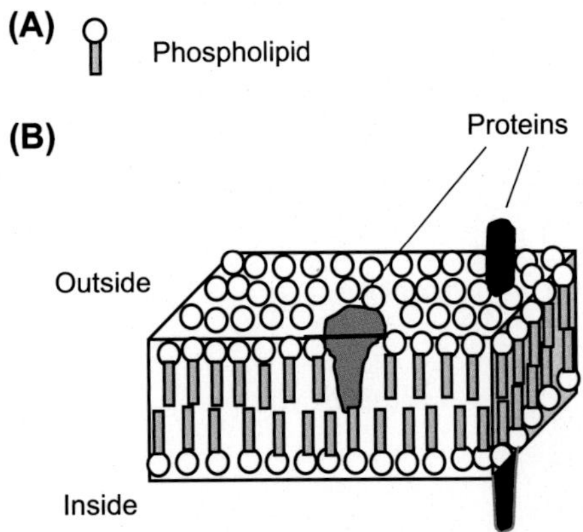

FIGURE 10.4
Phospholipid molecules (A) aggregate and form the bilayer cell membrane. The phospholipid (A) is composed of a hydrophilic glycerol-phosphate head (depicted as a white sphere) with long hydrophobic fatty acid tails (depicted as a gray bar connected to the sphere). (B) Modern cell membranes. The membrane is composed of bilayer of phospholipid. Functional proteins are embedded in the membrane. Some proteins penetrate the membrane to transport ions from inside to outside and vice versa. Early life should have a similar membrane but simpler functions.

10.4 THE PROTOCELL

The first life may have been a single cell that included the following three functional units:

1. Compartments that separate the inside of the cell from the outside physically and chemically;
2. Cytoplasm, which is the cell content. In the cytoplasm, the chemical reactions called "metabolism" work for production of energy and materials to maintain life;
3. Genetic information, which is stored safely and passed on to the next generation by RNA, DNA, etc.
   In addition to these three basic functions, the first life needs to perform:
4. Reproduction or duplication of life.

The first cell is called the "protocell." Many scientists propose that the protocell would have had compartments (membranes), including a self-reproducing compartment (Varela et al., 1974; Morowitz et al., 1988; Lopez-Garcia et al., 2006).

Then, what characteristics did these membranes have? Did they have amphiphilic or nonamphiphilic characteristics? The answer may be found in the characteristics of present-day cells. Modern cells are surrounded by membranes that keep their integrity, facilitate the material-exchange between the external environments, and protect ion gradients, which is advantageous when using energy sources, such as chemical reactions or visible lights (Lopez-Garcia et al., 2006).

The protocell membranes are thought to be similar to the modern cell. The cell membranes are essentially made of phospholipids, amphiphilic molecules composed of a hydrophilic glycerol-phosphate head (depicted as a sphere in Fig. 10.4A) bound to long hydrophobic fatty acid tails (depicted as a bar combined to the sphere in Fig. 10.4A). They form bilayers and carry proteins for energy-producing processes (Fig. 10.4B). The cell membranes define the self-boundary, and retain the cell's necessary contents. The early cell membrane had had few functions; as time passed, the number of functions increased.

When the cell was duplicated, the cell membrane was also duplicated (Fig. 10.5). The protocell is a cell-like compartment enclosed in a lipid vesicle containing nucleic acid and other biomolecules. The important problem is how they acquired the ability to self-replicate.

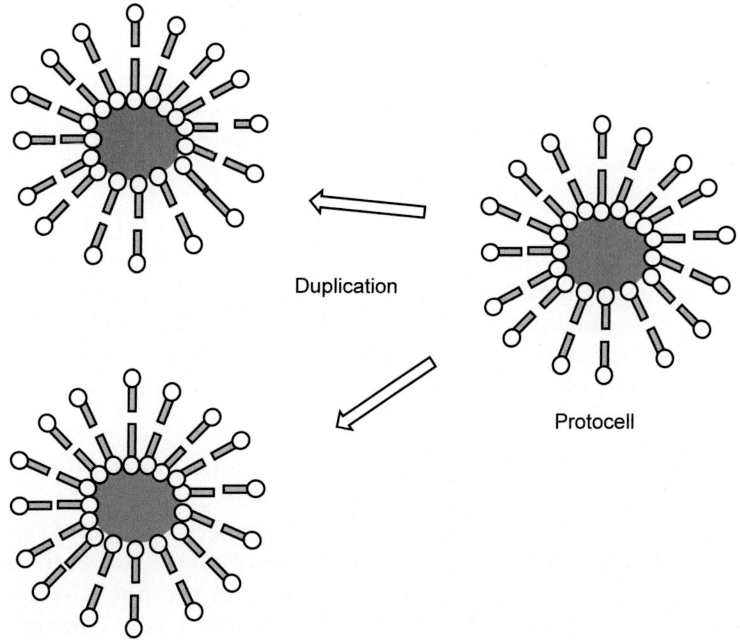

FIGURE 10.5
Duplication of the protocell. Inside and outside could be made of simple amphiphiles or phospholipids and phospholipids, respectively.

10.5 A REPRODUCTION MODEL OF THE PROTOCELL

Damer and Deamer (2015) proposed a "mille-feuille" reproduction model of the protocell made of lipid vesicles in a hydrothermal pool (see Fig. 10.6). The hydrothermal pool provides energy (heat), and bio-related organic molecules such as proteins, lipids, nucleic acids, and cations and anions. At the bottom of the pool, the lipids form sheets, perhaps double-layered sheets as shown in Fig. 10.4.

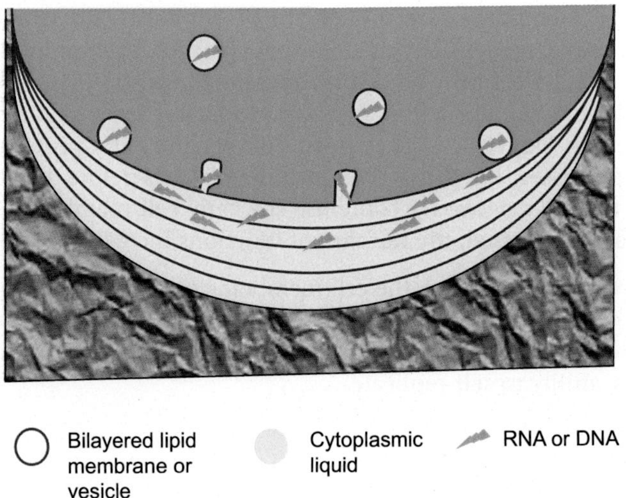

FIGURE 10.6
A reproduction model of protocell made of lipid vesicles at a hydrothermal pool.
Modified from Damer B, Deamer D. Coupled phases and combinatorial selection in fluctuating hydrothermal pools: a scenario to guide experimental approaches to the origin of cellular life. Life 2015;5:872–87.

The sheets overlay as "mille-feuille" in the hydrothermal pool. Between each sheet is cytoplasmic liquid in which biological metabolism can be performed. This liquid also contains RNA or DNA, which can regenerate themselves. From the uppermost sheet, vesicles containing cytoplasmic liquid with RNA/DNA are formed. The vesicle is the protocell, each of which contains one set of RNA/DNA.

The hydrothermal pool is very advantageous for the formation of the lipid bilayer membrane. When phospholipids are dispersed in water, they can aggregate together as micelle (Fig. 10.7A), which is a spheric aggregate (Fig. 10.7A), but does not form a double layered sheet (membrane). However, when the solution containing phospholipids is dried up in the pool, it can form a bilayer sheet (the left side of Fig. 10.7B) or a monolayer sheet (the right side of Fig. 10.7B). If the area is hydrothermal, the heat for evaporation is given geothermally and evaporation often occurs to form the bilayer membrane.

When the UV light is too strong, it is harmful for metabolic chemical reactions and storage of biological information by RNA/DNA, life in deep water is preferred. Otherwise, the metabolism requires sun shine light, and the formation of the protocell may not occur in the deep sea. When biological activities require visible light, shallow life by vesicles or sheets is preferred.

The advantage of this model is that replication of the bilayer can occur easily from the sheet, as shown in Fig. 10.7B. This model can simply overcome three criteria of the protocell shown in Section 10.4. Furthermore, the evolution of life can occur very rapidly in the sheet or the vesicles of the bilayered lipids, because one life can easily amalgamate with other lives through the similar bilayered lipids.

(A)

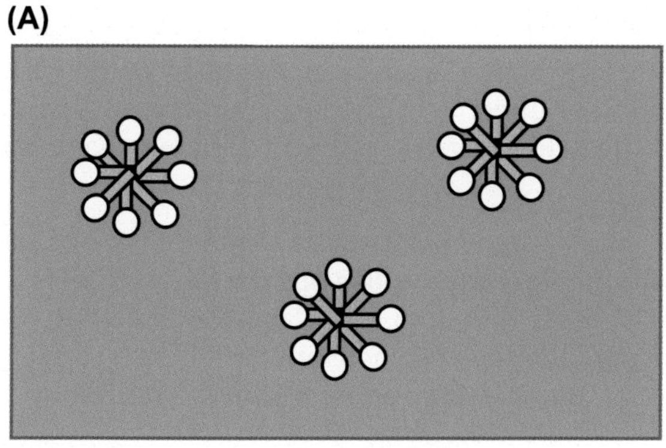

(B)

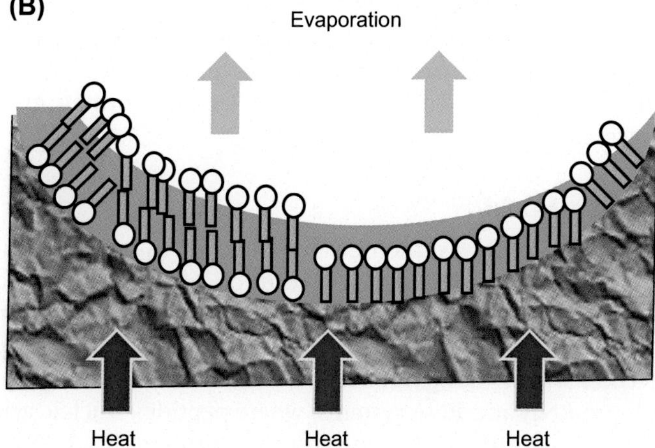

FIGURE 10.7
(A) Micelles of phospholipids in solution. The phospholipids do not form a sheet. (B) When the solution is drying up, the phospholipids can form a bilayer (left) or a monolayer (right).

10.6 HYPOTHETICAL EVOLUTION STEPS OF LIFE

It is still a long way from the protocell to modern life. The genetic information transfer system would have needed to evolve, and the present RNA/DNA system would have had to be formed and developed.

The building blocks of life are only thermodynamically synthesized, and many steps are required and hypothesized. At the beginning, polymers (nucleic acids and peptides; it is the prebiotic or prebiological world) occurred. Then, formation of supramolecular architectures (membrane and protocells) occurred.

Subsequently, three main hypotheses are proposed to replicate or translate complicated information to the next generation. They are:

1. Peptide world
2. Peptide + nucleic acid world
3. RNA world

Section 10.6 explains these three hypothetical evolution steps of life.

10.6.1 Peptide World

Prebiotic or prebiologic amino acids are synthesized by pathways explained in this chapter. By heating or under dehydrating conditions such as those in Figs. 10.6 or 10.7B, especially with a catalyst, amino acids are polymerized and form proteins.

The simplest life is made by only one peptide sequence (or one protein) of 10 amino acids. To make the story simple, this protein (or this life) is formed without duplication of 20 kinds of amino acids, and the possibility of the amino acid sequence (or the protein) is 20^{10}. Therefore, even if such simple life is reproduced, some kind of an information storing system is required.

10.6.2 Peptide + Nucleic Acid World

In order to encode amino acids, peptide nucleic acids (PNAs) were proposed (see Fig. 10.8A) with a peptide-like backbone and nucleic bases on the side chains (Bohler et al., 1995). The PNAs not only act as information carriers, they also work as catalysts in a pre-RNA world.

10.6.3 RNA World

In the RNA world hypothesis, the nucleic acid acted as an information storage and as the catalysts at the initial stage of evolution. However, the sudden appearance of self-replicating RNA is considered to be unlikely. The RNA world should have been preceded by the pre-RNA world, in which another system of replicating molecules was working. Or there is a possibility that life evolved using directly both pre-RNA and RNA systems, where peptide + nucleic acid world are working. The RNA world finally evolved to the protein, RNA and DNA world of today.

Cafferty et al. (2016) proposed the building blocks of melamine and barbituric acid nucleotides (see Fig. 10.9A). The three pairs of each nucleotide form a sheet of Hexad (Fig. 10.9B), which is combined by hydrogen bonds. These sheets are piled in water, and form a stacked tube superstructure (Fig. 10.9C). These are considered to be the model compounds of prebiotic RNA.

10.7 ASSEMBLY OF NUCLEIC ACIDS WITHOUT ENZYME

How nucleic acids first assembled and then started the earliest cellular life about 4 billion years ago is a fundamental question. RNA-like polymers can and must be synthesized nonenzymatically from mononucleotides in conditions simulating a prebiotic hydrothermal site undergoing cyclic hydration (Rajamani et al., 2008). The presence of a phospholipid enhanced the yield of polymeric products because the lipid matrix serves to concentrate and organize the mononucleotides.

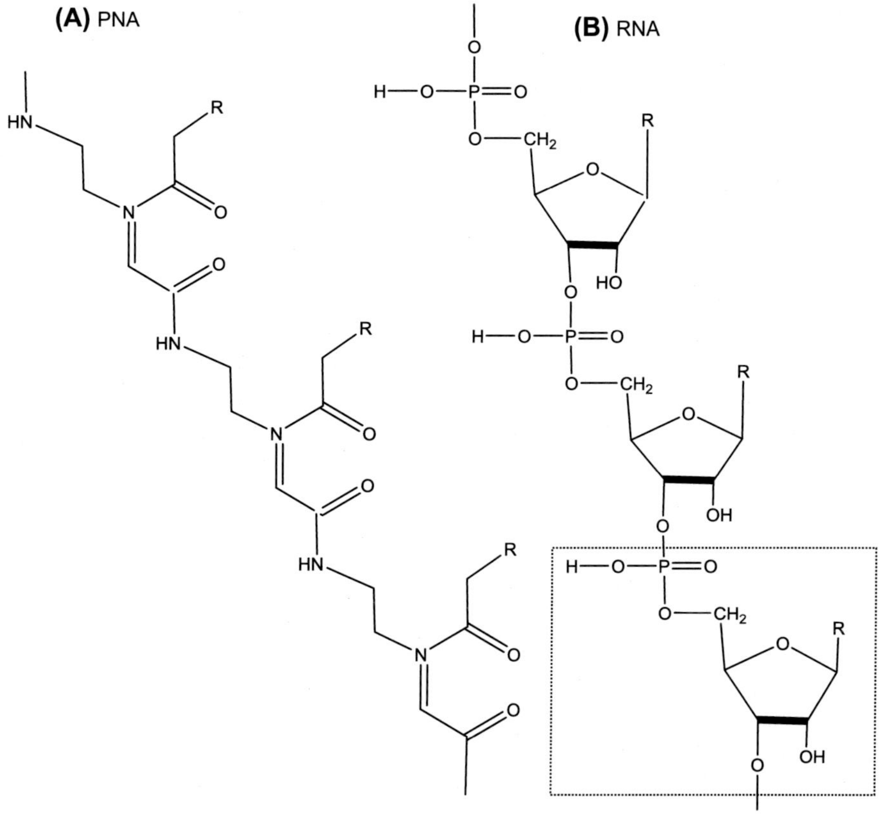

FIGURE 10.8
Comparison of (A) the peptide nucleic acid as a hypothetical information carrier with (B) ribonucleic acid. At R in each chain-like chemical formula, different molecules are bonded, and information is stored and transferred. The dotted molecule with R = adenine is adenosine monophosphate, and used in the experiment in Fig. 10.2.

The idea of guided polymerization dates back more than 40 years (Blumstein et al., 1971; Barrall and Johnson, 1979; Deamer, 2012), but there have been no studies of the arrangement of monomers within an organizing matrix.

Toppozini et al. (2013) experimentally determined whether in fact mononucleotides are captured and organized within a multilamellar structure that is produced when liposomes and solutes undergo dehydration (see Figs. 10.4B, 10.6, and 10.7B). The multiple dry-ups can cause lipid bilayers (e.g., Damer and Deamer, 2015; and see Figs. 10.4B, 10.6, and 10.7B). The results should also be a critical test of the proposed mechanism by which mononucleotides polymerize within the matrix without an enzyme condition. If it cannot be demonstrated that mononucleotides are captured in layers between lipid lamellae, the hypothesis would be excluded as a possible explanation. Toppozini et al. (2013) obtained a solution for that question.

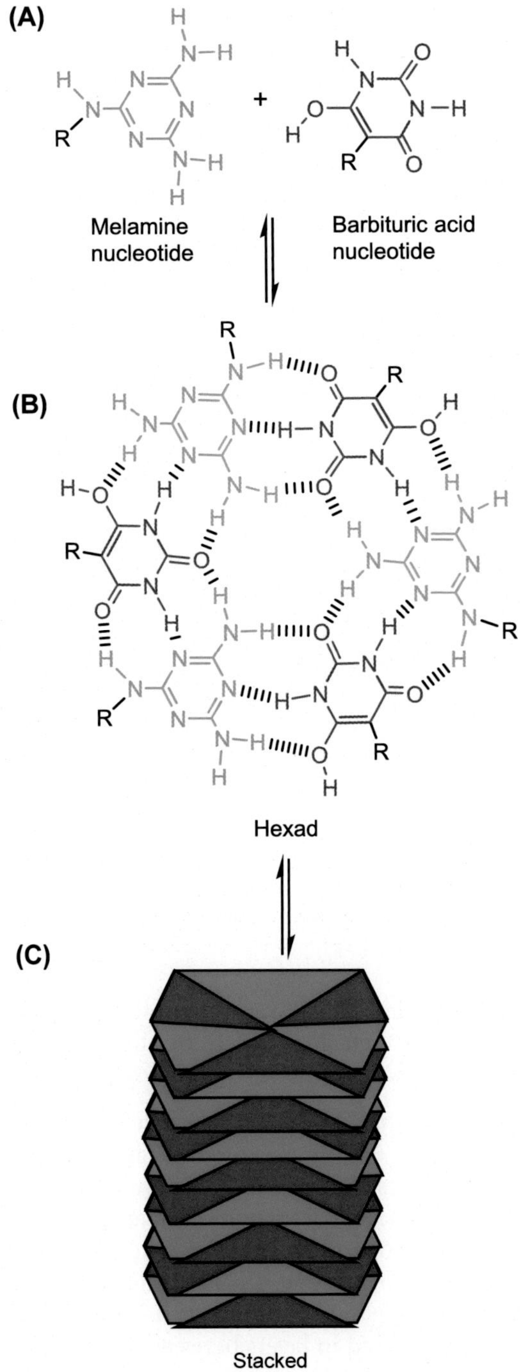

FIGURE 10.9
Model compounds of prebiotic RNA. (A) Melamine and barbituric acid nucleotides. (B) Three pairs form a sheet of Hexad. (C) The Hexad sheets form superstructure of stacked tube.

They used X-ray scattering to investigate adenosine monophosphate (AMP) molecules captured in a multilamellar phospholipid matrix composed of 1,2-dimyristroyl-*sn*-glycero-3-phosphocholine (DMPC) (see Figs. 10.10A and B). Instead of forming a random array, the AMP molecules are highly entangled, which may facilitate polymerization of the nucleotides into RNA-like polymers.

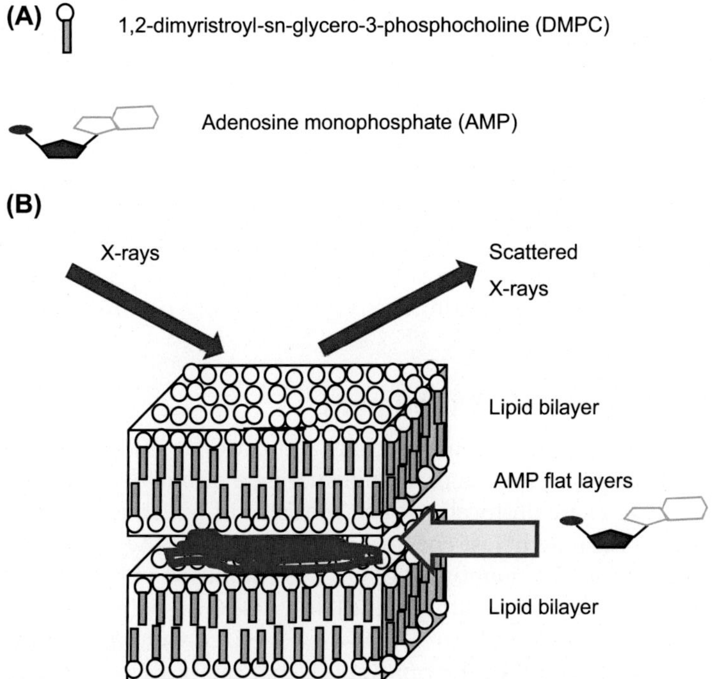

FIGURE 10.10
Experimental set up used by Toppozini et al. (2013). (A) 1,2-dimyristroyl-sn-glycero-3-phosphocholine (DMPC) and adenosine monophosphate (AMP). AMP can be a flat molecule and sandwiched between the bilayer of DMPC. (B) DMPC form lipid bilayers. These bilayers stack each other and form multilayers of the bilayers. AMP can be a flat molecule and stored between bilayers of AMP.

10.8 THE CYTOPLASM OF THE PROTOCELL

All modern cells contain more K^+, PO_4^{3-} and transition metals than those ions in modern oceans, lakes or rivers. Cells maintain ion gradients by using membrane enzymes (membrane pumps) that are embedded in cell membranes (see Fig. 10.4B). The first cell (early protocells) could not possess either ion-tight membranes or membrane pumps. Therefore, concentrations of inorganic ions between the protocell and the environment would equilibrate. In other words, the composition of inorganic ions in the modern cell reflect the "hatcheries" of the first cells (Mulkidjanian et al., 2012).

High concentrations of K^+, Zn^{2+}, Mn^{2+}, and PO_4^{3-} with a high K^+/Na^+ ratio of the modern cells are geochemically observed not in the marine settings but in the vapor-dominated zones of inland geothermal systems. Under the anoxic, CO_2-dominated primordial atmosphere, geothermal fields resemble the internal condition of modern cells. Therefore, the protocell might have been born in shallow ponds of cooled geothermal vapor that were lined with porous silicate minerals mixed with metal sulfides and enriched in K^+, Zn^{2+}, Mn^{2+}, and PO_4^{3-} compounds (Mulkidjanian et al., 2012).

This is consistent with the idea that the protocell was born in geothermal or hydrothermal systems introduced by Damer and Deamer (2015) (Figs. 10.6 and 10.7B).

10.9 THE CELL CYCLE

There are two famous Japanese teams, the Sugawara and Yomo teams, that study cell cycles. Both teams used membranes in modern cells and added enzymes, both of which might not have existed at the beginning of the protocell. However, such studies are important in protocell studies because nothing is known about the protocell.

10.9.1 Sugawara's Experiment

Tadashi Sugawara of Kanagawa University and coworkers (Kurihara et al., 2015) described protocells in four steps, demonstrating a primitive model of a conventional cell cycle. These protocells, which consisted of bacteria-sized lipid vesicles, enclosed DNA and polymerase enzymes to replicate DNA. They called the protocell Giant Vesicle (GV). The membrane of GV was composed of POPC (1-palmitoyl-2-oleoyl-*sn*-glycero-3-phosphocholine) and POPG [1-palmitoyl-2-oleoyl-sn-glycero-3-phospho-rac-(1-glycerol)], and synthesized cationic membrane lipid (V) containing an amphiphilic catalyst (C) that catalyzes the hydrolysis of the membrane lipid precursor (V*) to yield V and electrolyte (E). ($V^* + C \rightarrow V + E$).

The open question in the protocell study is how do we establish a protocell that can self-proliferate over multiple generations in a given environment? The protocell is like a cell-like compartment, not yet alive, but with many characteristic features of living cells (Chen, 2006).

Sugawara's model cell cycle is summarized in Fig. 10.11. The model has four stages. In the ingestion stage, vesicles containing nucleotides (dNTP) are fused with the DNA-containing protocells. Then in the replication stage, DNA is replicated using the nucleotides and DNA polymerase in the protocell. In the maturation phase, replicated DNA protrudes into the membrane to form the catalyst complex. With the addition of lipid precursors, this complex starts generating lipids, and the membrane of the protocell begins to grow (the maturation stage). Finally, the membrane becomes unstable, and the protocell divides into two daughter protocells (the division stage).

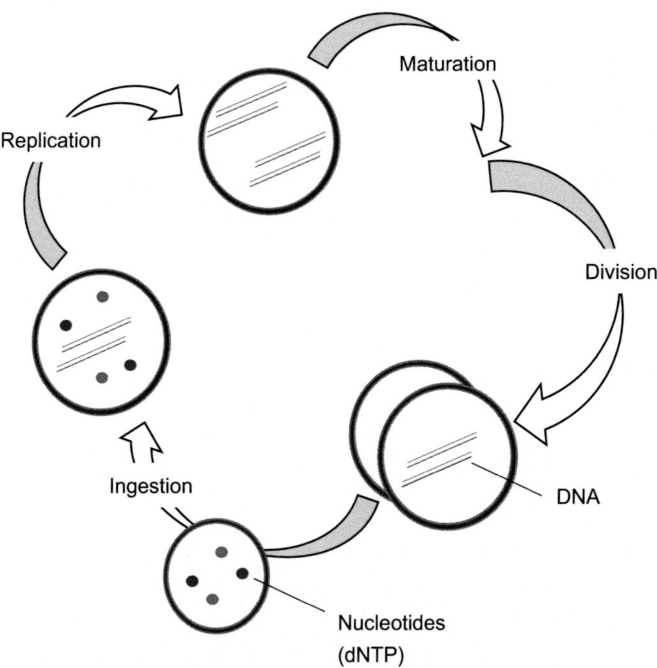

FIGURE 10.11
The primitive model cell cycle proposed by Sugawara's group (Kurihara et al., 2015).

10.9.2 Yomo's Experiment

The Yomo team at Osaka University focused on the sustainable proliferation of phospholipid vesicles because it has been difficult to achieve. They successfully achieved sustainable proliferation of liposomes that replicate RNA within them (Tsuji et al., 2016).

Nutrients for RNA replication and membranes for liposome proliferation were combined by using a modified freeze–thaw technique. The liposomes were mixed, frozen by liquid nitrogen and centrifuged. Then the supernatant was discarded and dissolved with a new buffer solution. The inner contents of liposomes could have leaked as a result of the freezing. These liposomes showed fusion and fission compatible with RNA replication and distribution to daughter liposomes. The schematic diagram of the experiment is shown in Fig. 10.12. The RNA in daughter liposomes was repeatedly used as the template in the next RNA replication and was distributed to granddaughter liposomes. Liposome proliferation was achieved by 10 cycles. Therefore, we propose the use of cultivable liposomes as an advanced protocell model with the implication that the simultaneous supplement of both the membrane material and the nutrients of inner reactions might have enabled protocells to grow sustainably.

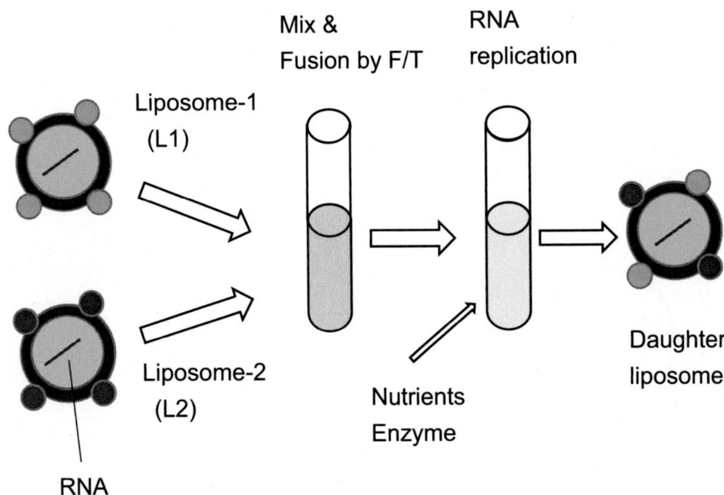

FIGURE 10.12
The schematic diagram of the cell cycle by Yomo's group (Tsuji et al., 2016).

10.10 ROAD FROM PROTOCELL TO ARCHAEA, BACTERIA, AND EUKARYOTA

If the protocell has evolved into the three present types of life (Archaea, Bacteria, and Eukaryota), the common ancestor is called the "cenancestor" (Fitch and Upper, 1987) or the last universal common ancestor (Forterre and Philippe, 1999). The cenancestor existed several billion years ago, and its diversification made three domains of life.

As nothing is known about the protocell, biologists are interested in how eukaryotic cells formed. Recently, the idea of the cenancestor has changed into the model that the eukaryotic cell was generated by endosymbiosis of a bacteria cell with an archaea cell. The origin of the eukaryote cell, which contains a nucleus and other membrane-bound compartments, is one of the conundrums regarding the evolution of life on Earth (Embley and Martin, 2006).

The most important evolution was the appearance of mitochondria. The mitochondria are thought to have formed when a bacteria cell began to live inside an archaeal cell as a form of endosymbiosis (a mutually beneficial relationship in which one organism lives inside another). The bacterium is considered to have provided its host cell with additional energy, and the interaction eventually resulted in a eukaryotic cell, which has the advantage that more energy can be produced. One question is when the evolution of the host cell occurred. There are three models: mitochondria is an early event ("mito-early"); the complexity of the eukaryotic cell was already established ("mito-late"); or after the endomembrane system was formed ("mito-intermediate") (see Fig. 10.13; Ettema, 2016).

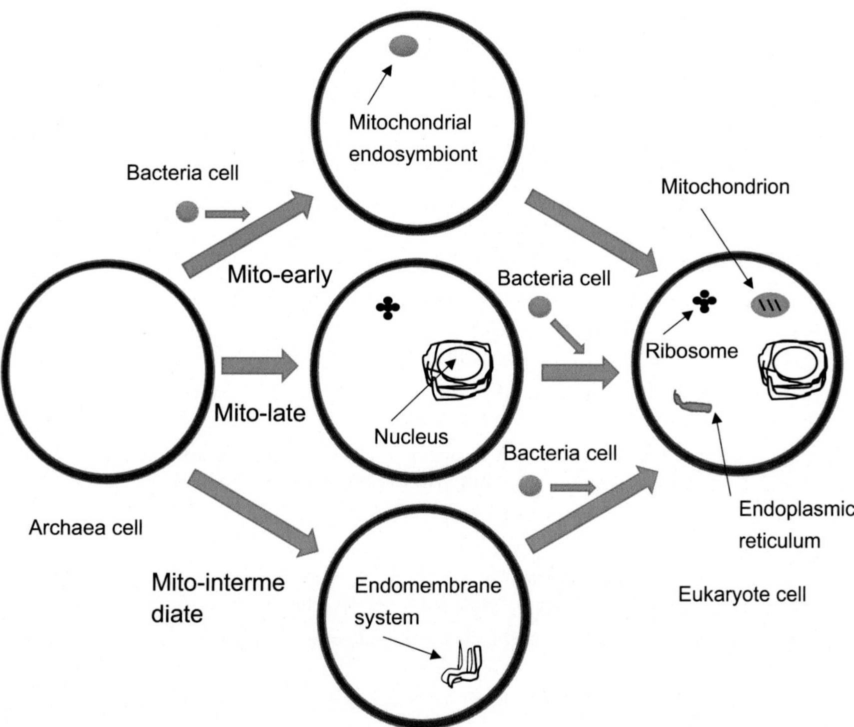

FIGURE 10.13
Origin of eukaryote cells and mitochondria (Ettema, 2016). The origin of eukaryote cells are divided into "Mito-early", "Mito-intermediate" and "Mito-late" models.

By tracing phylogenetic signals of proteins that were present in the last eukaryotic common ancestor (LECA), Pittis and Gabaldón (2016) identified different classes of protein according to the timing of their appearance in eukaryotes. The most recently acquired LECA proteins are dominated by bacterial proteins that are primarily located in mitochondria. Pittis and Gabaldón identified a bacterial LECA protein that was acquired before these mitochondrial proteins. Several of these proteins seem to have existed in intracellular membrane systems, such as the endoplasmic reticulum and the Golgi apparatus. The host cell was already chimeric before the mitochondrial endosymbiosis. Several bacterial proteins that predate the mitochondrial endosymbiosis already worked in intracellular membrane systems. This suggests that the host cell already had a considerable degree of complexity, which is supportive of a late mitochondrial origin ("mito-late") (Pittis and Gabaldón, 2016).

References

Barrall EM, Johnson JF. A review of the status of polymerization in thermotropic liquid crystal media and liquid crystalline monomers. J Macromol Sci C 1979;17:137–70.

Blumstein A, Kitagawa N, Blumstein R. Polymerization of p-methacryloyloxy benzoic acid within liquid crystalline media. Mol Cryst Liq Cryst 1971;12:215–27.

Bohler C, Nielsen PE, Orgel LE. Template switching between PNA and RNA oligonucleotides. Nature 1995;376:578–81.

Cafferty BJ, Fialho DM, Khanam J, et al. Spontaneous formation and base pairing of plausible prebiotic nucleotides in water. Nat Commun 2016. http://dx.doi.org/10.1038/ncomms11328/.

Chen IA. GE prize-winning essay. The emergence of cells during the origin of life. Science 2006;314:1558–9.

Damer B, Deamer D. Coupled phases and combinatorial selection in fluctuating hydrothermal pools: a scenario to guide experimental approaches to the origin of cellular life. Life 2015;5:872–87.

Deamer D. Liquid-crystalline nanostructures: organizing matrices for nonenzymatic nucleic acid polymerization. Chem Soc Rev 2012;41:5375–9.

Embley TM, Martin W. Eukaryotic evolution changes and challenges. Nature 2006;440:623–30.

Ettema TJG. Mitochondria in the second act. Nature 2016;531:39–40.

Fitch WM, Upper K. The phylogeny of transfer-RNA sequences provides evidence for ambiguity reduction in the origin of the genetic-code. Cold Spring Harbor Symp Quanti Biol 1987;52:759–67.

Forterre P, Philippe H. The last universal common ancestor (LUCA) simple or complex? Biol Bull 1999;196:373–5.

Kurihara K, Okura Y, Matsuo M, et al. A recursive vesicle-based model protocell with a primitive model cell cycle. Nat Commun 2015. http://dx.doi.org/10.1038/ncomms9352.

Lopez-Garcia P, Moreira D, Pereto J. Origins and evolution of compartments in Earth, Moon, and Planets. 2006;98:171–4.

Mulkidjanian AY, Bychkof AY, Dibrova DV, Galperin MY, Koonin EV. Origin of first cells at terrestrial, anoxic geothermal fields. Proc Natl Acad Sci U.S.A. 2012;109:E821–30.

Morowitz HJ, Heinz B, Deamer DW. The chemical logic of a minimum protocell. Orig Life Evol Biosph 1988;18:281–7.

Pittis AA, Gabaldón T. Late acquisition of mitochondria by a host with chimaeric prokaryotic ancestry. Nature 2016;531:39–40.

Rajamani S, Vlassov A, Benner S, et al. Lipid assisted synthesis of RNA-like polymers from mononucleotides. Orig Life Evol Biosph 2008;38:57–74.

Toppozini L, Dies H, Deamer DW, Rheinstaedter MC. Adenosine monophosphate forms ordered arrays in multilamellar lipid matrices: insights into assembly of nucleic acid for primitive life. PLoS One 2013;8:e62810.

Tsuji G, Fujii S, Sunami T, Yomo T. Sustainable proliferation of liposomes compatible with inner RNA replication. Proc Natl Acad Sci U.S.A. 2016;113:590–5.

Varela FG, Maturana HR, Uribe R. Autopoiesis: the organization of living systems its characterization and a model. Biosystems 1974;5:187–96.

CHAPTER 11

Possibility of Life on Other Planetary Bodies in Our Solar System

Chapter Outline

11.1 INTRODUCTION

In Table 2.3, the planetary bodies in our Solar System were summarized. Table 11.1 includes the planets and moons from Table 2.3 and dwarf-planets are added. These celestial bodies were recently studied by spacecraft in detail, and very interesting results were obtained. In particular, finding liquid water on various moons is very important, because it means life could exist. In this chapter, scientific results are shown for each celestial body that revealed the existence of liquid water. Movements of ice on the moons of the giant planets Jupiter and Saturn and on dwarf planet Pluto were also found.

11.2 MARS

11.2.1 Atmosphere of Mars

After Mars lost its magnetosphere 4 billion years ago, solar wind directly hit the Martian ionosphere and stripped away atoms from its outer layer, lowering the density of Mars' atmosphere. The Mars Global Surveyer (launched in 1996

Origins of the Earth, Moon, and Life. http://dx.doi.org/10.1016/B978-0-12-812058-3.00011-9

Table 11.1 The Selected Planets, Moons, and Dwarf Planets

		Orbital Characteristics		Physical Characteristics			
Target	Category	Semi-major Axis (AU)	Orbital Period (× Earth)	Radius (× Earth)	Mass (M_E)	Peculiar Characteristics	Space-Craft
Earth	Planet	(1.49×10^8 km)	(365.25 d)	(6.357×10^3 km)	(5.97×10^{24} kg)	Life	–
Mars	Planet	1.52	1.88	0.532	0.107	CO_2 air	Curiosity
Ceres	Dwarf planet	2.76	4.60	0.0744	0.00015	Bright spot	Dawn
Vesta	Dwarf planet	2.36	3.63	0.0413	0.00004	Eucrites	Dawn
Io	Moon, Jupiter	0.00283	1.77 d	0.286	0.0150	Sulfur-volcanoes	Pioneer 10/11, Voyager 1/2, Galileo, New Horizons
Europa	Moon, Jupiter	0.00450	3.55 d	0.246	0.0080	Ice-covered	Pioneer 10/11, Voyager 1/2, Galileo, New Horizons
Ganymede	Moon, Jupiter	0.0995	7.15 d	0.414	0.0248	Salty sea	Pioneer 10/11, Voyager 1/2, Galileo, New Horizons
Callisto	Moon, Jupiter	0.0722	16.7 d	0.379	0.0180	Ice-covered	Pioneer 10/11, Voyager 1/2, Galileo, New Horizons
Enceladus	Moon, Saturn	0.00160	1.37 d	0.040	1.8×10^{-5}	Salty sea, life?	Voyager 1/2, Cassini
Titan	Moon, Saturn				$\times 10^{-5}$	Atmosphere	Voyager 1/2, Cassini
Pluto	Dwarf planet	39.487	247.94	0.18	0.00218	Ice-covered	New Horizons
Charon	Moon, Pluto					Rocky	New Horizons

by the National Aeronautics ad Space Administration; NASA), Mars Express (launched in 2003 by the European Space Agency; ESA) and MAVEN (launched in 2013 by NASA) detected ionized atmospheric particles. The atmospheric pressure of Mars is only 600 Pa compared to the Earth of 1.0×10^5 Pa, indicating that the atmospheric pressure on Mars is only 0.6% that of Earth. The composition of the atmosphere was 96% CO_2, 1.9% Ar, 1.9% N_2, and traces of O_2 and water.

Traces of methane were detected and produced at two places by Mars Express. Methane in the atmosphere decomposes quickly by sunshine. Thus, the methane production event is a very recent one. The Mars rover, Curiosity (see Section 11.2.2), also found methane.

11.2.2 Mars Curiosity

The self-portrait of Curiosity is shown in Fig. 11.1. Curiosity has been moving on the surface of Mars for more than 3 years since 2012. The rover's observations confirmed that once there was plenty of water on the surface making lakes 3.8–3.3 billion years ago. There were streams delivering and depositing sediments as

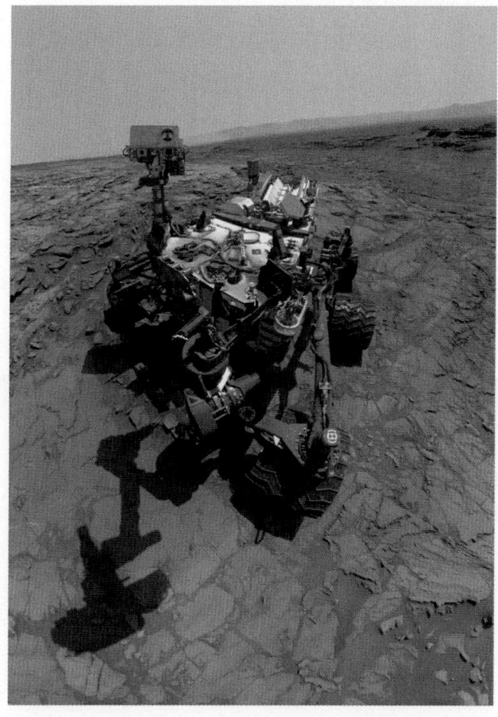

FIGURE 11.1
A self-portrait of the Mars Rover, Curiosity on Mars. For scale, the rover's wheel is 50 cm in diameter and 40 cm in width.
Image credit: NASA/JPL-Caltech/MSSS.
http://www.nasa.gov/sites/default/files/thumbnails/image/pia19920_bigsky-selfie.jpg.

FIGURE 11.2
A view of Mars by Curiosity. Such views confirmed that water filled this area, forming a lake. There were streams delivering sediments (mudstones) that deposited in layers.
Image credit: NASA/JPL-Caltech/MSSS.
http://www.nasa.gov/sites/default/files/thumbnails/image/pia19839-galecrater-main.png.

shown in Fig. 11.2. The rocks in Fig. 11.2 are finely laminated mudstones, which should be deposited as a lake (Clavin, 2015).

Before Curiosity, some scientists thought that wind-blown dust and sand accumulated in dry conditions. However, Curiosity revealed that conditions were wet. From the map data and Curiosity camera images, the water depth from the floor was estimated to be 800 m, which raises a new question: How did liquid water exist on Mars?

The rover is equipped with the Tunable Laser Spectrometer (TLS). In 2013, atmospheric methane was <1.3 ppbv, and it was concluded that current methanogenic microbial activity on Mars is less probable (Brown and Webster, 2013) because of the low methane concentration in the air.

11.3 THE DAWN MISSION FOR ASTEROIDS 4 VESTA AND 1 CERES

11.3.1 Introduction

Dawn was launched by NASA to study asteroids 4 Vesta (Vesta) and 1 Ceres (Ceres). The objectives of the Dawn mission were to acquire color images, topographic maps, elemental composition maps, and mineral composition maps, and to measure gravity fields in order to understand their formation, evolution, and their roles in forming the terrestrial planets.

First, the Dawn spacecraft went to Vesta, which is rocky, dry, and bright. It has an iron core, and a mantle and crust made of basalts, some of which came to Earth as the HED meteorites.

Ceres was icy, wet, and dark. It is expected to have a rocky core, icy mantle, and dusty clay surface. Why are they so different? This is one purpose of the mission.

Ceres is the largest asteroid, and it is also classified as a dwarf planet like Pluto. It has a metal-rich core and a clay surface. An ocean once existed. Now it is frozen or sublimated. Are there remnants of the ocean or organic materials? And does biological activity remain?

11.3.2 Vesta

Giant impact basins in the south were found (the bottom side of Vesta obtained by Dawn is shown in Fig. 11.3). This is the largest crater relative to body size in the Solar System. Scaled to Earth, it would stretch from Washington, DC, over the North Pole, to Tokyo. The central peak is more than twice as high as Mt. Everest, rivaling Olympus Mountain on Mars as the tallest mountain in the solar system. The massive impacts shook through Vesta, leaving giant scars across the surface (horizontal lines on the left and zigzag lines on the right are observed in Fig. 11.3) (Pirani and Turrini, 2016).

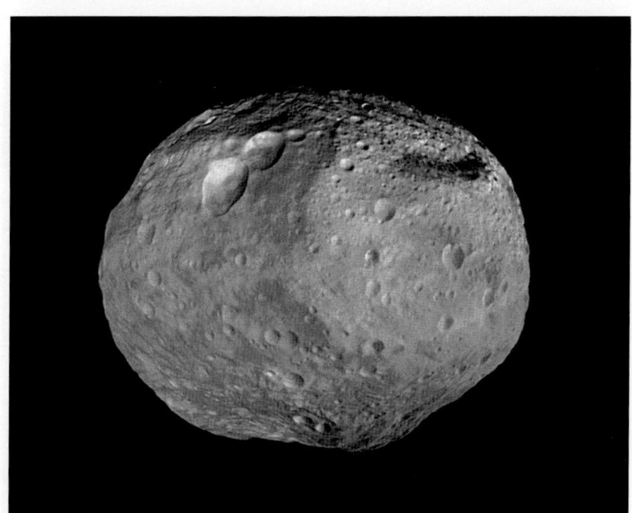

FIGURE 11.3
Full view of the giant asteroid Vesta taken by the Dawn spacecraft.
Image credit: NASA/JPL-Caltech/UCAL/MPS/DLR/IDA.
https://www.nasa.gov/mission_pages/dawn/images/index.html?id=337671.

Research of the HED meteorites found that the HEDs must have come from a large silicate-dominated, differentiated body like Vesta that must have at least one big crater. One of the purposes of visiting Vesta was to prove the connection of Vesta to the HED meteorites. After Dawn visited Vesta, we were sure we had found the parent body of HEDs.

Significant amounts of hydrogen on Vesta's surface were observed. This hydrogen should be in water or water-mineral forms, which seems to have come from impacters. Dark carbonaceous materials in the HED meteorites are found as water bound in the minerals. They are present on the surface of Vesta and are believed to have come from impacters composed of more primitive, icy materials. Marcia Crater, which is a large, fresh, 62-km diameter crater in the northern hemisphere, shows sharp rims, bright and dark layers, fluidized ejecta, and pitted terrains (The crater is on the top-left in Fig. 11.3.). Small pits away from a central structure are observed, which might be made by the release of volatiles from the sub-surface. Buried ice could exist in Vesta, having come from water rich impacters such as Ceres (Prettyman et al., 2012).

11.3.3 Mystery of the Bright Spots on Ceres

The surface of Ceres is cratered throughout. Bright spots were found on the surface of Ceres (see Fig. 11.4), both on some smooth and some chaotically fractured terrains. The bright spots may be ice or salts with high reflection. Brightness indicates recent activity, which could be vapor emission. The bright materials are distributed on the floor of the crater and wall, with a concentration of very bright materials in the center or along the fractures (Landau, 2015).

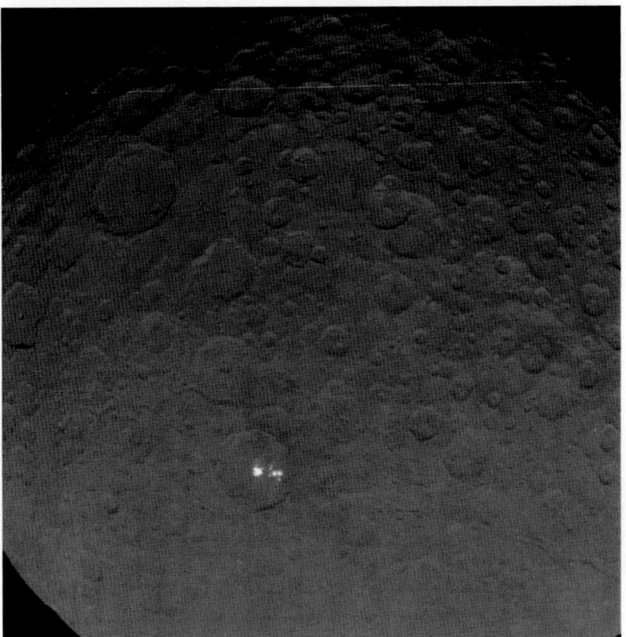

FIGURE 11.4
Ceres' bright spots. The intense brightness of these spots may be due to the reflection of sunlight by highly reflective materials on the surface, possibly ice.
Image credits: NASA/JPL-Caltech/UCLA/MPS/DLR/IDA.
https://www.nasa.gov/sites/default/files/thumbnails/image/pia19559-1041.jpg.

Vesta's craters are bowl-shaped. In contrast, Ceres' craters often have smooth floors, which may contain impact melt, and most are irregular in shape. They are similar to craters of a cold icy moon of Saturn, which contain no melt but are also irregular in shape (Krohn et al., 2016).

Dawn's exploration of Vesta and Ceres tested long-held paradigms and provided new theories about the early solar system. Protoplanets were forming very early during the hot phase of the solar nebula and delivered iron cores and rocky material to the forming terrestrial planets. Comets and wet asteroids delivered water to the inner solar system, as shown in the hydrated material on Vesta (Russel and Raymond, 2012).

11.4 FOUR CONTRASTING MOONS OF JUPITER

11.4.1 Overviews

The four largest moons of Jupiter (Io, Europa, Ganymede, and Callisto) are shown in Fig. 11.5. It is astonishing that the appearance of each Jovian satellite is so different, which became known only after visits by spacecrafts.

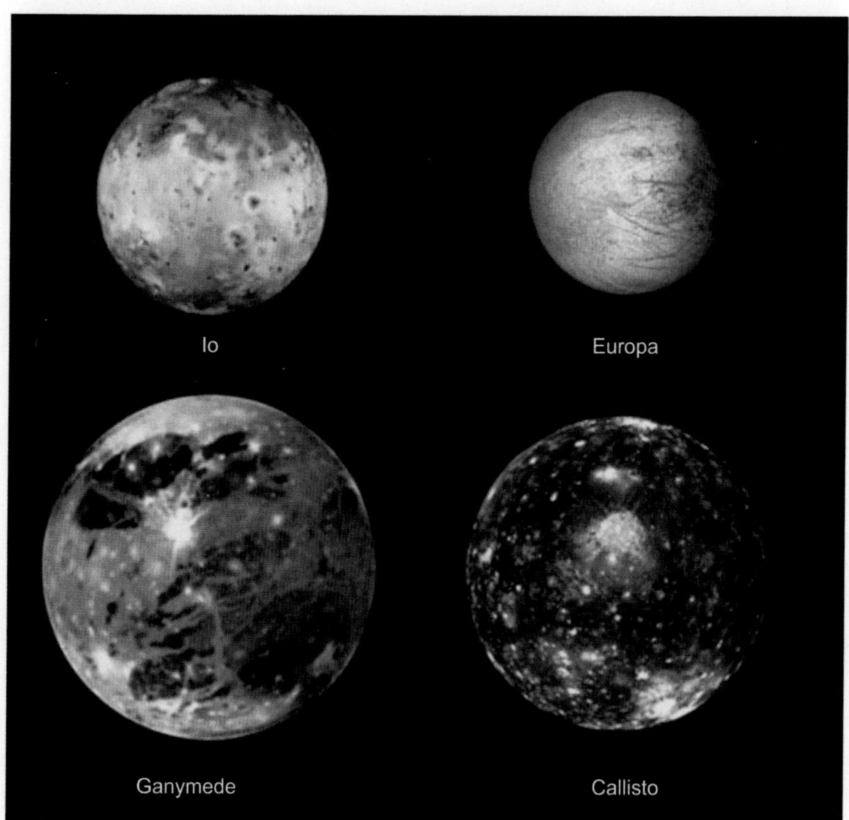

FIGURE 11.5
The four Galilean Satellites of Jupiter. The appearance of these neighboring satellites is amazingly different even though they are relatively close to Jupiter (0.35 Gm for Io; 1.8 Gm for Callisto). These images were acquired at very low "phase" angles (the sun-moon-spacecraft angle) so that the sun is illuminating the Jovian moons from behind the spacecraft. The colors have been enhanced to bring out subtle color variations of surface features. North is to the top of all the images, which were taken by the Galileo spacecraft. The original image of NASA was slightly rearranged.
Image credit: NASA/JPL/DLR.
http://photojournal.jpl.nasa.gov/jpegMod/PIA01400_modest.jpg.

Cutaway views of the possible internal structures of the Galilean satellites are shown in Fig. 11.6. With the exception of Callisto, all the satellites have metallic (iron, nickel) cores (shown in gray) drawn to the correct relative size, and all the cores are surrounded by rocky (shown in brown) shells. Io's rock or silicate shell extends to the surface, whereas the rock layers of Ganymede and Europa (drawn to correct relative scale) are in turn surrounded by shells of water in ice or liquid form (shown in blue and white and drawn to the correct relative scale).

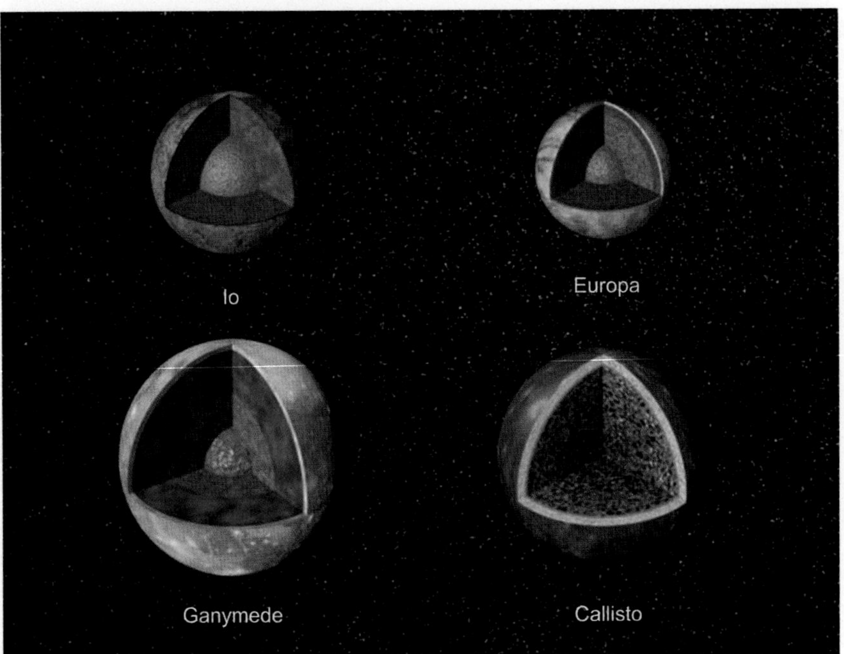

FIGURE 11.6
Cutaway views of the possible internal structures of the Galilean satellites.
Image credit: NASA/JPL.
http://photojournal.jpl.nasa.gov/jpegMod/PIA01082_modest.jpg.

Callisto is shown as a relatively uniform mixture of comparable amounts of ice and rock. Recent data, however, suggests a more complex core as shown here.

The surface layers of Ganymede and Callisto are shown as white to indicate that they may differ from the underlying ice/rock layers in the percentage of rock they contain. The white surface layer on Europa could be similar, although it could also suggest an ice layer overlying a liquid water ocean.

Europa might have a liquid water ocean under a surface ice layer several to 10 km thick; however, this evidence is also consistent with the existence of a liquid water ocean in the past. Therefore, it is not certain whether there is a liquid water ocean on Europa at present.

The four Jovian moons are emphasized not only because they have interesting characteristics but also because they have the possibility of life in their underlying "liquid" water (or in their hot environment in the case of Io). The possibility of life on these moons is higher than that on Mars! Here we learn more about these Jovian moons.

Why can liquid water exist on cold moons? It is said that the heat is generated by the tidal force of the giant mother planet. When the icy planets exist with internal water as shown in Europa or Ganymede in Fig. 11.6, the core can move inside the moon. Thus the energy can be transferred as the tidal force, friction, or potential energy from the giant mother planet, Jupiter. However, once all water becomes ice as on Callisto, it becomes difficult to transfer energy this way. The inner liquid in the moon is a prerequisite for energy transfer, but without energy transfer, the inner liquid does not exist. (It is a chicken-and-egg debate) (Box 11.1).

BOX 11.1 THE PHASE DIAGRAM OF WATER

The phase diagram of water is shown in Fig. Box 11.1A and B. There are many interesting points in the phases of water. One is that there are at least 10 polymorphs in ice. Many properties (viscosity, self-diffusion, Raman spectra, etc.) of cold liquid water change around at 200 MPa (we are living in 0.1 MPa). Pure liquid water cannot exist at any pressure at a temperature of −22°C.

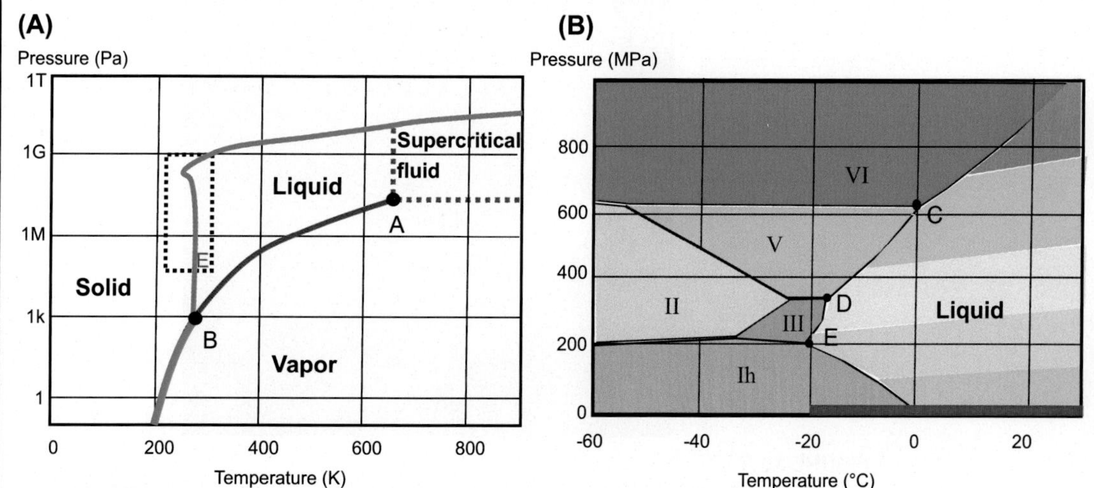

FIGURE BOX 11.1

The phase diagram of water.

(A) The phase diagram of water. The horizontal and vertical axes are temperature (°C) in linear-scale and pressure (Pa) in logarithmic-scale, respectively. The blue, green and red lines show boundaries between solid-liquid, solid-vapor and liquid-vapor, respectively. The area of right and upper side of dotted purple lines below the blue line is supercritical fluid. Points A and B are the critical and triple points, respectively. The temperature (°C) and pressure (MPa) of A and B are 647 and 22; and 273 and 0.6, respectively. The red "E" indicates the Earth's condition. The dotted area is enlarged in Fig. (B).

(B) The enlarged phase diagram of the dotted area of (A). The right side green areas indicate liquid water areas. The gradation change of colors are drawn along the density change. The left side blue areas indicate the ice (solid) areas. The ice phases of Ih, II, III, V and VI are included in this area. Points C, D and E are triple points where three phases coexist. The temperature (°C) and pressure (MPa) of C, D and E are 0.16 and 632; -17 and 350; and -22 and 210, respectively. The red line at the bottom indicates the typical habitable zone of the Earth.

11.4.2 Io

Io, which is slightly larger than the Earth's moon, is the most colorful of the Galilean satellites. Its surface is covered by deposits of actively erupting volcanoes and hundreds of lava flows. The volcanic vents are visible as small dark spots in Fig. 11.5. Several of these volcanoes are very hot, and at least one reached 2000 °C in 1997. Prometheus, a volcano located slightly to right of the center on Io's image, was active in 1979 and was still active in 1996. The active volcanic eruption of Io is shown in Fig. 11.7. The bright, yellowish, and white materials located at equatorial latitudes in Fig. 11.7 are believed to be composed of sulfur and sulfur dioxide. The polar caps are darker and covered by a redder material (Lopes, 2015).

FIGURE 11.7
Active volcanic plume of Io. This color image, acquired by Galileo, shows two volcanic plumes on Io. The plume is 140 km high. The second plume, seen near the boundary between day and night, is called Prometheus. The shadow of the airborne plume can be seen extending to the right of the eruption vent. The vent is near the center of the bright and dark rings. Plumes on Io have a blue color, so the plume shadow is reddish. It is possible that this plume has been continuously active for more than 18 years. North is to the top of the picture. The resolution is about 2 km per picture element.
Image credit: NASA/JPL/University of Arizona.
http://photojournal.jpl.nasa.gov/jpegMod/PIA01081_modest.jpg.

11.4.3 Europa

Europa has a very different surface from its rocky neighbor, Io. A Galileo image of Fig. 11.8 hints at the possibility of liquid water beneath the icy crust of this moon. The bright white and bluish parts of Europa's surface are composed of

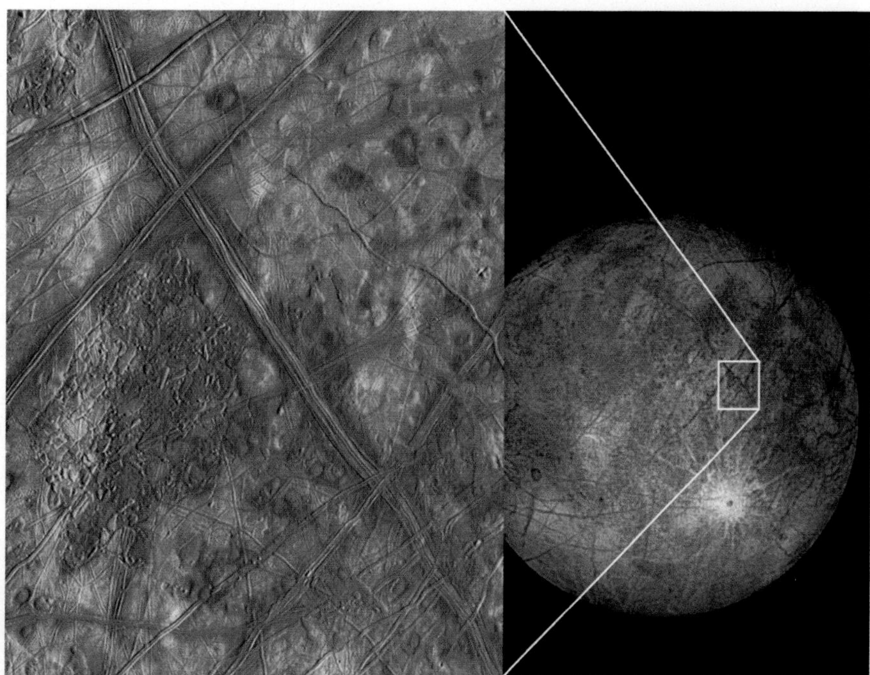

FIGURE 11.8
Jupiter's icy moon Europa. Jupiter's moon Europa has a crust made of blocks, which are thought to have broken apart, as shown in the image on the left. Europa may have had a subsurface ocean at some point. The presence of a magnetic field leads scientists to believe an ocean is present at Europa today. Reddish-brown areas represent non-ice material resulting from geologic activity. White areas are material ejected during the formation of the impact crater. Icy plains are shown in blue tones to distinguish coarse-grained ice (dark blue) from fine-grained ice (light blue). *Long, dark lines* are ridges and fractures in the crust. These images were obtained by the NASA's Galileo spacecraft.
Image credit: NASA/JPL/University of Arizona.
http://www.nasa.gov/sites/default/files/images/337344main_image_1339_full.jpg.

water ice. In contrast, the brownish regions on the right side of the image may be covered by salts (such as hydrated magnesium-sulfate) and an unknown red component. The yellowish terrain on the left side of the image is caused by some other unknown contaminant. This global view was obtained in 1997; the finest details that can be discerned are 25 km across (Castillo-Rogez, 2015).

Indications of possible plume activity were reported in 2013 by researchers using NASA's Hubble Space Telescope. NASA's Hubble Space Telescope observed water vapor above the frigid south polar region of Europa, providing the first strong evidence of water plumes erupting off the moon's surface.

However, NASA's Cassini spacecraft did not find the plume activity during its 2001 flyby of Jupiter. Either the plume activity was infrequent or the plumes are smaller than the plume on Enceladus (see Section 11.5.1).

Europa has become one of the most exciting destinations in the solar system for future exploration because it shows strong indications of having an ocean beneath its icy crust; thus, there is a possibility of life.

11.4.4 Ganymede

Ganymede, larger than the planet Mercury, is the largest Jovian satellite. Its distinctive surface is characterized by patches of dark and light terrain (see Fig. 11.5). Bright frost is visible at the north and south poles. The very bright icy impact crater, Tros, is near the center of the image in a region known as Phrygia Sulcus. The dark area to the northwest of Tros is Perrine Regio. The dark terrain to the south and southeast is Nicholson Regio. Ganymede's surface is characterized by a high degree of crustal deformation. Much of the surface is covered by water ice, with a higher amount of rocky material in the darker areas. The global view of Fig. 11.5 was taken in September 1997 when Galileo was 1.68 Gm from Ganymede; the finest details that can be discerned are about 67 km across (Cook, 2014).

Fig. 11.9 is an image map (right) and a geological map (left) of Ganymede. The image map was made by data from Voyager 1 and 2, and the Galileo spacecraft. The geology of Ganymede is very complex.

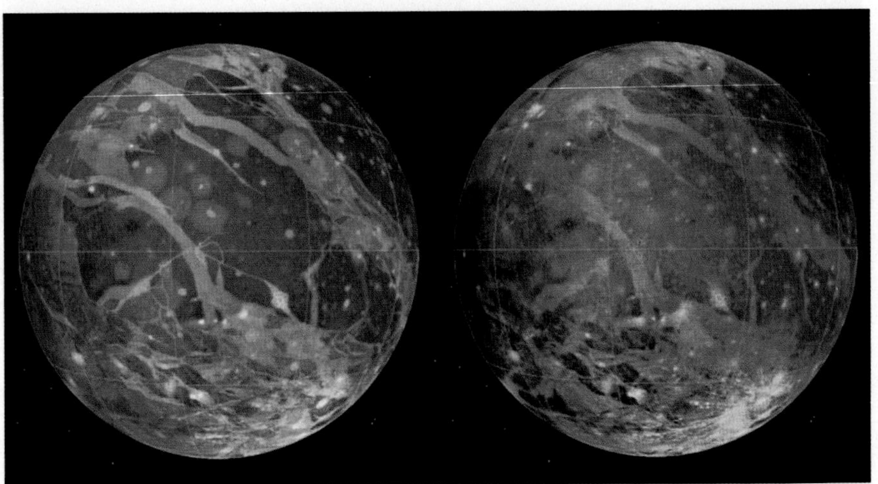

FIGURE 11.9
Ganymede. To present the best information of Jupiter's moon Ganymede, global images from NASA's Voyager 1 and 2 and NASA's Galileo spacecraft were assembled. The right map served as the base map for the left geologic map of Ganymede.
Image Credit: USGS Astrogeology Science Center/Wheaton/NASA/JPL-Caltech. http://photojournal.jpl.nasa.gov/jpegMod/PIA17901_modest.jpg.

In Fig. 11.10, an image of an aurora around Ganymede was taken by NASA's Hubble Space Telescope. If a saltwater ocean were present, Jupiter's magnetic field would create a secondary magnetic field in the ocean. This friction of the magnetic fields would suppress the reaction of two auroras. This ocean resists Jupiter's magnetic field so strongly that it reduces the aurora to 2 degrees, instead of 6 degrees if the ocean was not present. Thus, scientists estimate the ocean is 10 km thick—10 times deeper than Earth's ocean—and is buried under a 150 km crust of ice.

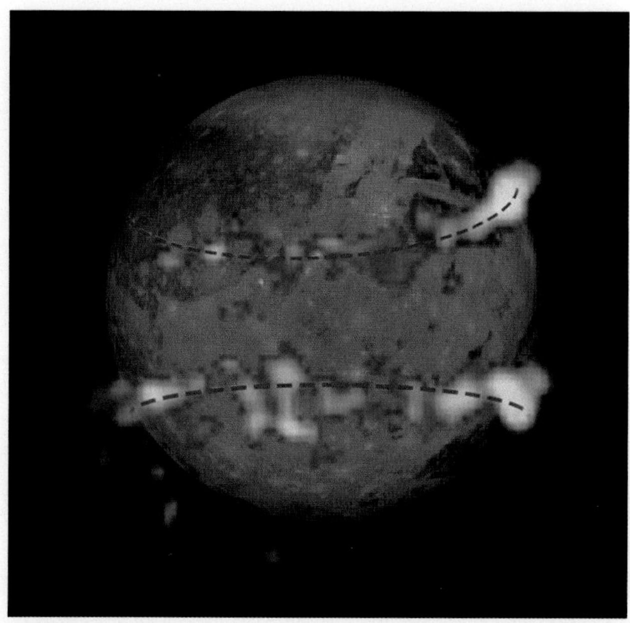

FIGURE 11.10
A NASA Hubble Space Telescope image of Ganymede's auroral belts (colored blue in the illustration) are overlaid on a Galileo orbiter image. The amount of rocking of the moon's magnetic field suggests that the moon has a subsurface saltwater ocean.
Image credit: NASA/ESA.
https://www.nasa.gov/sites/default/files/thumbnails/image/15-33i2.png.

Identifying liquid water is crucial in the search for life and for habitable worlds beyond the Earth. A deep ocean under the icy crust of Ganymede indicates the possibility of life beyond Earth. The subterranean ocean is thought to have more water than all the water on Earth's surface.

NASA's scientists made a sandwich model, shown in Fig. 11.11. This artist's conception of Ganymede illustrates its interior oceans. Ganymede's oceans may have 25 times the volume of those on Earth. This model, based on experiments in the laboratory that simulate salty seas, shows that the ocean and ice may be stacked up in multiple layers.

Ice comes in different forms depending on pressures. "Ice I," the least dense form of ice, floats on top (See Fig. Box 11.1 for the phase diagram of pure water. Note that it is not that of salty water.). As pressure increases, ice molecules become more tightly packed and thus denser. Because Ganymede's oceans are up to 800 km deep, they would experience more pressure than oceans on Earth. The deepest and most dense form of ice in Ganymede could be "Ice VI."

Furthermore, the model shows that a strange phenomenon might occur in the uppermost liquid layer, where ice floats upward. In this scenario, cold plumes

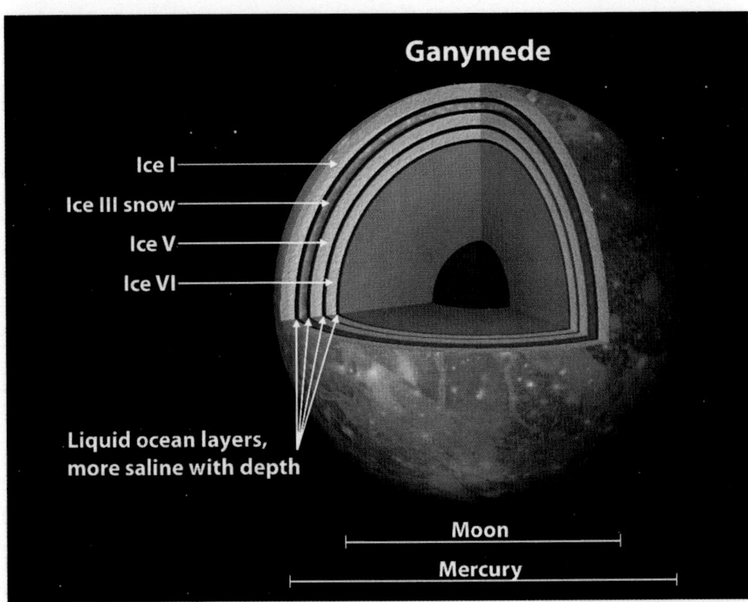

FIGURE 11.11
A sandwich model of ice and oceans of Ganymede (artist's concept).
Image credit: NASA/JPL-Caltech.
http://photojournal.jpl.nasa.gov/figures/PIA18005_fig1.jpg.

cause Ice III to form. As the ice forms, salt precipitates out. The salt then sinks down while the ice "snows" upward. Eventually, this ice would melt, resulting in a slushy layer in Ganymede's structure (Clavin, 2014).

The fact that salty water stays at the bottom of the rocky sea floor, rather than ice, is favorable for the occurrence of life. Researchers think that life emerges through a series of chemical interactions at water-mineral interfaces, so a wet sea floor on Ganymede might be a key ingredient for life there.

11.4.5 Callisto

Fig. 11.5 was taken in 1997 when Galileo was 684,500 km from Callisto; the finest details that can be discerned are about 27 km across. Callisto's dark surface is pocked by numerous bright impact craters (see Fig. 11.5). Of the four largest moons, Callisto's orbit is the farthest from Jupiter. The large Valhalla multiring structure (visible near the center of Fig. 11.5) has a diameter of about 4000 km, making it one of the largest impact features in the solar system. Although many crater rims exhibit bright, icy "bedrock" material, a dark layer composed of hydrated minerals and organic components is seen inside many craters and in other low-lying areas. Evidence of tectonic and volcanic activity, seen on the other Galilean satellites, appears to be absent on Callisto (Dodd, 2015).

11.5 MOONS OF SATURN

In this section, Saturn's moons, Enceladus and Titan, are discussed and explained. In particular, Enceladus emits water vapor, which means liquid water exists inside the moon. As Saturn is farther from the Earth than Jupiter and thus colder, it is amazing that clear images of Enceladus are obtained, and that liquid water exists on Saturn's moon.

11.5.1 Enceladus: Saturn's Moon of Moving Ice

Cassini was launched in 1997 and entered into orbit around Saturn in 2004. Enceladus was found to be an icy moon of Saturn. Cassini first discovered the "tiger stripes" on Enceladus, indicating the young and active surface. Then Cassini confirmed that Enceladus has a towering plume of ice (see Fig. 11.12). Furthermore, it has water vapor and organic molecules spraying from its south polar region (see Fig. 11.12). Cassini determined that the moon has a global ocean and likely hydrothermal activity, meaning it could support simple life (Platt et al., 2014).

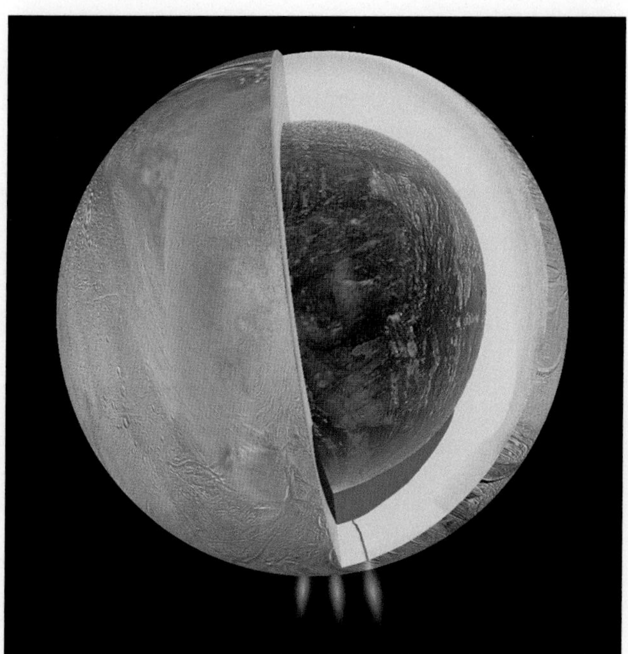

FIGURE 11.12
This diagram illustrates the possible interior of Saturn's moon Enceladus based on a gravity investigation by NASA's Cassini spacecraft and NASA's Deep Space Network, reported in 2014. The gravity measurements suggest an ice outer shell and a low density, rocky core with a regional water ocean sandwiched in between at high southern latitudes. Views from Cassini's imaging science subsystem were used to depict the surface geology of Enceladus and the plume of water jets gushing from fractures near the moon's south pole. Enceladus is 504 km in diameter.
Image credit: NASA/JPL-Caltech.
http://www.nasa.gov/sites/default/files/pia18071_enceladus-interior.jpeg.

The Enceladus plume is thought to come from the ocean below (the blue region depicted between the core and the white ice layer in Fig. 11.12). The spacecrafts (Voyagers 1 and 2) have flown closer to the surface of Enceladus before, but never observed such an active plume.

The flyby was not intended to detect life, but will provide powerful new insights about how habitable the ocean environment is within Enceladus. Cassini scientists are hopeful that the flyby will provide insights about how much hydrothermal activity—that is, chemistry involving rock and hot water—is occurring on Enceladus. This activity could have important implications for the potential habitability of the ocean for simple forms of life. Scientists also expect to better understand the chemistry of the plume.

The flyby will help to solve the mystery of whether the plume is composed of column-like, individual jets, icy curtain eruptions, or a combination of both. The answer would clarify how material is getting to the surface from the ocean below. Researchers are not sure how much icy material the plumes are actually spraying into space. The amount of activity has major implications for how long Enceladus might have been active.

Scientists expected the north polar region of Enceladus to be heavily cratered, based on low-resolution images from the Voyager mission, but high-resolution Cassini images show a landscape of stark contrast (see Fig. 11.13). Thin cracks

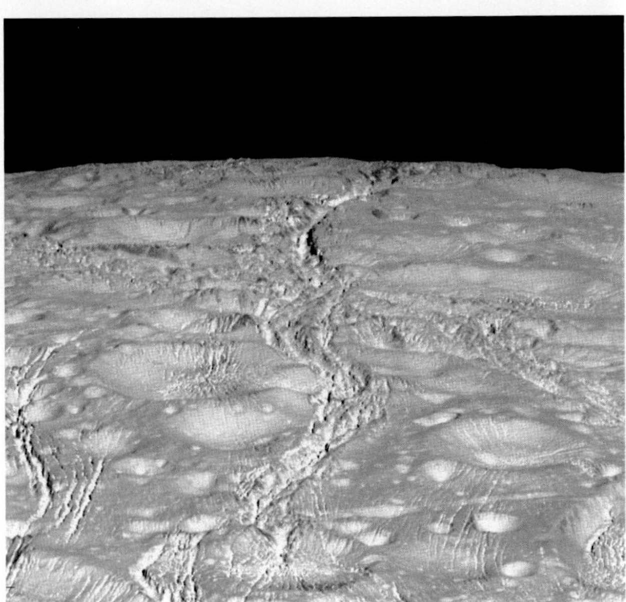

FIGURE 11.13
NASA's Cassini spacecraft zoomed past Saturn's icy moon Enceladus on Oct. 14, 2015, capturing this image of the moon's north pole. North on Enceladus is up. The view was acquired at a distance of about 6000 km from Enceladus and at a Sun-Enceladus-spacecraft angle of 9 degrees. Image scale is 35 m per pixel.
Image credit: NASA/JPL-Caltech/Space Science Institute.
http://www.nasa.gov/sites/default/files/thumbnails/image/pia19660.jpg.

cross over the pole—the northernmost extent of a global system of such fractures. Before this Cassini flyby, scientists did not know if the fractures extended so far north.

11.5.2 Titan

Saturn's moon Titan has seas and lakes filled with liquid hydrocarbons (see Fig. 11.14). Apart from Earth, Titan is the only body in the Solar System to possess surface lakes and seas, which have been observed by the Cassini spacecraft. As Titan's surface temperature is roughly −180 °C, liquid methane and ethane, rather than water, dominate Titan's hydrocarbon equivalent of Earth's water. With plenty of organic materials and a relatively high temperature, there could be life on Titan.

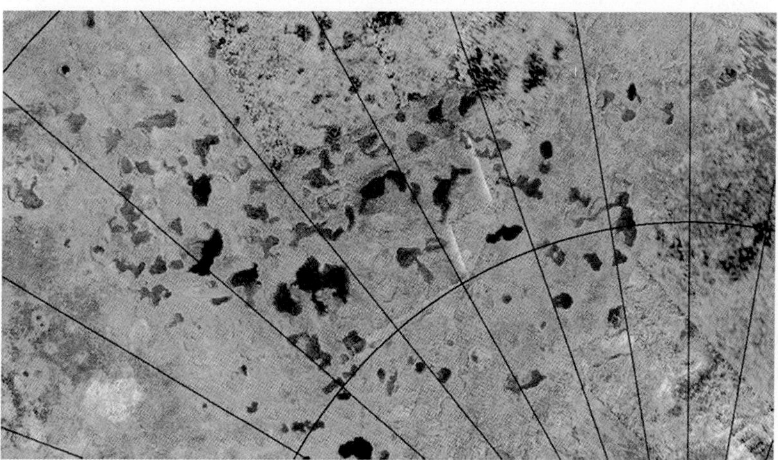

FIGURE 11.14
Saturn's moon, Titan. Radar images from NASA's Cassini spacecraft reveal many lakes on Titan's surface, some filled with liquid and some appearing as empty depressions. It is like karstic landforms on Earth. *Image credit: NASA/JPL-Caltech/ASI/USGS. http://www.jpl.nasa.gov/spaceimages/images/largesize/PIA17655_hires.jpg.*

Cassini has identified two forms of methane- and ethane-filled depressions that create distinctive features near Titan's poles. There are vast seas several hundred kilometers across and up to several hundred meters deep, fed by branching, river-like channels. There are also numerous smaller, shallower lakes, with rounded edges and steep walls that are generally found in flat areas. Cassini has also observed many empty depressions.

The lakes are generally not associated with rivers; thus, they are thought to fill up by rainfall and liquids feeding them from underground. Some of the lakes fill and dry out again during the 30-year seasonal cycle on Saturn and Titan. But exactly how the depressions hosting the lakes came about in the first place is poorly understood.

Recently it was found that Titan's lakes are reminiscent of what are known as karstic landforms on Earth. These are terrestrial landscapes that result from the erosion of dissolvable rocks, such as limestone and gypsum, in groundwater and rainfall percolating through rocks. Over time, this leads to features like sink-holes and caves in humid climates, and salt pans where the climate is more arid. The rate of erosion creating such features depends on factors such as the chemistry of the rocks, the rainfall rate, and the surface temperature. While all of these aspects clearly differ between Titan and Earth, it is surprising that the underlying process may be surprisingly similar.

It was proposed that it would take around 50 Myr to create a 100-m depression at Titan's relatively rainy polar regions, consistent with the youthful age of the moon's surface. The dissolution of the surface was a major cause of landscape evolution on Titan and could be the origin of its lakes. Scientists calculated how long it would take to form lake depressions at lower latitudes where the rainfall is reduced. The much longer timescale of 380 Myr is consistent with the relative absence of depressions in these geographical locations (Baldwin et al., 2015).

11.6 PLUTO AND CHARON

In 2015, NASA's New Horizons spacecraft approached the Pluto system, comprising Pluto with its moons: Charon, Nix, Hydra, Styx, and Kerberos.

11.6.1 Pluto

Close-up images of a region near Pluto's equator reveal a great surprise: a range of young mountains rising as high as 3500 m above the surface of the icy body. The mountains likely formed no more than 100 Myr ago relative to the 4.56 Gyr of the Solar System, and may still be in the process of building (see Fig. 11.15). The youthful age estimate is based on the lack of craters.

Unlike the icy moons of giant planets, Pluto cannot be heated by gravitational interactions with a much larger planetary body. Some other process must be generating the mountainous landscape. Thus we must rethink what powers geological activity on many other icy worlds. The mountains are probably composed of Pluto's water-ice "bedrock."

Although methane and nitrogen ice covers much of the surface of Pluto, these materials are not strong enough to build the mountains. Instead, a stiffer material, most likely water-ice, created the peaks. At Pluto's temperatures, water-ice behaves more like rock (Green and Yeomans, 2015).

Even though Pluto is extremely far away from the sun, its surface shows movement of ice, which might indicate the existence of liquid water, meaning subsistence of some kind of life.

FIGURE 11.15
Heart on Pluto. NASA's New Horizons spacecraft took this photo on Jul. 13, 2015 when the spacecraft was 768,000 km from the surface. This is the last and most detailed image sent to Earth before the spacecraft's closest approach to Pluto on Jul. 14. This view is dominated by the large, bright feature informally named the "heart," which measures approximately 1600 km across. Even at this resolution, much of the heart's interior appears remarkably featureless—possibly a sign of ongoing geologic processes.
Image credit: NASA/APL/SwRI.
http://www.nasa.gov/sites/default/files/thumbnails/image/tn-p_lorri_fullframe_color.jpg.

11.6.2 Charon

Many scientists expected Charon to be a monotonous, crater-battered world; instead, they're finding a landscape covered with mountains, canyons, landslides, surface-color variations, and more.

It was discovered that the plains south of Charon's canyon have fewer large craters than the regions to the north, indicating that they are noticeably younger. The smoothness of the plains, as well as their grooves and faint ridges, are clear signs of wide-scale resurfacing (see Fig. 11.16).

One possibility for the smooth surface is a kind of cold volcanic activity, called cryovolcanism. There is the possibility that an internal water ocean could have frozen long ago, and the resulting volume change could have led to Charon cracking open, allowing water-based lavas to reach the surface at that time (Talbert, 2015).

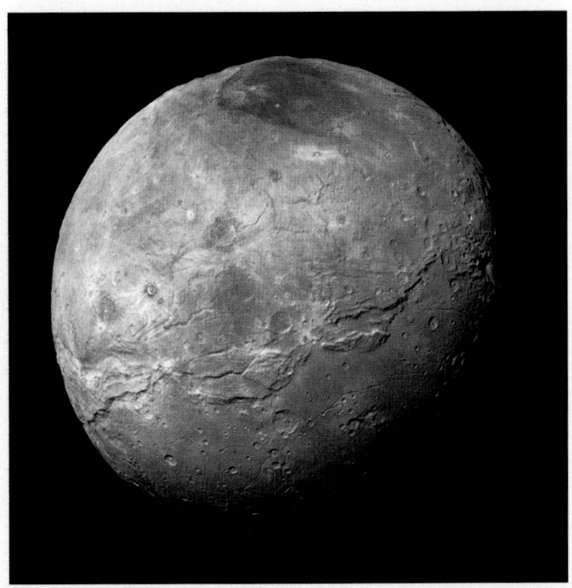

FIGURE 11.16
Pluto's big moon, Charon. At half the diameter of Pluto, Charon is the largest satellite relative to its planet in the solar system.
Image credit: NASA/JHUAPL/SwRI.
http://www.nasa.gov/sites/default/files/thumbnails/image/nh-charon-neutral-bright-release.jpg.

References

Baldwin E, Landau E, Dyches P. The mysterious 'lakes' on Saturn's moon Titan. Cassini at Saturn. 2015. Available from: http://www.nasa.gov/feature/the-mysterious-lakes-on-saturns-moon-titan.

Brown D, Webster G. NASA Curiosity rover detects no methane on Mars. 2013. Release 13–288. Available from: http://www.nasa.gov/press/2013/september/nasa-curiosity-rover-detects-no-methane-on-mars/#.VvnFI5Vf2ck.

Castillo-Rogez J. Solar System exploration. 2015. Available from: http://solarsystem.nasa.gov/people/castillo-rogezj.

Clavin W. Solar System exploration. 2014. Available from: http://solarsystem.nasa.gov/news/2014/05/01/ganymede-may-harbor-club-sandwich-of-oceans-and-ice.

Clavin W. NASA's Curiosity rover team confirms ancient lakes on Mars. 2015. Available from: http://www.nasa.gov/feature/jpl/nasas-curiosity-rover-team-confirms-ancient-lakes-on-mars.

Cook J-R. Solar System exploration. 2014. Available from: http://solarsystem.nasa.gov/news/2014/02/12/largest-solar-system-moon-detailed-in-geologic-map.

Dodd S. Solar System exploration. 2015. Available from: http://solarsystem.nasa.gov/planets/callisto/indepth.

Green J, Yeomans D. Solar System exploration. 2015. Available from: http://solarsystem.nasa.gov/planets/pluto/basic.

Krohn K, Jaumann R, Tosi F, Nasu A, et al. Geologic mapping of the Ac-H-6 quadrangle of Ceres from Nasa's Dawn mission: compositional changes. Geophys Res Abstr 2016;18:EGU2016–7848.

Landau L. New clues to Ceres' bright spots and origins. 2015. Available from: http://www.jpl.nasa.gov/news/news.php?feature=4785.

Lopes L. Solar System exploration. 2015. Available from: http://solarsystem.nasa.gov/people/lopesr.

Pirani S, Turrini D. Asteroid 4 Vesta: dynamical and collisional evolution during the Late Heavy Bombardment. Icarus 2016;271:170–9.

Platt J, Brown D, Bell B. NASA space assets detect ocean inside Saturn moon. 2014. Available from: http://solarsystem.nasa.gov/news/2014/04/03/nasa-space-assets-detect-ocean-inside-saturn-moon.

Prettyman TH, Mittlefehldt DW, Yamashita N, et al. Elemental mapping by Dawn reveals exogenic H in Vesta's regolith. Science 2012. http://dx.doi.org/10.1126/science.1225354.

Russel C, Raymond C. The Dawn mission to minor planets 4 Vesta and 1 Ceres. New York: Springer Science & Business Media; 2012.

Talbert T. Pluto's big moon Charon reveals a colorful and violent history. 2015. Available from: http://www.nasa.gov/feature/pluto-s-big-moon-charon-reveals-a-colorful-and-violent-history.

CHAPTER 12
Conclusions

Chapter Outline

We have so far learned the present (2015–16) state of astroscience regarding the origins of the Earth, the Moon, and life in 11 chapters by using interdisciplinary approach. For the origin of the Earth and the Moon, refinements of astrophysical models and cosmochemical parameters for the giant impact are required. The search for remnants of the giant impact on the Moon's surface by astrogeology is especially important. Sampling of the far side of the Moon should give us new insight into the origin of the Moon, and polar samples are also important and interesting when investigating the Moon's origin.

The oldest geological samples on Earth exist in the core of zircons. The oldest traces of life are chemofossil graphite inclusions in such zircons. Thus, improvements of spot analytical techniques need to be refined. Techniques of secondary ion mass spectrometry (SIMS) should be improved to obtain higher sensitivity and stability.

With regard to the origin of life, life-related molecules were synthesized by collisions between icy materials in space or on Earth at high velocity. Even nucleic acids, which are the main component of genes, can be made in such conditions. However, it is still not known how the building blocks made life or the protocell, which has a cell membrane and cytoplasm, and stores genetic information with the ability to reproduce. How the protocell appeared is enigmatic, but it is probable that it began to "live" in geothermal or hydrothermal areas.

Some researchers can make self-reproductive materials, but for the time being this has not been called life. However, nobody doubts that more sophisticated protocells will be created within 10 years, given that bacteria has been recently synthesized (Hutchison et al., 2016). Ethical problems can occur in this type of research.

Some moons of Jupiter and Saturn have oceans of liquid water. Life could be found in these celestial bodies. We cannot imagine the characteristics of such life; even in the deep ocean of the Earth, we find unexpected life. Future observations will reveal new life.

Reference

Hutchison III CA, Chuang R-Y, Noskov VN, et al. Design and synthesis of a minimal bacterial genome. Science 2016;351(6280). http://dx.doi.org/10.1126/science.aad6253.

Origins of the Earth, Moon, and Life. http://dx.doi.org/10.1016/B978-0-12-812058-3.00012-0

Index

'*Note*: Page numbers followed by "f" indicate figures, "t" indicate tables, and "b" indicate boxes.'

M

N

Printed in the United States
By Bookmasters